Integrated Water Resources Management in a Changing World

Integrated Water Resources Management in a Changing World

Lessons Learnt and Innovative Perspectives

Editors
Dietrich Borchardt and Ralf Ibisch

IWA Publishing
London • New York

Published by **IWA Publishing**
Alliance House
12 Caxton Street
London SW1H 0QS, UK
Telephone: +44 (0)20 7654 5500
Fax: +44 (0)20 7654 5555
Email: publications@iwap.co.uk
Web: www.iwapublishing.com

First published 2013. Reprinted 2013

British Library Cataloguing in Publication Data
A CIP catalogue record for this book is available from the British Library

Library of Congress Cataloging-in-Publication Data
A catalog record for this book is available from the Library of Congress

ISBN: 9781780405261 (Paperback)
ISBN: 9781780405278 (eBook)

Contents

About the Editors and Contributors

EDITORS

Dietrich Borchardt holds a master's degree in fisheries biology, a PhD in hydrobiology and qualified as a professor in water resources management. His research fields are freshwater ecology, water quality modeling and ecosystem management. He currently holds a full professorship at Technical University of Dresden, Germany and is head of department Aquatic Ecosystem Analysis and Management at the Helmholtz Centre for Environmental Research – UFZ in Magdeburg, Germany. In 2011 he chaired the scientific steering committee of the international conference on Integrated Water Resources Management in Dresden, Germany.

Ralf Ibisch holds a PhD in biology and has expertise in aquatic ecology and water resources management. He is currently working as a senior scientist at the Helmholtz Centre for Environmental Research – UFZ in Magdeburg, Germany. In 2011 he chaired the organizing committee of the international conference on Integrated Water Resources Management in Dresden, Germany.

AUTHORS

Jan Adamowski, McGill University, Quebec, Canada

Emad Al-Karablieh, University of Jordan, Amman, Jordan

Usman Khalid Awan, Center for Development Research, Bonn, Germany & University of Agriculture, Faisalabad-Pakistan

Ilona Bärlund, Helmholtz Centre for Environmental Research – UFZ, Germany

Gero Bauser, Institute of Environmental Engineering, ETH Zurich, Switzerland

Jürgen Benndorf (†), Technical University of Dresden, Germany

Anik Bhaduri, Center for Development Research, Bonn, Germany

Thomas Bonn, University of Heidelberg, Germany

Dietrich Borchardt, Helmholtz Centre for Environmental Research – UFZ, Germany

Marian Brenda, Technical University of Darmstadt, Germany

Wolfgang Britz, University of Bonn, Germany

Ana Buxo, Technical University of Lisbon, Portugal

Christopher Conrad, University of Würzburg, Germany

Jutta Deffner, Institute for Social-Ecological Research (ISOE), Frankfurt/Main, Germany

Shuhuai Duan, Beijing Soil and Water Conservation Center, Beijing, China

Hartmut Eckstädt, University of Rostock, Germany

Martina Flörke, Center for Environmental Systems Research, University of Kassel, Germany

Stephen Foster, World Bank/Global Water Partnership GW-MATE (Groundwater Management Advisory Team)

Tim Fröhlich, Institute for Technical and Scientific Hydrology GmbH, Hannover, Germany

Lothar Fuchs, Institute for Technical and Scientific Hydrology GmbH, Hannover, Germany

Markus Funke, SYDRO Consult GmbH, Darmstadt, Germany

Héctor Garduño, World Bank/Global Water Partnership GW-MATE (Groundwater Management Advisory Team)

M. Ghanem, Palestinian Hydrology Group, West Bank/ Palestine

Thomas Grischek, University of Applied Sciences Dresden, Germany

Jens Grundmann, Technical University of Dresden, Germany

Joseph Guttman, MEKOROT Water Company Ltd, Tel Aviv, Israel

Johannes Halbe, Institute of Environmental Systems Research, University of Osnabrück, Germany

Fred Hattermann, Potsdam Institute for Climate Impact Research (PIK), Germany

Björn Helm, Technical University of Dresden, Germany

Anna-Katharina Hornidge, Center for Development Research, Bonn, Germany

Ayman Jarrar, Palestinian Water Authority, Ramallah, Palestine

Kurt Mørck Jensen, Danish Institute for International Studies (DIIS), Copenhagen, Denmark

Alexander Jokisch, Technical University of Darmstadt, Germany

Ralf Junghanns, Adensis GmbH, Dresden, Germany

Hans-Peter Kaiser, Stadt Zurich Wasserversorgung, Zurich, Switzerland

Inka Kaufmann Alves, University of Kaiserslautern, Germany

Wolfgang Kinzelbach, Institute of Environmental Engineering, ETH Zurich, Switzerland

Hagen Koch, Potsdam Institute for Climate Impact Research (PIK), Germany

Peter Krebs, Technical University of Dresden, Germany

Christina Kröger, University of Rostock, Germany

Uli Kuhlmann, TK Consult AG, Zurich, Switzerland

Arnim Kuhn, University of Bonn, Germany

Ellen Kynast, Center for Environmental Systems Research, University of Kassel, Germany

John Lamers, Center for Development Research, Bonn, Germany

Franz Lennartz, Technical University of Dresden, Germany

Stefan Liersch, Potsdam Institute for Climate Impact Research (PIK), Germany

Pia Lipp, Water Technology Center (TZW), Karlsruhe, Germany

Hubert Lohr, SYDRO Consult GmbH, Darmstadt, Germany

Stefan Lorenz, Leibniz-Institute of Freshwater Ecology and Inland Fisheries (IGB), Berlin, Germany

Ahmad M. Manschadi, Center for Development Research, Bonn, Germany

Amer Marei, Al-Quds University, Jerusalem

Beatrice Sabine Marti, Institute of Environmental Engineering, ETH Zurich, Switzerland

Muhammad Mehmood-Ul-Hassan, Center for Development Research, Bonn, Germany

Ralph Meissner, Helmholtz Centre for Environmental Research – UFZ, Germany

M. Teresa Condesso de Melo, Technical University of Lisbon, Portugal

Maria Paula Sofio Silva Mendes, Technical University of Lisbon, Portugal

Volker Mohaupt, Federal Environment Agency, Dessau-Rosslau, Germany

Jacinto Nascimento, Technical University of Lisbon, Portugal

Claudia Pahl-Wostl, Institute of Environmental Systems Research, University of Osnabrück, Germany

Stefania Paris, Huber SE, Berching, Germany

Lothar Paul, Technical University of Dresden, Germany

Thomas Petzoldt, Technical University of Dresden, Germany

Hoang Mai Phan, Technical University of Dresden, Germany

Martin T. Pusch, Leibniz-Institute of Freshwater Ecology and Inland Fisheries (IGB), Berlin, Germany

Helle Munk Ravnborg, Danish Institute for International Studies (DIIS), Copenhagen, Denmark

Luis Ribeiro, Technical University of Lisbon, Portugal

Sandra Richter, Helmholtz Centre for Environmental Research – UFZ, Germany

Susanne Rolinski, Potsdam Institute for Climate Impact Research (PIK), Germany

Amer Salman, University of Jordan, WERSC, Amman, Jordan

Cornelius Sandhu, University of Applied Sciences Dresden, Germany

Nicole Scheifhacken, Technical University of Dresden, Germany

Natalie Schmidt, Water Technology Center (TZW), Karlsruhe, Germany

Christof Schneider, Center for Environmental Systems Research, University of Kassel, Germany

Niels Schütze, Technical University of Dresden, Germany

Naief Seder, Jordan Valley Authority, Amman, Jordan

Bertram Skibinski, Technical University of Dresden, Germany

Irene Slavik, Technical University of Dresden, Germany

Fritz Stauffer, Institute of Environmental Engineering, ETH Zurich, Switzerland

Catalin Stefan, Technical University of Dresden, Germany

Tibor Stigter, Technical University of Lisbon, Portugal

Ali Subah, Ministry of Water and Irrigation, Jordan

Abdel Rahman Tamimi, Palestinian Hydrological Group, Ramallah, Palestine

Ellen Teichert, Center for Environmental Systems Research, University of Kassel, Germany

Tatyana Terekhanova, Technical University of Dresden, Germany

Andreas Tiehm, Water Technology Center (TZW), Karlsruhe, Germany

Bernhard Tischbein, Center for Development Research, Bonn, Germany

Jens Tränckner, Technical University of Dresden, Germany

V. N. Tran, Hanoi University of Civil Engineering, Hanoi, Vietnam

Wolfgang Uhl, Technical University of Dresden, Germany

Wilhelm Urban, Technical University of Darmstadt, Germany

Markus Venohr, Leibniz-Institute of Freshwater Ecology and Inland Fisheries (IGB), Berlin, Germany

Paul L.G. Vlek, Center for Development Research, Bonn, Germany

Jeanette Völker, Helmholtz Centre for Environmental Research – UFZ, Germany

Peter Werner, Technical University of Dresden, Germany

Leif Wolf, Commonwealth Scientific and Industrial Research Organisation (CSIRO), Queensland, Australia

Heinz-Peter Wolff, Office for Quantitative Analyses, Stuttgart, Germany

A. Xu, University of Rostock, Germany

Claudia Zawadsky, Water Technology Center (TZW), Karlsruhe, Germany

Moritz Zemann, University of Karlsruhe, Germany

B. Zhang, Beijing Soil and Water Conservation Center, Beijing, China

Martin Zimmermann, Technical University of Darmstadt, Germany

Preface

The present volume contains a selection of the main contributions made to the international conference on Integrated Water Resources Management (IWRM) entitled 'Management of Water in a Changing World: Lessons Learnt and Innovative Perspectives' that was held from 12 to 13 October 2011 in Dresden, Germany. The conference was funded by the German Federal Ministry of Education and Research (BMBF), supported by the International Water Association (IWA), the Global Water Systems Project (GWSP), and organised by the Helmholtz Centre for Environmental Research (UFZ). This book contains papers which were presented during the conference, either as keynote lectures in plenary sessions or as submitted papers in one of the thematic sessions.

All the papers were peer-reviewed by international experts before publication in journals issued by the International Water Association (IWA). Neither all the submitted papers nor all the conference presentations could be included in the book: Several key presentations were not accompanied by written papers and thus could not be included. However, all presentations and abstracts and the corresponding PowerPoint presentations as well as a video recording of the panel discussion are available at the conference website http://www.bmbf.iwrm2011.de/. Readers are encouraged to complete their review of the conference and its messages by consulting this interesting on-line source of accompanying scientific material.

The present publication has two major objectives: 1) to document major contributions to the conference by providing background reading for those wishing to engage in the issues addressed and debates held in October 2011 at the IWRM conference, and 2) to summarise the main messages issuing from the conference.

It was organised around six main topics reflecting important issues addressed during the IWRM conference.

1. Water resources in changing environments

The papers in this section of the book deal with the potential of quantification methods, especially modelling studies, as aids for estimating water resources and water use, management evaluation and decision making processes under changing environmental and/or socio-economic conditions. The contributions concentrate on large scale, conceptual studies and give examples of selected case studies.

2. Groundwater management

This part of the book highlights modelling and decision-support tools for sustainable groundwater management strategies. Groundwater provides much of the water used in irrigation and industrial production and is of great importance for drinking water supplies. However, groundwater resources throughout the world are being subjected to over-abstraction and an increasing threat of pollution from urbanisation, industrial development, agricultural activities and mining enterprises.

3. Technologies and implementation

This section aims to identify and evaluate promising technologies that allow for efficient use of resources in the pursuit of two aims: meeting growing needs and demand for water and limiting the negative consequences of over-exploitation. The contributions draw attention to new approaches that could be of service to authorities and those responsible for urban water services in satisfying growing demands. Also, this part of the book touches upon the question of appropriate frameworks (socio-economic, governance, participation) that are prerequisites for the successful implementation of new technologies.

4. Water management indicators at different scales

A variety of indicators has been developed in recent decades to assess water quantity and quality status, vulnerability and threats at catchment and global scales. Water quality indicators and water management indicators are indispensable tools for water management staff and policy-makers and they have also received considerable public attention. The papers in this part of the book discuss key water management indicators and highlight the question of their respective suitability in connection with core management decisions.

5. Information and decision support systems

Decision Support Systems (DSS) help in structuring and solving decision-related problems in IWRM. They are designed to assist decision makers in identifying relevant information for the decision(s) to be made, to predict the impact of different management options and to evaluate alternatives. DSS underpin efficient water management approaches and make decision-making processes more transparent. The papers included here present various approaches for the development of decision support tools for intra-disciplinary and interdisciplinary issues.

6. Water governance: actors and institutions

Water management under changing conditions requires appropriate governance systems. This part deals with issues concerning the design of water governance systems that are capable of responding to increasing degrees of complexity and uncertainty. The criteria and indicators that are and will be needed for the analysis of water governance schemes and institutional performance are discussed.

The editors take full responsibility for the selection and collation of the papers. The respective author(s) take full credit and responsibility for the content of the published contributions. There are several papers which could appropriately be included in more than one of the main subject areas described above; in such cases our aim has been to place the paper in question in the section which best reflects its main thrust. Taken together, the papers provide an overview on important issues concerning the conceptual framework of Integrated Water Resources Management (IWRM).

In retrospect, two aspects of the IWRM conference held in Dresden in 2011 appear especially clear: On the one hand it proved possible to demonstrate that state-of-the-art research solutions can help to solve real-world problems. On the other hand it is apparent that our present knowledge concerning the interdependencies between different areas of activity is still limited, so forecasts of what may lie ahead of us are rendered less useful on account of inaccuracies and uncertainties. The discussions during the conference highlighted the lack and inadequacy not only of data, but also of conducive governance frameworks. The latter set the constraints within which we have to deal with our present and future water challenges.

Conditions are changing fast and this has already led to irreversible damage to water resources and aquatic ecosystems in many regions of the world. Therefore, the important message from the conference is that we have to speed up the implementation of IWRM and the enforcement of the resultant programmes of measures.

Dietrich Borchardt and Ralf Ibisch

Magdeburg, June 2013

Message from the Dresden International Conference on Integrated Water Resources Management

Management of Water in a Changing World: Lessons Learnt and Innovative Perspectives

12th–13th October 2011, Dresden, Germany

In the light of the global challenges caused by climate change, land use and demographic changes the sustainable use and the protection of natural resources are top priorities for sustainable development. Enormous efforts will be necessary to ensure the supply of clean and safe water to the world population and to protect vital aquatic ecosystems. To meet this challenge the concept of Integrated Water Resources Management (IWRM) was introduced under the Agenda 21 and aimed at the coordinated development and management of water, land and related resources in order to maximize the resultant economic and social welfare in an equitable manner without compromising the sustainability of vital ecosystems. Ten years later, at the World Summit for Sustainable Development in Johannesburg in 2002, all countries agreed to "develop IWRM and Water Efficiency Plans".

Part of this ongoing process was the Dresden International Conference on IWRM "Management of Water in a Changing World: Lessons learnt and innovative perspectives", which took place from 12th–13th October 2011 and was funded by the German Federal Ministry of Education and Research under a priority research programme on IWRM. The event attracted an audience of more than 350 participants, representing 40 countries worldwide. Hereby the following conclusions were drawn:

1. The concept of IWRM has gained wide acceptance in the majority of countries worldwide in the last 20 years. However, while considerable progress has been made to include IWRM in national policies, strategies and laws, the actual implementation of IWRM is lagging behind.
2. There are strong linkages but also substantial trade-offs between water security, food security and energy security. IWRM should be seen as a pathfinder process for the implementation of an Integrated Resource Management.
3. Besides economy, energy and food the environment with its vital ecosystems should be treated with high relevance.
4. Successful IWRM works across sectors and levels: horizontally across sectors such as economy, energy, agriculture, environment, science, vertically from international over national, regional, basin to local levels. It works with an intense dialogue between governmental institutions, science, NGO's and society in order to achieve more sustainable solutions.
5. Successful IWRM includes targeted and coordinated Capacity Development on different levels (in particular academic, administrative, technical, stakeholder).
6. Economics plays a key role in effective water resources management. Not the natural resource as such, but the water services should be treated as an economic good.
7. IWRM based infrastructures typically serve multi-purpose schemes (e.g. wastewater management for protecting the environment and human health, water storage schemes for producing energy or food and mitigation of extreme events such as floods and droughts).
8. IWRM provides a framework for the necessary integration of all the sectors involved. However, there are

limitations due to insufficient knowledge about their interactions.

9. It is necessary to further strengthen the scientific basis of water resources management. Required research approaches have to be interdisciplinary including development and innovation, action oriented and transdisciplinary with a substantiated science policy interface.
10. The implementation of IWRM and the realisation of the respective programs have to be accelerated. Dynamics of change are fast and already lead to irreversible damages of water resources in many regions of the world.

AUTHORS

Prof. Dr. Dietrich Borchardt, Helmholtz Centre for Environmental Research – UFZ, Germany (Chair of the conference)

Dr. Peter Koefoed Bjørnsen, UNEP-DHI Centre for Water and Environment, Denmark

Prof. Dr.-Ing. Dr. h.c. Janos J. Bogardi, Global Water System Project, Bonn, Germany

Prof. Torkil Jønch Clausen, Water Policy Adviser, DHI Group, Senior Adviser, Global Water Partnership, Denmark

Dr. Ines Dombrowsky, German Development Institute, Germany

Héctor Garduño, International Consultant, Mexico

Prof. Dr.-Ing. Norbert Jardin, Ruhrverband, Essen, Germany

Prof. Dr. Alan Jenkins, CEH, Natural Environment Research Council, Oxfordshire, Great Britain

Dr. Stefan von Keitz, Hessian Ministry of Environment, Agriculture, Rural Areas and Consumer Protection, Germany

Dr. Rivka Kfir, South African Government, Water Research Commission, South Africa

Prof. Dr. Peter Krebs, University of Technology, Dresden, Germany

Prof. Dr. Helmut Kroiss, Vienna University of Technology, Austria

Prof. em. Dr. Christian Leibundgut, University of Freiburg, Germany

Prof. Dr. Wolfram Mauser, Ludwig-Maximilians University Munich, Germany

Dr. Timothy Moss, Leibniz Institute for Regional Development and Structural Planning, Germany

Dayanand Panse, Indian Water Works Association, India

Prof. Dr. Peter Reichert, Eawag, Duebendorf, Switzerland

Prof. Dr. Seppo Rekolainen, Finnish Environment Institute, Helsinki, Finland

Prof. Dr. Dr. Karl-Ulrich Rudolph, University Witten/Herdecke, Germany

Prof. David L. Rudolph, University of Waterloo, Ontario, Canada

Dr. Per Stålnacke, Norwegian Institute for Agriculture and Environmental Research, Norway

Bai-Mass M. Taal, African Ministers Council on Water, Nigeria

Prof. Dr. Min Yang, Chinese Academy of Sciences, China

Report from the Dresden International Conference on Integrated Water Resources Management

Management of Water in a Changing World: Lessons Learnt and Innovative Perspectives

12th–13th October 2011, Dresden, Germany

Sabrina Kirschke, Ralf Ibisch, Christian Stärz and Dietrich Borchardt

1. INTRODUCTION

On October 12 & 13, the international conference on Integrated Water Resources Management (IWRM) "Management of Water in a Changing World: Lessons Learnt and Innovative Perspectives" took place in Dresden, Germany. The conference was sponsored by the German Federal Ministry of Education and Research (BMBF), supported by the International Water Association (IWA) and the Global Water Systems Project (GWSP) and organised by the Helmholtz Centre for Environmental Research (UFZ). 365 participants from science and practice from 40 countries discussed lessons learnt and innovative perspectives of IWRM concepts and their implementation in more than 20 technical sessions, international key note speeches, a high level panel discussion and a two-day poster session.

This report aims to summarize the most important results from the conference based on six sources: Firstly, junior rapporteurs participated in the thematic sessions and made notes on the session results with respect to leading questions. Secondly, the leading rapporteurs of the thematic sessions each gave a five minute statement at the end of the conference, including a personal key message on the respective topic. Thirdly, more than 80 conference participants (approx. 25% of all participants) took part in an online survey and subsequently formulated a community statement on lessons learnt and innovative perspectives. Important information has also been drawn from the key note speakers presentations and from interviews conducted with eight key contributors such as Wilfried Kraus, the Deputy Director General of "Sustainability, Climate, Energy" from the German Federal Ministry of Education and Research (BMBF) and Bai-Mass Taal from AMCOW, the African Ministers' Council on Water. Finally, one important source was a statement prepared by the steering committee for the international conference in Bonn on "The Water, Energy and Food Security Nexus".

This report is highly subjective: The discussions and opinions are often subjective interpretations of facts, the relationship between science and practice which was discussed at the conference being an example. The leading and junior rapporteurs were also asked to make personal statements on the innovative lessons learnt and future perspectives rather than making sophisticated summaries of the conference contents. The extract of ideas made in this report is also based on a subjective view of different information sources and thus does not form an objective factual analysis.

We would like to present the conference results using the following structure, showing a consecutive level of generalization: On a first level, forming a low degree of generalization, we present topic-related insights based on the reports from both junior and leading rapporteurs (Chapter 2). On a second level, we attempt to link the respective topics to one another, a task which is based on the observations made by the junior rapporteurs (Chapter 3). On a third level, representing the highest degree of generality, we show some general challenges for IWRM such as energy supply and climate change which were emphasized in various sources (Chapter 4). Concluding, we look at the often mentioned challenge of implementation and particularly at the science-policy interfaces which takes all previously mentioned levels into consideration (Chapter 5).

2. CONFERENCE TOPICS

Sustainable water resources management is based on integrated knowledge from different disciplines and sectors. Against this background, we present some core insights on seven identified scientific topics that were discussed during the conference (see Figure 1). These lessons learnt and future perspectives, mentioned below, are based on information given by the junior and leading rapporteurs.

2.1 Technologies and Implementation

Technologies and their implementation represented a core area of the discussions during the IWRM conference. In the sessions the talks ranged from i) adapted technologies for specific IWRM-problems such as water supply in climatically extreme conditions and ii) the process of implementation to iii) the challenges faced when maintaining technologies in the long term in developing and emerging nations.

Figure 1 | IWRM Topics at the conference on "Management of Water in a Changing World: Lessons.*Learnt and Innovative Perspectives", 12 – 13 October 2011, Dresden, Germany.*

On the topic of technological innovation, Univ.-Prof. Dr.-Ing. Jörg Londong stated that only a few new technologies were presented whereas more traditional technologies were demonstrated that had been applied in a new environmental context such as adapted water treatment and supply technologies. In this context, it was stressed that all water components, particularly the reusing of wastewater must be given due consideration. It was also emphasized that flexible rather than fixed technologies are needed in order to achieve sustainable solutions. On the topic of implementation in the short and particularly in the long term, it was repeatedly pointed out that technologies cannot stand on their own but must be embedded in an enabling environment associated with extensive stakeholder participation and adequate Capacity Development measures such as workshops, on-the-job trainings and education. International cooperation with long term perspectives has also been highly recommended for the maintenance of technical solutions.

It was suggested at the conference that in the future, existing technologies must be adjusted and adapted to new purposes and local conditions, in parallel to the development of new technologies. Improved monitoring and evaluation of technical measures are also required, whereby a technical evaluation is just a first step in the evaluation process and must be complemented by surveys on acceptance in local communities. Finally, it was highlighted that more efficient and flexible technological solutions are needed for today's fast growing megacities and populations.

2.2 Water Resources in Changing Environments

The topic of water resources in changing environments was discussed in two sessions. Core issues addressed during the sessions related particularly to the effects of climate change, urbanization and human water uses such as agricultural intensification, irrigation and the storage of water in different spatial entities (from the basin to continental scale).

The following aspects were particularly emphasized: Firstly, the gathering of data is a key factor for successful IWRM, however data is still scarce in many regions and often particularly so in areas where IWRM is most urgently required. Secondly, it has been highlighted that there is a strong need for a better understanding of processes. Thirdly, essential methods were discussed, for example integrated computer modelling in order to depict the various effects of changing environments on water resources and scenario techniques for evaluating different future pathways. It was pointed out that a range of methods must be used as water-related problems vary over different regions and scales.

The development of better models (e.g. with links between socio-economic and ecosystem models) and environmental change scenarios remains a core challenge for the future. Further, some participants questioned how the disparity between models designed for larger spatial scales and real world water problems found on the small scale can be overcome. Thus the fundamental question – raised by the leading rapporteur of the topic, Dr.-Ing. Martina Flörke – still remains: Are models really capable of representing the reality of IWRM and if they are able to do so, how can this information be communicated and translated into practice?

2.3 Information and Decision Support Systems

The topic of information and decision support was highly relevant at the IWRM conference. In two sessions, scientists and practitioners discussed different information management systems and decision support tools as well as their practical use in achieving IWRM. Special attention was paid to questions on the information required by such a system and on the structure of the developing process for such tools.

The following tools were presented and discussed: GIS databases, manuals and planning maps, the MONERIS model, the Water Evaluation and Planning System (WEAP), a semantic IWRM Wiki, a digital multimedia atlas in different languages and a technology tool-kit with information material and tools inside a physical box. Prof. Dr. Stefan Kaden, the leading rapporteur of this topic, also underlined that "there is no unique approach for decision support systems" and that the choice of system is highly dependant on the study area. It was further emphasized that decision support tools often do not need to be developed from scratch; existing tools can and should often be adapted to local conditions. Water managers frequently fail to use decision support tools and information management due to knowledge deficits and a lack of participation in the tool design process. Practical problems such as limited internet access for stakeholders are also common.

Looking to the future, the online questionnaire respondents rather agreed that decision support tools require further development in order to achieve IWRM. In this context, some individuals at the conference particularly highlighted the following challenges for the future: On the technical side, they demanded better models based on more reliable and improved data. Future models should have an improved processing of uncertainties in global change such as climate, socio-economic and land use change. More complex models should serve the increased demand for an integrated approach and integrate interacting processes in catchments, and should also be capable of processing contradicting management objectives. On the other hand, one participant suggested that decision support systems "need simplification for pre-planning phases where data availability is scarce". It has also been noted that the development of more meaningful indicators for decision support is required. Generally it will require some effort in order to implement operational management rules in a generalised and user-friendly form.

2.4 Capacity Development

Capacity Development (CD) in IWRM and social learning theory was discussed in two sessions. Central questions that were addressed were related to the role and design of CD processes in IWRM, with a special focus on the experiences made by projects funded by the Federal Ministry of Education and Research (BMBF) in the funding priority IWRM.

The following conclusions are highlighted as a result of the discussion: Firstly, the participants stressed that there is a need for targeted and coordinated CD on all levels (in particular academic, administrative and technical) and for all age groups. Secondly, cultural specificities must be respected when designing and implementing CD measures. Thirdly, the scientific evaluation of specific CD measures needs further development. CD measures for the settlement of disputes must also be boosted in order to support IWRM.

Conference participants also stressed the importance of CD for researchers for the development and implementation of innovative IWRM solutions. Specialists with a well-founded interdisciplinary education are necessary as a lack of these skills may lead to an insufficient handling of complexity. This interdisciplinary IWRM education could already start at school level and was also strongly supported by those participating in the online questionnaire.

Dr. Steffen Niemann, the leading rapporteur on CD, emphasized that despite many good initiatives on CD, there is a need for better cooperation and coordination between different actors e.g. universities and associations, and that this represents an essential requirement for the future. More attention must thus be paid to new media and tools which are often underestimated, such as trade fairs.

2.5 Water Governance

What can be learned from examples such as the EU Water Framework Directive? How do political and institutional changes effect the solving of water-related problems? And what causes conflict and cooperation over water resources? These and other questions were discussed during three governance related sessions.

The results are highlighted in the following points: First and foremost it was stressed that there is no one-size-fitsall approach, as there are for example major differences between developed and developing nations. The EU Water Framework Directive is thus not easily globally applicable, especially in some Eastern and Southern Non-EU countries. Secondly, the concept of participation was particularly highlighted by the online survey participants, in addition to concepts and theories such as new institutional economics, power relationships, liberal functionalism, polycentric governance approaches and political economy. The participation of all stakeholders, including the local population, in the design and implementation of IWRM measures and the decentralization of funds and functions etc. was demanded. The leading rapporteur on governance, PD Dr. Heike Walk suggested that the BMBF should establish more control mechanisms for good performance in participation and coordination in addition to sanctions with the aim of supporting IWRM solutions. Thirdly, a number of specific governance constraints for IWRM solutions were highlighted by some individual conference attendees: Even if legally binding rules exist, they are often hindered to implementation due to overlapping competencies between different administrative bodies, e.g. from financial, environmental and agricultural sectors. Further administrative constraints that were mentioned relate to institutional separations between ground and surface waters and between water quantity and water quality issues. The online questionnaire respondents also rather agreed that existing institutions are often overburdened with the complex task of efficiently coordinating different sectors of society in order to achieve IWRM.

The conference participants identified several challenges for the future in both scientific and practical issues. On the scientific side, a better understanding is required of the role of third parties in water cooperation and of EU experiences in developing and transition countries. Multilevel-governance, which also includes the transboundary dimension, remains an underdeveloped field of research in the water sector. Comparative analysis is an important methodological approach here. In practise, overlapping responsibilities and weak, inappropriate governance structures were highlighted in the online survey. The establishment of adapted institutions thus remains a major challenge for the future.

2.6 Groundwater Management

The topic of groundwater management was discussed in three sessions. A focal point of the discussions was the issue of sustainable groundwater management in areas where groundwater resources are overexploited by irrigated agriculture. A second major point in the discussions was the link between ground and surface waters and the combined management of both.

For the sustainable management of groundwater resources, the following points were highlighted: (1) managed aquifer recharge is a tool for solving problems around shrinking and low-quality groundwater resources; (2) in order to manage karst aquifers, extensive hydrogeological basic research work is necessary in order to understand the system, in addition to a combination of active and passive management strategies; (3) in order to maintain the quality and quantity of groundwater resources, precise criteria must be determined and vulnerability assessments conducted prior to the implementation of further measures in stressed ecosystems. On the topic of conjunctive use and management, coupled groundwater- surface water models were presented which simulate interactions between resources.

For science, the development of adaptive groundwater management strategies which define groundwater allocations using a combination of groundwater level measurements, numerical groundwater models and water balance approaches remains a central challenge. Aspects of water quality and quantity must also be taken into consideration more often. Moreover, the implementation of different management approaches remains a challenge: The problem is that tools such as groundwater flow models lack the capability for implementing operational management rules in a generalised and user-friendly form. Political, institutional and economical restrictions also hinder the implementation of groundwater strategies. In this context, Héctor Garduño underlined that the "hidden resource" groundwater is often overlooked by political decision makers.

2.7 Economic Instruments

Economic instruments for achieving financially sustainable water resources management were a highly relevant topic at the IWRM conference. Issues that were raised during the three sessions on the topic concerned the financing of IWRM measures, in particular the payment of water ecosystem services and the contribution of businesses to IWRM, but also the question of quantifying the economic value of ecosystem services.

It was clear that investments and financing must be sustainable, so the refinancing of investments is an essential consideration. This aspect was strongly supported both by the leading rapporteur Prof. Dr. Karl-Ulrich Rudolph and the participants of the online survey. Furthermore, even when businesses already contribute to water-related solutions, it has been noted that improvement is required when combining suitable water management and business strategy approaches. On the issue of payment of ecosystem services (PES), developing countries can also learn from experiences gained with institutional prerequisites in developed countries. Here, it has been shown that the durability of PES is rather given if a small group of stakeholders is involved in its design and leadership.

As for methodological approaches, combined hydro-economic modelling in river basins, multi-criteria analysis and cost-benefit-analysis remain vital tools for answering questions on financially sustainable solutions but these require further development. The development and implementation of adequate financing solutions remains a further challenge for the future. The annual worldwide investment in water infrastructure required for solving water-related problems currently stands at about 400–500 billion Euro. Although financial resources may often be available, the problem remains that the threshold for financing projects is quite high so that many smaller projects ultimately lose out on funding. A further challenge is to improve on the extent to which the actual refinancing potential of local populations is taken into account, as infrastructure refinancing must be guaranteed in order to support financially sustainable solutions. Finally, societal problems with IWRM financing have been underlined: In some regions of the world, the willingness to pay for water is less pronounced due to historically conventional behavior. These mindsets must also be taken into consideration.

3. LINKS BETWEEN THE CONFERENCE TOPICS

The conference participants made it clear that no single IWRM topic alone can solve water-related problems such as bad water quality or water shortages. Technical innovations will not persist if institutional prerequisites, financial sustainability and the relevant capacities are not in place. Technical innovations will not be sustainable if they have negative effects on water resources such as groundwater, or on the general public – hidden impacts that are predictable with adequate decision support tools. This example illustrates what one anonymous conference participant noted in response to the governance issue, that "we need to think (water governance) out of the box".

The topics that were interlinked at the conference have been identified below, generally based on the information given by the junior rapporteurs. Some overlaps between the different topics of the IWRM conference are apparent: The topics of Water Resources in Changing Environments (WRCE) and Information and Decision Support Systems (IM/DSS) both include aspects of modelling and decision support. Overlapping is also apparent in relation to groundwater and other topics such as WRCE, IM/DSS and technologies as technologies and decision support tools can be used in both ground and surface water management. The topic Technologies and Implementation also highly overlaps with implementation research and practices that were presented during the sessions on Capacity Development, Governance and Economic Instruments as all of these are part of the implementation process. Finally, the topics Governance and Economics may highly interlock depending on the understanding of the term 'governance'.

Despite these overlaps, many links and the necessity for cooperation between the different topics were identified and are presented below. The focal point of the discussions was the respective topic as it is perceived by representatives of other disciplines. This information has been sporadically complemented by links made by representatives of the respective topic. The following shows that thinking outside the box is generally the rule, rather than the exception.

3.1 Technologies and Implementation

Many links between Technologies and Implementation and other IWRM conference topics have been identified. Firstly, predication & simulation software was seen as relevant for the modelling of salt dynamics, and a link to WRCE but also to economic modelling was thus established. Economic links have been made whilst highlighting the fact that all technologies must be embedded in a socio-economic context, and that financing methods must be considered even in the R&D phase. This aspect seems to be commonly underestimated in current technological developments. Strong links to governance issues were also put forward by those participating in the technical sessions as well as in the online survey, both highlighting that governance ultimately sets the general scene for the implementation of innovative technologies. Also, a high degree of stakeholder participation in the designing process has been stressed in order to support sustainable solutions. Here, stakeholders from local but also from higher levels must be involved and their objectives identified. Finally, the relevance of Capacity Development has been underlined in order to guarantee long term solutions. The majority of those participating in the online survey strongly agreed with the assumption that the absence of capacities is one of the main reasons for the unsatisfactory implementation of IWRM-related measures.

3.2 Water Resources in Changing Environments

The topic of Water Resources in Changing Environments has been predominantly linked to modelling (including economic modelling). The model results are often the basis for technical solutions, and information management and decision support tools for both ground and surface waters. Furthermore, a strong link to Capacity Development has been established as the need for the training of trainers has been stressed in order to support adequate information management.

3.3 Information and Decision Support Systems

Information and decision support tools rely on (improved) data as well as information and modelling techniques provided by ground and surface water specialists (see WRCE, GW, Economics). When a broader perspective of the term 'economics' is used, decision support tools become necessary when choosing optimal management strategies. The importance of participatory modelling and management paradigms was stressed in relation to governance, to be of use for the analysis of complex social-economical systems and to guarantee decision acceptance. Participants in the sessions on decision support tools also lamented the effects of inadequate legislation and ineffective administration. Scientists also deemed adequate user knowledge to be an essential aspect, as insufficient knowledge also leads to the misuse of DS-systems.

3.4 Capacity Development

Capacity Development was mentioned in almost every session at the IWRM conference but the perspectives on CD were quite different, depending on the topic of the session. Asked about the necessity of CD, participants in the technological sessions emphasized the importance of CD in the long-term maintenance of technologies. Those attending the sessions on WRCE and IM/DSS stressed that CD is necessary for the adequate management of information and the subsequent efficient use of project results in the short and long term. Economists maintained that CD is necessary for local governments in order to be able to provide the necessary political framework for businesses. Moreover, it was noted in technological sessions that CD measures can be undertaken by different people, and it was also emphasized that scientists are an important group in the organization and implementation of CD measures. Further, different target groups for CD measures were discussed: individual users (technological session), beneficiaries (economic session), trainers (WRCE), local governments (economic session) and institutions (governance session) were named in particular. Workshops and on-the-job-training courses were suggested among others as tools during the technological session.

3.5 Water Governance

Governance issues were also very prominent in various sessions. The conference participants discussed both general governance issues and specific governance-related aspects. The point was made that governance sets the general scene for implementation (during the technological session), that it must be improved through institutional Capacity Development (CD session) and that there is a strong connection between politics and economics (economic session). The relevance of participation was particularly highlighted (technological session, IM/DSS, Groundwater Management and Economic Instruments). In technological sessions, a high degree of e.g. local and regional stakeholder participation in the designing process was seen as crucial in supporting sustainable solutions. The relevance of participatory modelling (IM/DSS) and participation in the design of measures (economics), e.g. for the acceptance of decisions, were also stressed. Absent legislation and ineffective administration (IM/DSS) in addition to differences in the legal frameworks of different countries (groundwater) were also lamented. Groundwater experts clarified that they do not support "top-down provisions" whereas economists search for the necessary political framework for business involvement.

3.6 Groundwater Management

The resource groundwater was interlinked to all other sessions. A special focus during the groundwater sessions was placed on modelling and technological innovations. There is also an important link to information and decision support systems. Groundwater modelling, the simulation of river-groundwater interactions and the striving for adaptive groundwater management strategies are important links to the topic of IM/DSS. Groundwater specialists also highlighted economic issues such as the financial risks of certain methods and governance issues such as legal frameworks. In his key note speech and report on groundwater governance, Héctor Garduño underlined the importance of participation for sustainable groundwater management, referring to the enabling rather than organisation of society. He simultaneously warned that "participation is not the magic bullet", as it must be based on sound hydrogeology, complemented by adequate regulation and economic incentives. Groundwater experts also highlighted the importance of Capacity Development, e.g. in the form of training for trainers in order to address the 'SUR' challenge which stands for sustainability, upscalability and replicability. Generally a multidisciplinary approach is desired for solving ground water-related problems. Finally, participants in the online survey stressed the necessity of linking further research on ground and surface waters.

3.7 Economic Instruments

Economic instruments were also well interlinked with the other conference topics. It was stressed that technologies must be embedded in the current and future socio-economic context and that the issue of financing must be considered early in the research & development phase. In the sessions on Water Resources in Changing Environments and Information and Decision Support Systems the importance of integrating economic variables into ecosystem modelling and decision support tools was underlined. Groundwater specialists pointed out the financial risks of certain technologies. Governance specialists discussed the structuring of payment for watershed ecosystem services and drew the general conclusion that there is a high degree of connection between politics and economics. Economic experts also stressed the importance of Capacity Development for beneficiaries when implementing measures and also for local governments by providing the necessary political framework for businesses. Finally, the leading rapporteur of the topic economics, Prof. Dr. Dr. Karl-Ulrich Rudolph, stated that "economics is not the exclusive property of economists" and thus requested an intensive exchange between economists and other disciplines.

4. FUTURE CHALLENGES FOR THE MANAGEMENT OF WATER

Beside the above mentioned issues, the conference participants discussed core challenges for the future – or "icebergs" using the conference chair Prof. Dietrich Borchardt's metaphor. Some of these core challenges for sustainable water resources management will be highlighted below. As indicated by the IWRM conference steering committee, they demonstrate the relevance of closer cooperation between all water relevant sectors and levels.

4.1 Land Use

Land management issues are particularly challenging with regard to a sustainable water resources management. The Steering Committee placed particular emphasis on the "strong linkages but also substantial trade-offs between water security" and food security and that "IWRM should be seen as pathfinder process for the implementation of an Integrated Resource Management."

The following links between sustainable water and land management were mentioned. In the session on Water Resources in Changing Environments questions were raised about the effects of environmental changes such as land use change, agricultural intensification and irrigation on water resources and models were presented that integrate hydrological and land use aspects, amongst others. In DSS-sessions, the challenge of uncertainty under land use change was particularly highlighted. Participants in these sessions also stressed that multicriterial optimisation algorithms are necessary for contradicting objectives such as profit-oriented agriculture versus aquifer sustainability. The high relevance of land use dynamics with regard to groundwater was particularly stressed. A core issue was the sustainable extraction of groundwater for agriculture, under the consideration that the largest abstractions of groundwater are made for irrigated agricultural purposes. Further, the Conference Steering Committee stated the technological perspective on the links between water and land management: "IWRM based infrastructures typically serve multi-purpose schemes (e.g. water storage schemes for producing energy or food and mitigation of extreme events such as floods and droughts)." Finally, it has been emphasized that the polluter-pays principle must be applied to land management issues as this is not currently the case.

4.2 Climate Change

Climate change is a core challenge for Integrated Water Resources Management. Participants in the sessions on Water resources in Changing Environments were particularly interested in how environmental changes such as climate change effect water resources in different spatial entities from the basin to the continental scale. In sessions on decision support systems, the relevance of uncertainty under climate change was stressed. Participants in the sessions on governance also discussed how governance features can contribute to adequate adaptation to climate change.

4.3 Energy

Energy supply is another critical challenge for sustainable water resources management. On this topic, the Conference Steering Committee's position was that there are "strong linkages but also substantial trade-offs between water security" and energy security. Questions raised in this context referred particularly to the effects of dams and power plants on water resources. The negative effects on water resources were highlighted in addition to the positive aspects of "IWRM based infrastructures that typically serve multi-purpose schemes", e.g. "water storage schemes for producing energy or food and mitigation of extreme events such as floods and droughts". A higher efficiency of energy producing technologies was also desired. In general, the participants in the online questionnaire strongly agreed that cooperation between experts from the water and energy sectors needs to be intensified.

4.4 Urbanization

Urbanization is one of the most significant future challenges mentioned at the IWRM conference. One conference attendee pointed out that by 2050, more than 80% of the world's population (more than 6 billion people) will live in cities. Participants asked how it will be possible to facilitate water supply and wastewater disposal in fast growing cities and especially in mega cities. The effects of urbanization on water resources were also discussed, particularly during the sessions on Water Resources in Changing Environments.

4.5 Demographic change

Demographic change in developing, emerging and industrialised countries is another important challenge that was addressed during the conference. In developing and emerging countries, continued population growth increases the demand for water and food. Prof. Dr. Dietrich Borchardt noted that "water is running out for food production" which raises the question of how water-efficient food production can be achieved (see also the challenge of land use change). In industrialized countries, it was emphasized that technologies must be adjusted in order to a) reduce water demand, particularly in rural areas, and b) remove pharmaceutical residues.

5. IMPLEMENTATION OF IWRM SOLUTIONS AND SCIENCE-POLICY INTERFACES

During the IWRM conference, research results on the main IWRM topics were presented, links between these topics were established and core challenges for the future were formulated. Nonetheless, the implementation of scientific results into practice remains a core requirement for the future. The IWRM Conference Steering Committee stated that "while considerable progress has been made to include IWRM in national policies, strategies and laws, the actual implementation of IWRM is lagging behind". The IWRM conference chair, Prof. Dietrich Borchardt asked in his key note speech: "Do we really have so much time for planning? Or do we have to speed up with action?" The Steering Committee answered this question clearly stating that the realisation of IWRM solutions must be accelerated as the dynamics of change "are fast and already lead to irreversible damage to water resources in many regions of the world". The participants of the online survey also tended to agree that the implementation of measures is slow in comparison with the dynamics of driving forces and pressures resulting from land-use change, demographic change and resource use. Aspects of water supply and sanitation in developing and emerging countries and water quality aspects in industrialised countries are core issues that must be addressed.

In order to strengthen the implementation of IWRM solutions, the Steering Committee stressed that IWRM research approaches must be "transdisciplinary with a substantiated science policy interface". In other words, successful IWRM "works with an intense dialogue between governmental institutions, science, NGOs and society in order to achieve more sustainable solutions", but must especially bridge the gap between science and political decision makers in order to support the transfer of results to applied IWRM. This desire was expressed in plenary discussions, in the special sessions and by the participants of the online survey.

PD Dr. Heike Walk referred to transdisciplinary projects in particular when she suggested that the BMBF should further support the strategic exchange between scientists and people outside the scientific community such as authorities and NGOs by supporting networks and workshops. Several methods of integrating science into practical work have been proposed in order to foster the exchange between politicians and scientists. Adapted decision support tools are, amongst others, vital tools for bringing IWRM strategies to governmental decision makers. These systems facilitate the transfer of scientific results to decision makers and can thus inform and influence political decisions. According to Dr.-Ing. Martina Flörke, scientists must consequently make it clear that they cannot provide predictions and they must continuously communicate the uncertainties of developed models. In the online survey it was further agreed that there is a strong need to clarify the conditions under which decision support tools are used in decision-making processes. Implementation research and governance analysis in particular also remain essential to the implementation of IWRM solutions.

The interaction between scientists and politicians remains a challenge due to the different logic prevalent amongst the two societal groups. Whilst politicians have governmental targets and restrictions to consider, scientists must still work for scientific credibility that is not given for good management consulting. Scientists also tend to think in the long-term whereas politicians need direct impacts with immediate economic benefits within short periods. A further challenge is that scientists and politicians need to find a common language for communicating scientific results and transferring knowledge. Referring to this interaction between science and policy, Peter Reichert appealed to scientists in his key note speech, asking them to "stimulate integrative thinking (…) even if sectoral management prevails". Finally, at the conference it was underlined that IWRM is a philosophy rather than a template to be followed step by step. The IWRM approach is not too complex but too general and abstract to be easily implemented. The best approach, according to the participants seems to be to focus on local conditions, and to support long-term reforms which encompass the different topics, interactions and challenges addressed in this conference and beyond.

ACKNOWLEDGEMENTS

This document has been produced by the organising committee of the IWRM conference. We would like to thank all of the conference participants for their presentations and for contributing to the discussions. Special thanks go to the leading and junior rapporteurs involved in each of the topics, namely Univ.-Prof. Dr.-Ing. Jörg Londong, Kerstin Matthies and Jörg Seegert for "Technologies and Implementation"; Dr.-Ing. Martina Flörke, Tim aus der Beek, Michael Schäffer and Michael Strauch for "Water Resources in Changing Environments"; Prof. Dr. Stefan Kaden, Stefan Lorenz and David Riepl for "Information and Decision Support"; Dr. Steffen Niemann for "Capacity Development"; PD Dr. Heike Walk, Dr. Ross Beveridge and Nina Hagemann for "Water Governance"; Hector Garduño, Felix Grimmeisen and Sebastian Schmidt for "Groundwater Management" and Prof. Dr. Dr. Karl-Ulrich Rudolph, Michael Harbach and Dr. Paul Lehmann for "Economic Instruments". Last but not least, we would like to thank Dr.-Ing. Ilona Bärlund for having chaired the rapport session at the end of the IWRM conference.

Theme I

Water resources management in changing environments

Pan-European freshwater resources in a changing environment: how will the Black Sea region develop?

M. Flörke, I. Bärlund, C. Schneider and E. Kynast

ABSTRACT

Climate change and socio-economic driving forces will affect Europe's future freshwater resources. A large-scale water model is used to analyse these effects and to identify 'hot spots' of water stress in the Black Sea region, as an example of an area where future water demand is expected to exceed the available water resources. Two scenarios are analysed, describing different developments of water withdrawals. Depending on the scenario, water stress increases or decreases due to changing water withdrawals which are identified as the principal cause of additional water stress in the future. According to the 'economic-oriented' pathway, water withdrawals are expected to increase by 58%. In Turkey and Bulgaria where water is already scarce, a further decrease in water availability will exacerbate the situation. By contrast, the 'quality of life oriented' scenario, assuming raised awareness to save water, results in a reduction of water withdrawals by approximately 59%. The situation of decreasing availability and increasing demand leads to growing competition between users and may finally end in cross-sectoral conflicts. This type of modelling study helps to prepare and foresee which kind of management options (in which sectors especially, and where) would be required to reduce ecological, economic and social consequences.

Key words | Black Sea region, climate change, scenarios, SCENES, water stress, water use

M. Flörke (corresponding author)
C. Schneider
E. Kynast
Center for Environmental Systems Research,
University of Kassel,
Wilhelmshöher Allee 47,
34117 Kassel,
Germany
E-mail: *floerke@usf.uni-kassel.de*

I. Bärlund
Helmholtz Centre for Environmental Research – UFZ,
Brückstrasse 3a,
39114 Magdeburg,
Germany

INTRODUCTION

European freshwater resources are expected to change over the next decades due to climate and socio-economic impacts as well as political developments. Regarding climate change, the magnitude and direction of change differs between regions and seasons of the year and may have both positive and negative impacts (Bates *et al.* 2008). The distribution of freshwater resources in Europe shows a distinct north-south gradient (EEA 2004) and several studies indicate that Southern Europe will be more severely affected by climate change. However, freshwater resources are not only influenced by climate change but also by changing water withdrawals (water use). The assessment of future developments in water use is extremely important to foresee whether 'enough water for all' will be available or whether water will lead to conflicts. This in turn triggers the development of adaptive management strategies, like integrated water resources management, which need to be implemented in order to ensure a sustainable use of the available resource. A prolonged period of water shortage or even a water deficit leads to shortfalls in domestic water supply and economic losses in the industrial and agricultural sectors. An increasing and water demanding economy as well as population developments, e.g. population growth and urbanisation, will amplify the situation in the future and exacerbate the impact on water availability and freshwater ecosystems. Whether water is or will become a scarce resource depends on the balance of water demand versus water availability, and in addition on timing.

But what will Europe's freshwaters look like in the future? They will be influenced by a combination of

doi: 10.2166/ws.2013.027

environmental and socio-economic drivers as well as by policy and technology responses affecting them. In the 'Water Scenarios for Europe and Neighbouring States' (SCENES) project a set of both qualitative and quantitative scenarios were developed to describe freshwater futures up to 2050 (Kämäri *et al.* 2008). The overall scenario approach is based on the Story-and-Simulation Approach (Alcamo 2008) linking storyline development and modelling work in an iterative process involving stakeholders during consecutive workshops (Kok *et al.* 2011).

The main aim of this paper is to promote an understanding of Europe's freshwater resources by analysing the impact of climate change and changing patterns of water use, both in quantity and in sectoral diversity, on water scarcity. Water scarcity can result from intensive water use, low (climate driven) water availability or a combination of these. Water stress is a measure indicating the pressure put on water resources and aquatic ecosystems by the users of these resources, including municipalities, industries, thermal power plants and agricultural users. One of the most important indicators for quantifying water stress is the withdrawals-to-availability ratio (w.t.a.) on a river basin scale. This indicator has the advantage of being transparent and computable for all river basins and it has been used in several studies (Cosgrove & Rijsberman 2000; Vörösmarty *et al.* 2000; Alcamo *et al.* 2007). Higher water stress means stronger competition between society's users and between society and ecosystem requirements (Raskin *et al.* 1997). A drainage basin is assumed to be under low water stress if w.t.a. ≤ 0.2; under medium water stress if $0.2<$ w.t.a. ≤ 0.4 and under severe water stress if w.t.a. >0.4.

Climate change impacts are expected to intensify existing competition for water resources between the sectors. Bringing these factors explicitly into a European analysis of water stress can provide new insight into the relative importance of different elements of change and especially on future water management. We analyse two SCENES scenarios, which project partially contrasting assumptions on the development of driving forces. In particular, we identify and compare future 'hot spots' of change that should be monitored and studied further in small-scale studies.

METHODS

SCENES pan-Europe covers 52 countries which were grouped into seven regions according to the UN World regions: NA (Northern Africa), WE (Western Europe), NE (Northern Europe), SE (Southern Europe), EEc (Eastern Europe, central), EEe (Eastern Europe, eastern), and WA (Western Asia). Future trends in and magnitude of change of selected key drivers were retrieved for these regions and model results were summed up to this scale. For this study model results were extracted from the pan-European view and prepared for the area of interest, i.e. the Black Sea region. The assessment was carried out in several steps. First, two SCENES scenarios and the climate input were selected. Second, future annual average water resources were simulated on a monthly basis taking into account climate change and water use. Third, using time series on future socio-economic and other drivers from the SCENES scenarios, water withdrawals were calculated for five different water-related sectors in pan-Europe. Fourth, using the results of water withdrawals and water availability, 'hot spots' were identified where future water demand may not be fulfilled and where cross-sectoral conflicts may occur.

Overview of scenarios applied

For this paper, two SCENES scenarios that span a broad variety of how the future may unfold were chosen to visualise possible futures for Europe's freshwater resources. 'Economy First' ('EcF') is an economic-oriented scenario developing towards globalisation and liberalisation characterised by intensified agriculture and a slow diffusion of water-efficient technologies. Global demand for food and biofuels drives the intensification of agriculture with increasing need for irrigation and new cultivation area. Slow adoption of water-efficient technologies and low water-saving consciousness lead to higher water use. 'Sustainability Eventually' ('SuE') is a scenario that sketches the transition from a globalising, market-oriented Europe to environmental sustainability where quality of life becomes a central point. Economic aspects are important but growth is slow. Improvements in technology lead to increases in water use efficiency and investments in water-related R&D activities are initiated to share technological benefits within Europe. Water demand is strongly reduced by water savings and behavioural changes.

In order to take into account climate change, the IPCC SRES A2 scenario (IPCC 2007) was selected covering the whole time horizon up to the 2050s. The SRES A2 scenario describes a very heterogeneous world with high population growth, slow economic development and slow technological change. Within SCENES, climate change input from two different global circulation models (GCMs) was used to take into consideration uncertainties arising from climate modelling.

Modelling future European water resources

For the quantification of the pan-European SCENES scenarios and to compute the impact of climate change and other important driving forces on future water resources, the WaterGAP (Water – Global Assessment and Prognosis) water model was used (Alcamo *et al.* 2003; Döll *et al.* 2003; Flörke & Alcamo 2004). WaterGAP, developed at the Center for Environmental Systems Research, is designed for large-scale grid-based applications and its capabilities to simulate water availability and water use in scenario assessments is well tested (e.g. Global Environment Outlook GEO-4 (Rothman *et al.* 2007), State of the European Environment (EEA 2005), Millennium Ecosystem Assessment (Alcamo *et al.* 2005)). The model version applied in SCENES, WaterGAP3, computes both water availability and water uses on a 5 by 5 arc minutes grid (longitude and latitude; grid cell sizes of 6 × 9 km in Central Europe).

WaterGAP3 consists of two main components, a Global Hydrology Model (Verzano 2009) to simulate the terrestrial water cycle and a Global Water Use Model (Flörke & Alcamo 2004; aus der Beek *et al.* 2010; Flörke *et al.* 2011) to estimate water withdrawals and water consumption of the domestic, thermoelectric, manufacturing and agricultural sectors. Domestic, thermoelectric, manufacturing and livestock water uses are calculated annually; irrigation requirements were calculated on a monthly time step and then summed up to annual values. The aim in using the Global Hydrology Model was to simulate the characteristic macro-scale behaviour of the terrestrial water cycle to derive the annual average water availability for pan-Europe. Herein, water availability is defined as the total river discharge, which is the sum of surface runoff and groundwater recharge. In a standard model run, river discharges in approximately 180,000 grid cells (approximately 2,000 river basins >140 km^2 drainage area) in Europe are simulated. Natural cell discharge is reduced by the consumptive water use as calculated by the Water Use Model. For most water use sectors, except irrigation, only a small part of water is consumed, whereas most of the water withdrawn is returned to the environment for subsequent use. Water use for the agricultural and electricity production sectors are calculated on a 5 by 5 arc minutes grid scale, but for domestic and manufacturing sectors on a country scale. These country-scale estimates are then downscaled to the grid size within the respective countries by using generic downscale algorithms.

Main driving forces

Climate input

The baseline climate input including monthly information on precipitation and temperature covered the time frame 1961–1990. For the model simulations a combination of the datasets CRU TS 2.1 (Mitchell & Jones 2005) and CRU TS 1.2 (Mitchell *et al.* 2004) was used. To take into account the uncertainty in climate modelling, the climate input from two different GCMs was analysed: (1) the IPSL-CM4 model from the Institute Pierre Simon Laplace, France ('IPCM4') representing an A2 scenario – this scenario indicates high temperature increase over large parts of Europe, i.e. up to 5 °C in Northern and Eastern Europe, and low precipitation increase or decrease in Europe; and (2) the MIROC3.2 model from the Center for Climate System Research, University of Tokyo, Japan ('MIMR') representing an A2 scenario – in accordance with the IPCM4 model, the MIMR model projects a high temperature increase over Europe but in combination with a high precipitation increase or low decrease.

Population and GDP

Future trends in population and economic activity show significant differences in the rate and direction of change between the scenarios. In pan-Europe only NA and WA have a significant increase in population while a contrary trend occurs over the whole scenario period in EEc and

EEe. Changes in population are highest in EcF and lowest in SuE. A further increase of population by 8.5% (67.5 million people) is projected for the EcF scenario in Europe until 2050. On the other hand, after a short period of growth (until 2015) population is expected to decrease by 6.1% (36 million people) between 2005 and 2050 in SuE. In Europe, economic activity is highest in EcF and continues to grow over the whole scenario period resulting in a doubling of GDP. SuE is generally characterised by a slowdown in economic activity between 2005 and 2050 with an increase of 9.9% between 2005 and 2050.

Total and thermal electricity production

Both total and thermal electricity production are projected to increase twofold in Europe under the EcF scenario between 2005 and 2050. Thermal electricity production is remarkably expanded and increases further as the use of fossil fuels and nuclear energy sources becomes entrenched. At the end of the period, 87% of the total electricity production is intended to be generated by thermal power plants. A contrary picture is drawn by the SuE scenario where total electricity production increases by 14% over the entire time period and thermal electricity by 3.8%, respectively. Between 2015 and 2030 the shift towards renewable energy begins, resulting in a declining share of thermal generation from fossil fuels. Fossil fuel production still moves to nuclear, meaning the share of thermal production changes little. However, renewable energy sources start to take over in this scenario, so that thermal shares drop until the end of the time period (2050). Finally, 70% of total electricity production is generated by thermal power plants.

Irrigated area

Using the IPCM4 climate change input, an increase in irrigated area was simulated for the EcF and SuE scenarios in pan-Europe, but with a large discrepancy between the two (+73 and +8.5%, respectively). In EcF, irrigated area is increasing in all regions with hot spots being located in WE (+277%), NA (+80%) and EEc (+75%). In contrast, in SuE the extent of irrigated area is declining in EEe, SE, and WA. The MIMR climate change input in combination with the EcF socio-economic scenario resulted in a similar but less pronounced extension in irrigated area (+54% in total). Compared to the previous model runs, a slight decrease of irrigated area in Southern Europe can also be observed. The MIMR input together with the SuE scenario resulted in area increases in NA and WE while irrigated area is shrinking in all other regions.

Structural and technological developments

Additional driving forces determine how water use intensities alter due to structural and technological changes. Here, the concept of technological change accounts for the effect of improving technology which makes appliances more water efficient in the future, hence contributing to reductions in water use. In EcF, technological change rates vary between 0.3 and 0.6% per year, depending on the time frame and regional developments. Future estimates on the development of technological change are more consistent in SuE, assuming 0.6% per year between 2000 and 2025 and 1.2% per year after 2025 until 2050 in Europe. For the EcF scenario, project efficiencies for irrigation water withdrawals are constant or decreasing over the time period (in NA and WA), except for SE, EEc and EEe where an increase was considered. A contrary picture is drawn for SuE, where project efficiency is likely to increase in all regions.

Structural changes also have an influence on water use, leading to further reductions. In the domestic sector, structural changes are related to people's commitment to save water. Overall, the commitment to save water is higher in SuE compared to EcF, reflecting the uptake of new technologies and reductions in personal water use through educational programmes and raising awareness about environmental impacts. For the electricity production sector it was assumed for the SuE scenario that older power stations with once-through cooling systems were replaced by new ones with a tower cooling system. The lifetime of a power plant was 40 years in water-poor countries and 50 years in water-rich countries. No structural changes were assumed in the EcF scenario.

Identification of hot spots

Hot spots were identified on a river basin scale by investigating whether average annual water availability would satisfy the average annual water demand in the future. Here the concept of 'water stress' was used. Due to variations in water availability and water withdrawals within a year, water stress is likely to vary during the year, too; this momentary water stress was not considered in the approach applied.

RESULTS AND DISCUSSION

Future hot spots (water stress)

The water stress maps for the two SCENES scenarios under the IPCM4-A2 and MIMR-A2 climates show stressed areas for all scenarios (Figure 1) but the area is largest under EcF covering large parts of Europe and the Northern African coastline. The EcF scenario combined with the IPCM4-A2 climate, partially also under MIMR-A2, shows some stress in the more northern parts of pan-Europe such as the Baltic region even if water shortage is not perceived as an issue there. It is interesting to notice that water stress is highest under EcF but decreases tremendously in SuE under both climates. The main reason for this decline is the decreasing water withdrawal due to technological, structural and behavioural changes as well as socio-economic developments. This means that adaptation measures and policies are required throughout the scenarios and throughout all regions to reduce water stress.

It is obvious from Figure 1 that water is most likely to become a scarce resource in the Mediterranean region. This is in agreement with other studies, focussing e.g. on droughts. According to hydrological model simulations,

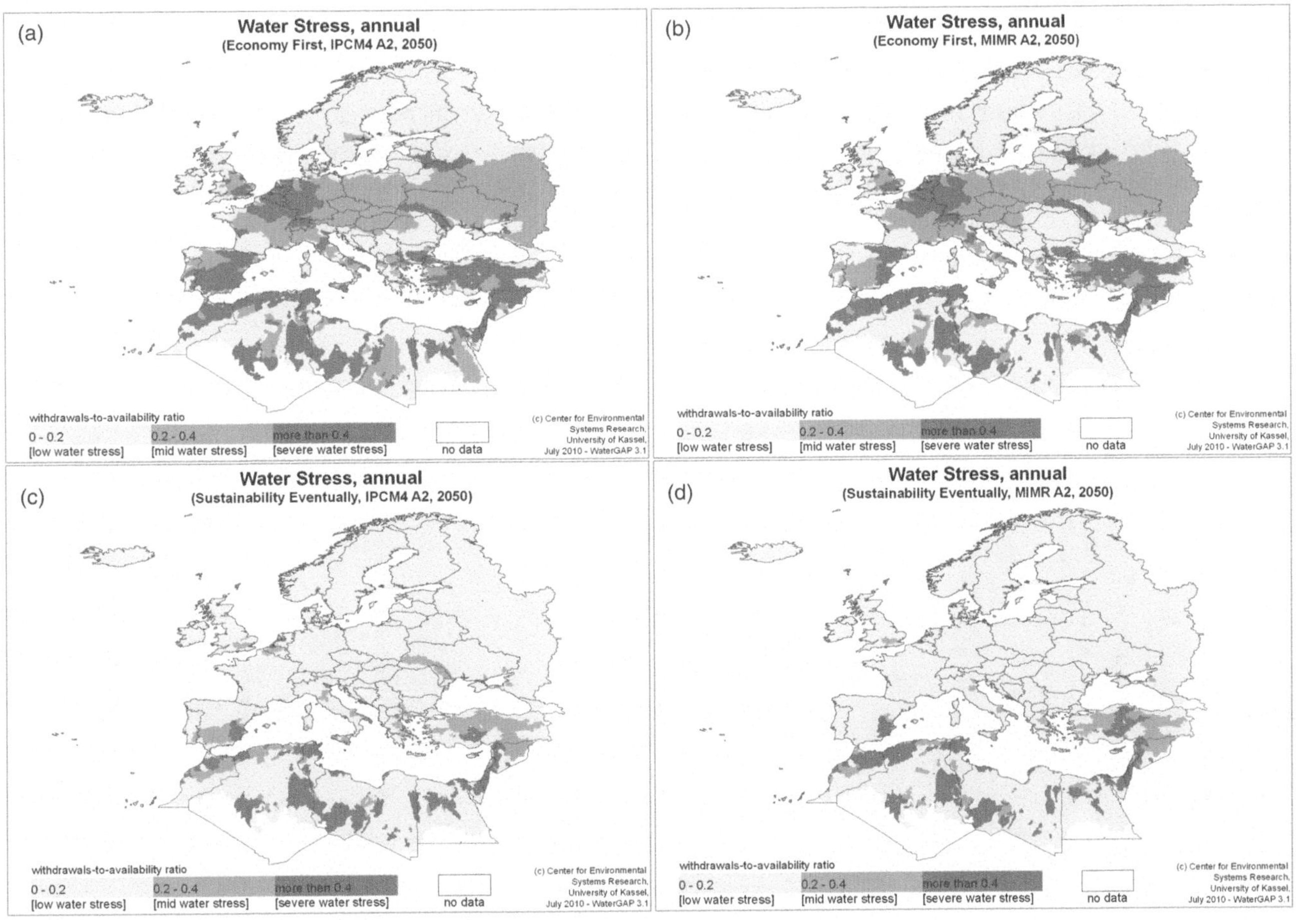

Figure 1 | Water stress in the 2050s based on the withdrawals-to-availability ratio in pan-Europe for the climate change and socio-economic scenario combinations: (a) EcF scenario and IPCM4-A2 climate, (b) EcF scenario and MIMR-A2 climate, (c) SuE scenario and IPCM4-A2 climate, and (d) SuE scenario and MIMR-A2 climate.

more frequent, severe and persistent stream flow droughts are to be expected especially in Southern Europe in the coming century (Weiß *et al.* 2007; Feyen & Dankers 2009). Next to the Mediterranean, the Black Sea region also appears as a water stressed area in the future. Thus we focus next on this region in order to perform a more detailed investigation.

Future water availability

Changes in precipitation will increase or decrease the average river runoff volume. Meanwhile, the expected increase in air temperature intensifies evapotranspiration nearly everywhere, and hence reduces runoff. These two effects interact differently at different locations and produce a net increase or decrease in water availability. Over most of the Black Sea rim countries, water availability is decreasing but the intensity and direction of change depends on the GCM (Figure 2). IPCM4-A2 and MIMR-A2 based calculations result in a decrease in annual water availability, especially in Turkey, up to more than 30% compared to the baseline (1961–1990). In contrast to IPCM4-A2, MIMR-A2 results show no change or even an increase in annual water availability in the northern rim countries of the Black Sea region. The results obtained are in agreement with other studies (Milly *et al.* 2005; IPCC 2007; Feyen & Dankers 2009) where decreases in water availability and increases in water scarcity are reported.

Future water use

Water withdrawals and consumption were calculated for the base year (2005) and the two scenarios (2050). Figure 3 shows the development in sectoral water uses (left diagram) and the percentage share of each sector of the total water withdrawals (right diagram) for the Black Sea region. In the base year approximately 130 km^3 of water was withdrawn from the region's freshwater reservoirs for domestic, manufacturing, thermoelectric and agricultural purposes. The agricultural sector is the major water user in the study region accounting for about 47% of the total water withdrawn in the base year, followed by water withdrawn for cooling purposes (26%) and the domestic (18%) and manufacturing (9%) sectors. Looking into the future, the EcF and SuE scenarios will result in remarkably different water withdrawals in the Black Sea region. Under the EcF scenario an increase in total water withdrawals of 58% is estimated whereas the SuE leads to a decrease by more than 59% mainly due to water savings in the energy sector. The direction of change is the same for each water-related sector within the scenario: sectoral water uses either increase (EcF) or decrease (SuE). However, sectoral

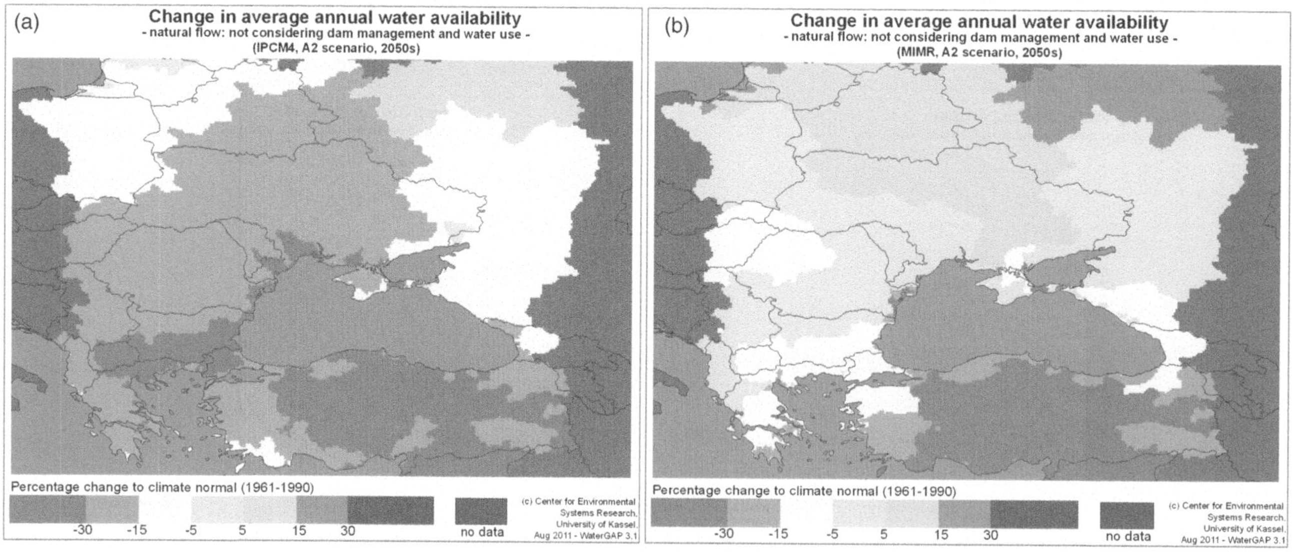

Figure 2 | Annual average renewable water availability for the 2050s under different climate conditions on a river basin scale. Map (a) climate input based on IPCM4-A2, and (b) climate input based on MIMR-A2.

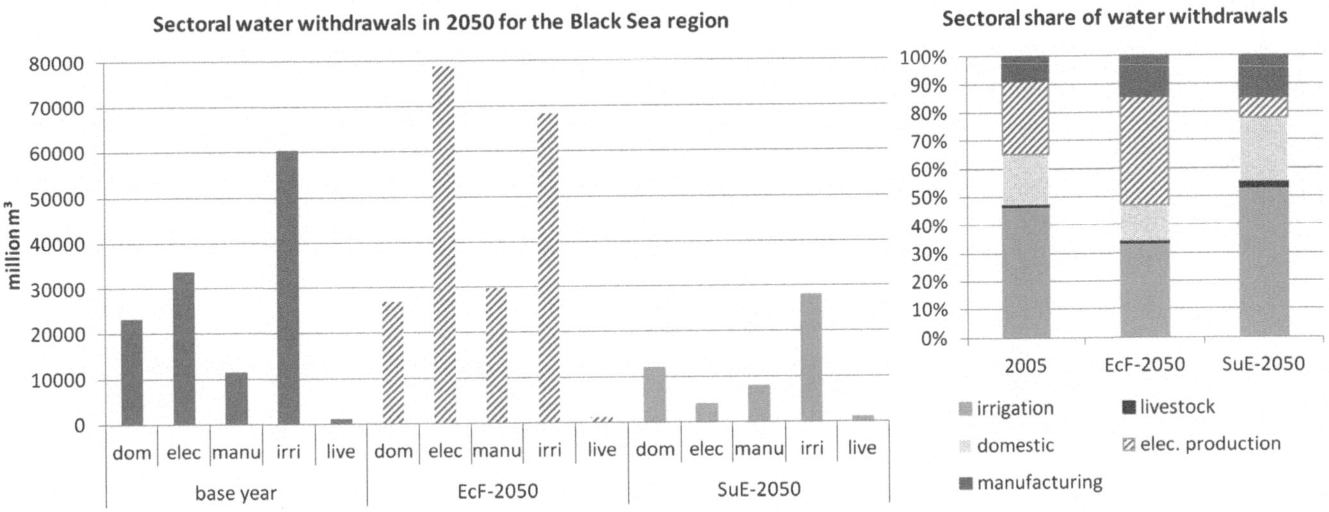

Figure 3 | Computed sectoral water withdrawals (left) and sectoral shares (right) in the Black Sea region for the years 2005 and 2050 for the EcF and SuE scenarios.

water withdrawals evolve over time in different directions depending on the scenario. Under the EcF scenario, water withdrawals of the manufacturing and thermal electricity production sectors will more than double and account for more than 50% of the total water withdrawn. Moderate changes are simulated for the domestic (+15%), livestock (+14%) and irrigation water withdrawals (+13%) by 2050. The increases under EcF result from an assumed water use behaviour that follows traditional patterns accompanied by reluctantly adopted scientific and technological innovations and slow improvements in water use efficiency. Irrigation water withdrawals are mainly driven by changes in irrigated area, which is assumed to increase in the Black Sea rim countries except Greece, accompanied by climate change impacts. Moreover, a strong growth in the generation of thermal electricity leads to a significant increase in cooling water. In SuE the highest reduction in water withdrawals is calculated for the industrial sector, which is contrary to the EcF scenario. Here, water abstractions are expected to decrease by 89% in the thermal electricity production sector, which is mainly driven by the assumption of changing technologies, i.e. cooling systems, and a reduction in thermal electricity production. The results for the manufacturing sector (−31%) are mainly influenced by a slow economic development leading to a reduced manufacturing output. Water withdrawals are expected to decrease for irrigation purposes by 54% and in the domestic sector by 48%, respectively, due to a combination of behavioural and technological changes with the promotion of technology transfer and efficiency improvements as well as changes in the extent of irrigated areas.

Although sectoral water withdrawals may increase or decrease in the future, the profile of water use is likely to change, too. In the EcF scenario, the industrial sector (thermoelectric and manufacturing sectors) is expected to be the major water user in 2050 and may account for 53% of the total water abstracted. Analysing the picture in more detail, it becomes obvious that cooling water needs of the energy sector are highest, accounting for 38% (manufacturing 15%). This presumably happens at the expense of other sectors like agriculture (sum of irrigation and livestock) which accounts for around 34%. The share of the domestic sector decreases slightly reaching 13% of the total water abstracted in EcF. By contrast, the results for SuE show that the agricultural sector will be the main water user (53%) in 2050, followed by the domestic (23%), manufacturing (15%) and electricity production (7%) sectors.

Cross-sectoral conflicts

Cross-sectoral conflicts occur if different water-related sectors compete for the same (scarce) resource, in particular freshwater. Climate change impacts might intensify existing competition for water resources between the sectors in the future. Figure 4 indicates river basins that are expected to

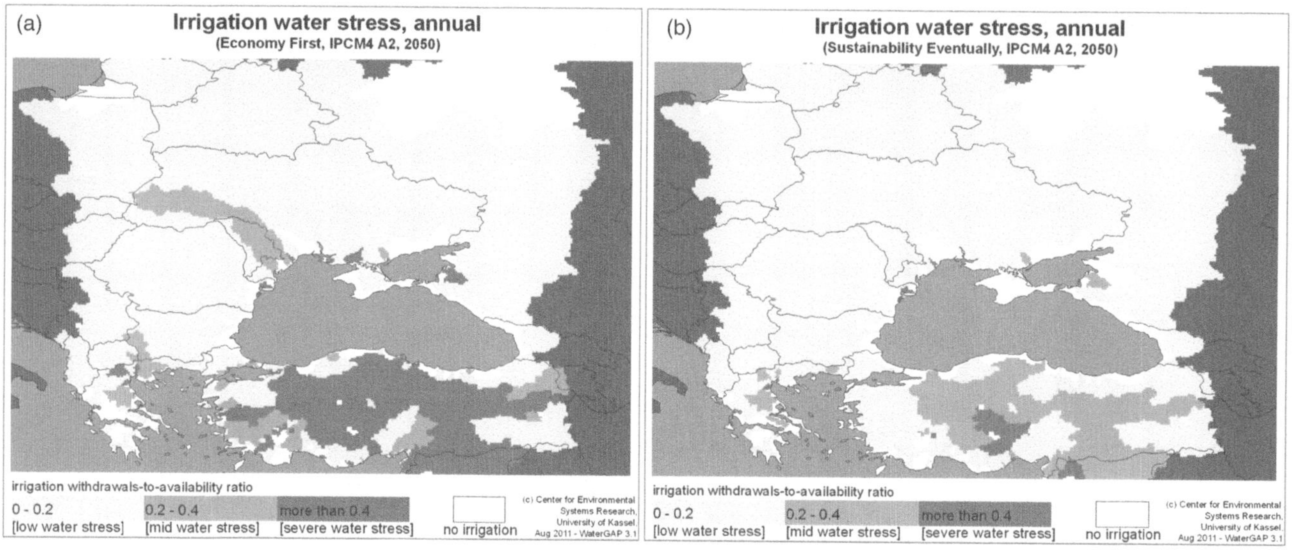

Figure 4 | Irrigation water stress in 2050 as calculated for the Black Sea region based on IPCM4-A2 climate input. Map (a) EcF scenario, and (b) SuE scenario.

be water stressed due to irrigation water withdrawals alone. Here, we define irrigation water stress as the irrigation water withdrawals-to-availability ratio. As can be seen from Figure 4(a) most of Turkey's river basins are under severe irrigation water stress in the EcF scenario, i.e. more than 40% of the water resources are abstracted for irrigation purposes. Other river basins, like the Dniester, are in the medium irrigation stress class but appear in the severe water stress class if total water withdrawals are considered (see Figure 1 for comparison). An increase in water stress is mainly caused by an increase in industrial water withdrawals which are most likely in the region. On the other hand, a decrease in irrigated area together with further improvements in efficiency leads to a reduction in water stress as calculated for the SuE scenario (Figure 4(b)). Although total water withdrawals are expected to decrease in SuE in the future, water stress appears due to declining water availability (see Figure 2(a)).

Yet drinking water will have priority over other uses particularly during severe water scarcity. In the Black Sea region, water use is dominated by agricultural water use, and this will remain the same in the future. However, this tendency will be dampened somewhat by continuing improvements in the water use efficiency of irrigation. Even under the SuE scenario in which a decrease in irrigated area is assumed, the region remains vulnerable to water scarcity caused by irrigation water requirements alone. This is especially true for Turkey (Figures 1 and 4). On the other hand, the agricultural sector is not the only water use sector, domestic and industrial sectors as well as water for millions of tourists are required. Considering a good ecological status of the water bodies, as required by the Water Framework Directive, water must be allocated to nature (ecosystems) which seems to be out of reach in some parts of this region.

CONCLUSIONS

The global water model WaterGAP3 was used to simulate future freshwater resources under two different scenarios and to identify future hot spots where water may become a scarce resource. In this case various water-related sectors may compete for the available water leading to cross-sectoral conflicts. This study provides an assessment of how climate change and changes in future freshwater demand affect water scarcity (water stress) in the Black Sea region.

Using a scenario approach helps to evolve different pathways into the future and to be prepared for various developments. Here we analysed two different scenarios. The first one, EcF, follows an economic-oriented way into the future accompanied by less consciousness for water resources, while the second is characterised by increasing awareness to save water. The scenarios may overstate the development of future water withdrawals but they span a variety of possibilities that can be used as a basis for flexible management.

Future water withdrawals are expected to increase following the EcF scenario pathway, but may decrease in a SuE future where especially less economic activity but further technological development and a changing behaviour result in a decrease in water withdrawals. In the EcF scenario it is very likely that the industrial sector grows, provided by a vision of market expansion. In contrast, this development is not in the focus of the SuE scenario. Here the proportion of agricultural water use accounts for 53%, and goes hand in hand with a reduction in industrial and agricultural water uses. Because of these developments in the industrial and agricultural sectors, the share of domestic water use will increase up to 23% of total water withdrawn.

Our model results show that particularly the Black Sea rim countries may suffer from climate change impacts since annual water availability is likely to decrease to 2050 by up to more than 30%. Especially in Turkey and Bulgaria, decreasing water availability exacerbates water stress.

In the Black Sea region, hot spots of water stress mainly occur in areas under intensive irrigation. Increasing irrigation efficiency can to some degree reduce irrigation water withdrawals; however, technological changes will not be sufficient to prevent this region from water stress. A more profound change in agricultural practices is needed as already indicated by the SuE scenario. Moreover, hot spots indicate locations where cross-sectoral conflicts may arise in the future and where the implementation of integrated water resources management may be important to avoid an overexploitation of water resources and hence potential conflicts. Short-term water stress is likely to appear within a year but is not covered by the approach applied.

Ideally, in impact studies a broad range of possible future climate scenarios simulated by an ensemble of existing GCMs should be used. In SCENES, climate output from two GCMs following the SRES A2 emission pathway were selected for reasons of resource constraints. It is well known that the choice of the emission scenario is of less importance for the early decades of the 21st century than for the later ones since they start to diverge in the second half of the century. However, for impact, vulnerability and adaptation analysis focussing on short and medium time scales, it is important to provide a meaningful combination of the climate change pathways with short- to medium-term scenarios for socio-economic development.

ACKNOWLEDGEMENTS

The financial support from the EU's Research Framework Programme (FP6) under contract no. 036822 (SCENES) is gratefully acknowledged. The authors acknowledge Dr David A. Wiberg from the International Institute on Applied Systems Analysis (IIASA) for providing time series of future socio-economic developments and electricity production.

REFERENCES

Alcamo, J. 2008 The SAS Approach: combining qualitative and quantitative knowledge in environmental scenarios. In: *Environmental Futures: The Practice of Environmental Scenario Analysis. Developments in Integrated Environmental Assessment* (J. Alcamo, ed.), Vol 2. Elsevier, Amsterdam, pp. 123–150.

Alcamo, J., Döll, P., Henrichs, T., Kaspar, F., Lehner, B., Rösch, T. & Siebert, S. 2003 Development and testing of the waterGAP 2 global model of water use and availability. *Hydrological Sciences Journal* **48** (3), 317–337.

Alcamo, J., van Vuuren, D., Ringler, C., Cramer, W., Masui, T., Alder, J. & Schulze, K. 2005 Changes in nature's balance sheet: model-based estimates of future worldwide ecosystem services. *Ecology and Society* **10** (2), 19. See: http://www.ecologyandsociety.org/vol10/iss2/art19/.

Alcamo, J., Flörke, M. & Märker, M. 2007 Future long-term changes in global water resources driven by socio-economic and climatic changes. *Hydrological Science Journal* **52** (2), 247–275.

aus der Beek, T., Flörke, M., Lapola, D. M. & Schaldach, R. 2010 Modelling historical and current irrigation water demand on the continental scale: Europe. *Advances in Geosciences* **27**, 79–85.

Bates, B. C., Kundzewicz, Z. W., Wu, S. & Palutikof, J. P. (eds) 2008 *Climate Change and Water*. Technical Paper of the Intergovernmental Panel on Climate Change, IPCC Secretariat, Geneva, 210 pp.

Cosgrove, W. & Rijsberman, F. 2000 *World Water Vision: Making Water Everybody's Business*. World Water Council, Earthscan Publications, London, p. 108.

Döll, P., Kaspar, F. & Lehner, B. 2003 A global hydrological model for deriving water availability indicators: model tuning and validation. *Journal of Hydrology* **270**, 105–134.

EEA (European Environment Agency) 2004 Impacts of Europe's changing climate: An indicator-based assessment. *Technical Report No. 2/2004*. EEA, Copenhagen.

EEA (European Environment Agency) 2005 *The European Environment – State and Outlook 2005*. EEA, Copenhagen.

Feyen, L. & Dankers, R. 2009 Impact of global warming on streamflow drought in Europe. *Journal of Geophysical Research* **114**, D17116.

Flörke, M. & Alcamo, J. 2004 European Outlook on Water Use, Technical Report prepared for the European Environment Agency. Kongens Nytorv. 6. DK-1050. Copenhagen, DK. See: http://scenarios.ewindows.eu.org/reports/fol949029.

Flörke, M., Bärlund, I. & Teichert, E. 2011 Future changes of freshwater needs in European power plants. *MEQ* **22** (1), 89–104.

IPCC (Intergovernmental Panel on Climate Change) 2007 Summary for Policymakers. In: *Climate Change 2007: The Physical Science Basis*. Contribution of Working Group I to the Fourth Assessment Report of the Intergovernmental Panel on Climate Change (S. Solomon, D. Qin, M. Manning, Z. Chen, M. Marquis, K. B. Averyt, M. Tignor & H. L. Miller, eds). Cambridge University Press, Cambridge, UK and New York, NY, USA.

Kämäri, J., Alcamo, J., Bärlund, I., Duel, H., Farquharson, F., Flörke, M., Fry, M., Houghton-Carr, H., Kabat, P., Kaljonen, M., Kok, K., Meijer, K. S., Rekolainen, S., Sendzimir, J., Varjopuro, R. & Villars, N. 2008 Envisioning the future of water in Europe: the SCENES project. *E-WAter (online journal of the European Water Association)*, **2008**.

Kok, K., van Vliet, M., Bärlund, I., Dubel, A. & Sendzimir, J. 2011 Combining participative backcasting and explorative scenario development: experiences from the SCENES project. *Technological Forecasting and Social Change* **78** (5), 835–851.

Milly, P. C. D., Dunne, K. A. & Vecchia, A. V. 2005 Global pattern of trends in streamflow and water availability in a changing climate. *Nature* **438**, 347–350.

Mitchell, T. D., Carter, T. R., Jones, P. D., Hulme, M. & New, M. 2004 A Comprehensive Set of High-resolution Grids of Monthly Climate for Europe and the Globe: the Observed Record (1901–2000) and 16 Scenarios (2001–2100), Working Paper 55. Tyndall Centre for Climate Change Research, UK.

Mitchell, T. D. & Jones, P. D. 2005 An improved method of constructing a database of monthly climate observations and associated high-resolution grids. *International Journal of Climatology* **25** (6), 693–712.

Raskin, P., Gleick, P., Kirshen, P., Pontius, G. & Strzepek, K. 1997 *Water Futures: Assessment of Long-Range Patterns and Problems*. Comprehensive assessment of the freshwater resources of the world. Stockholm Environment Institute, Stockholm, Sweden.

Rothman, D. S., Agard, J., Alcamo, J. 2007 Chapter 9. The future today. In: United Nations Environment Programme (UNEP) (eds), *Global Environment Outlook – GEO-4*. UNEP, Nairobi, Kenya, pp. 397–456. See: http://www.unep.org/geo/geo4/.

Verzano, K. 2009 Climate change impacts on flood related hydrological processes: Further development and application of a global scale hydrological model. *Reports on Earth System Science. 71-2009*. Max Planck Institute for Meteorology, Hamburg, Germany.

Vörösmarty, C. J., Green, P., Salisbury, J. & Lammers, R. B. 2000 Global water resources: vulnerability from climate change and population growth. *Science* **289**, 284–288.

Weiß, M., Flörke, M., Menzel, L. & Alcamo, J. 2007 Model-based scenarios of Mediterranean droughts. *Advances in Geoscience* **12**, 145–151.

First received 2 January 2013; accepted in revised form 20 March 2013

Integrating water resources management in eco-hydrological modelling

H. Koch, S. Liersch and F. F. Hattermann

ABSTRACT

In this paper the integration of water resources management with regard to reservoir management in an eco-hydrological model is described. The model was designed to simulate different reservoir management options, such as optimized hydropower production, irrigation intake from the reservoir or optimized provisioning downstream. The integrated model can be used to investigate the impacts of climate variability/change on discharge or to study possible adaptation strategies in terms of reservoir management. The study area, the Upper Niger Basin located in the West African Sahel, is characterized by a monsoon-type climate. Rainfall and discharge regime are subject to strong seasonality. Measured data from a reservoir are used to show that the reservoir model and the integrated management options can be used to simulate the regulation of this reservoir. The inflow into the reservoir and the discharge downstream of the reservoir are quite distinctive, which points out the importance of the inclusion of water resources management.

Key words | eco-hydrological modeling, integrated water resources management, reservoir management, Soil and Water Integrated Model

H. Koch (corresponding author)
S. Liersch
F. F. Hattermann
Department Climate Impacts & Vulnerabilities,
Potsdam Institute for Climate Impact Research,
P.O. Box 601203,
14412 Potsdam,
Germany
E-mail: *hagen.koch@pik-potsdam.de*

INTRODUCTION

To investigate the effects of land use and climate variability or change on water resources, models integrating the most important hydrological processes are needed. These models include natural processes at different spatial and temporal scales. Today, there are very few large river basins not affected by human intervention and regulation. The development and implementation of Integrated Water Resources Management at the river basin scale has to consider water management, e.g. reservoir regulation or water withdrawals. Natural processes and impacts, e.g. of climate change and variability, must be included too. Some hydrological models contain routines to integrate the effects of water withdrawal or reservoir management in one model setup. These routines are often relatively simple tank modules, where water is released if the tank is full. Sometimes minimum discharges for water releases can also be defined, where water is released as long as it is available in the reservoir. However, these simple routines cannot be used to reflect more sophisticated reservoir management rules. Some hydrologic and water balance models containing options to include reservoir management are presented in the following.

The widely used model SWAT (Soil and Water Assessment Tool; Neitsch *et al.* 2002) contains routines for the simulation of water balance of reservoirs. The first option is to deliver measured outflow series for reservoirs. These quantities are then used to calculate reservoir storage and release. The second option is an uncontrolled reservoir, where a specified quantity of water is released as long as water is available in the reservoir. If the maximum reservoir storage is reached, excess water is released from the reservoir. As a third option, target releases for controlled reservoirs can be simulated as a function of the desired target storage. The target storages are given on a monthly basis. If the simulated release does not meet predefined criteria, i.e. minimum or maximum discharge, the outflow is altered to meet these criteria.

The Hydrologic Modeling System (HEC-HMS; US ACE 2010) also provides three options for simulating reservoirs. For the first option the user has to provide a relationship between storage and discharge. This relationship must be monotonically increasing with storage. The second option is the release of water according to predefined, i.e. measured

doi: 10.2166/wst.2013.022

or externally specified, release rates. The last option is comparable to the first one, but here a number of outlet structures, e.g. a spillway and low-level outlet pipes, can be defined.

Arnold (2006) couples a reservoir module with the grid-based Water Balance Simulation Model (WaSiM-ETH) developed at the ETH Zürich. This module simulates the reservoir management as a standalone programme. Water abstractions from or injections to the river reach are then transferred to the WaSiM-ETH sub-catchment where the reservoir is located. However, due to the model concept, the model is usually applied to small or medium sized basins.

In the water balance model, the Large Area Runoff Simulation Model (LARSIM; Ludwig & Bremicker 2006), different options for reservoir management can be simulated. While the first option is a simple constant outflow rule, i.e. constant water release as long as sufficient water is in the reservoir, a second option enables the simulation of a reservoir outflow controlled by a downstream gauge at a moderate distance. The release is calculated in such a way that reservoir release and the discharge between reservoir and the control gauge do not exceed the desired discharge at the gauge. Using a third option, reservoir release can be simulated with seasonally changing release targets. For this option, target storage volumes and maximum allowable release volume must be defined.

This paper describes the development of a novel reservoir management model and the incorporation of this model as a module into the eco-hydrological model SWIM (Soil and Water Integrated Model; Krysanova *et al.* 1998, 2000). Up to now, this model is lacking explicit routines for water management. A more realistic simulation of the study area required the development of a new module for reservoir management.

Simulation results in terms of water management are presented and discussed. It can be shown that the new reservoir model and the management options implemented can be used to simulate reservoir regulation including hydropower production. Therefore, the integrated model can be applied to investigate the impacts of climate change on discharge and to study possible adaptation strategies in terms of reservoir management, i.e. changing release rules or including new reservoirs. While this paper deals mainly with the technical implementation of the reservoir module, the application to and results for the Upper Niger Basin including the Inner Niger Delta (IND) are presented by Liersch *et al.* (2012).

STUDY AREA

With a length of 4,200 km, the Niger River is the third longest river in Africa. The study area is the Upper Niger Basin in West Africa which covers an area of 350,000 km^2 (see Figure 1). Mali and Guinea have the largest share in the Upper Niger Basin, the shares of Ivory Coast, Burkina Faso, and Mauretania are small. The flow regime has a strong seasonal character and is driven by a unimodal monsoon-type of climate. During the 3–5 month rainy season (June–October) the headwaters of the basin receive up to 2,000 mm of rainfall whereas the IND in the northern part receives only 200–550 mm per year. The inter-annual variability of climate is high. The ecological integrity of the IND as well as human well-being depend largely on seasonal flooding of huge areas and thus on the magnitude and duration of flood peaks of the rivers Niger and Bani. The IND is threatened by decreasing water availability due to upstream water management (dams and large irrigation areas) as well as climate change and variability. Two major dams, the Sélingué dam and the Markala barrage, have been constructed and others are in the phase of planning. The Sélingué reservoir changes the natural flow regime by reducing flood peaks and increasing river flow during the dry season. The novel reservoir model is used to simulate these effects.

MODEL DESCRIPTION

Eco-hydrological model SWIM

The model SWIM (Krysanova *et al.* 1998, 2000) is a continuous-time spatially semi-distributed eco-hydrological model. It was developed from SWAT version '93 (Arnold *et al.* 1993) and MATSALU models (developed in Estonia for the agricultural basin of the Matsalu Bay, which belongs to the Baltic Sea; Krysanova *et al.* 1989) for climate and land use change impact assessment. The distribution of the SWIM model is restricted mainly to project partners. SWIM simulates hydrological processes, vegetation growth, erosion, and nutrient dynamics at the river-basin scale. It is a process-based model, combining physics-based processes and empirical approaches. Hydrotopes or hydrological response units (HRUs) are the core elements in the model. These elements are generated by overlaying geographic information system (GIS)-maps of land use/cover, soil, and sub-basins. The latter are derived

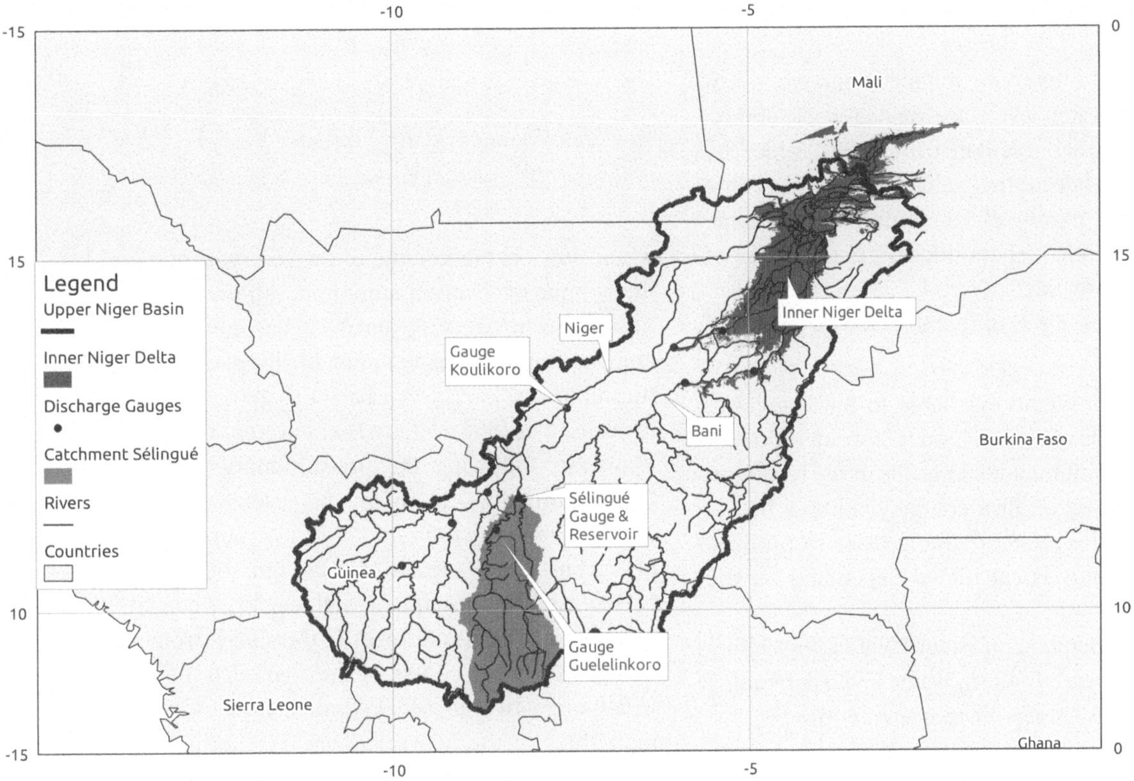

Figure 1 | Upper Niger Basin with location of gauges and reservoir used in this study.

from digital elevation models. The HRUs are considered as units with the same properties regarding bio-physical processes. It is important to understand that there is no lateral connection between HRUs. All processes are calculated at this spatial level using daily time-steps. Up to now in SWIM, only processes affecting the natural discharge as described are considered. Effects of water management, e.g. the management of reservoirs or water transfers, are not included in the model.

The eco-hydrological model, SWIM, requires spatial and temporal input data, which are described in the following. A global daily climate dataset in 0.5 degree resolution provided precipitation, air temperature, radiation, and humidity data for the years 1970 to 2001. The dataset was produced within the EU FP6 WATCH project (http://eu-watch.org/) and is based on monthly Climate Research Unit, Global Precipitation Climatology Centre, and sub-daily re-analysis data (ERA40). A 270 m resolution digital elevation model was constructed from the Shuttle Radar Topography Missions' (SRTM) 90 m resolution digital elevation model data. Soil parameters were derived from the Digital Soil Map of the World (FAO). Land use (cover) data were reclassified from Global Land Cover (GLC2000). River discharge data were obtained from the Global Runoff Data Centre (GRDC).

To estimate the quality of the SWIM model, besides annual mean discharges, the goodness of fit using the Nash–Sutcliffe efficiency (NSE) was used (Nash & Sutcliffe 1970). An NSE equal to 1 represents a perfect fit. The root mean squared error (RMSE), the absolute value of the bias (BIAS) and the percentage of the bias (PBIAS) are used to show the quality of fit.

Reservoir model

For reservoirs the active storage, sometimes called active capacity or carryover storage, and the dead storage, sometimes called permanent storage, must be delivered. The dead storage can be established using different measures. It can be set by the reservoir volume below the base outlet or due to biological or environmental requirements. The active storage is the reservoir volume that is used in day-to-day management of the reservoir, i.e. the volume that is filled in times of high flows and where water is released from during times of low flows. Also the gross capacity, i.e. active storage plus dead storage, has to be defined. If no dead storage is given, the active storage is equal to the gross capacity. Flood control storage is not considered up to now. Therefore, all water surpassing the gross capacity is released from the reservoir.

All input data for the reservoir management, e.g. maximum and minimum water level, are delivered as end-of-month values. The model calculates the daily values by linear interpolation between the value of the preceding month and the current month. By this interpolation, large leaps between the last day of 1 month and the first day of the next month can be avoided.

The model provides three different reservoir management options:

(i) Variable daily minimum discharge to meet environmental targets downstream under consideration of maximum and minimum water levels in the reservoir.
(ii) Daily release based on firm energy yield by a hydropower plant at the reservoir (the release to produce the required energy is calculated depending on the water level).
(iii) Daily release depending on water level (rising/falling release with increased/falling water level, depending on the objective of reservoir management).

In the first step, the current filling of the reservoir is calculated:

$$V_d = V_{d-1} + V_{\text{infl},d} + V_{\text{prec},d} - V_{\text{evap},d} - V_{\text{seepg},d} \quad (1)$$

where V is the volume of the reservoir (in million m^3), d is the current day, V_{infl} is the inflow volume of water entering the reservoir from upstream sub-basin(s), V_{prec} is the volume of precipitation over the reservoir surface area, V_{evap} is the volume lost by evaporation, and V_{seepg} is the volume lost by seepage (all in million m^3/d). In this calculation inflow from the upstream basin and precipitation over the reservoir surface are accounted for as a positive part of the reservoir balance, while potential evaporation is included as a negative part of the reservoir balance. Seepage losses can be included depending on reservoir filling and actual reservoir area. If the active storage is filled, excess water is seen as outflow from the reservoir. This quantity has to be accounted for in the calculation of outflow from the reservoir as described in the following equations. To keep the equations as simple as possible this part is omitted in the given equations.

Reservoir release option i

The reservoir release for option (i) is calculated:

$$V_{\text{outfl},d} = V_d - V_{\text{min},d} \qquad \text{if} \quad V_d - V_{\text{Qmin},d} < V_{\text{min},d} \quad (2a)$$

$$V_{\text{outfl},d} = V_{Q\,\text{min},d} \qquad \text{if} \quad V_{\text{min},d} < V_d - V_{Q\,\text{min},d} < V_{\text{max},d} \quad (2b)$$

$$V_{\text{outfl},d} = V_{Q\,\text{min},d} + (V_d - V_{Q\,\text{min},d} - V_{\text{max},d}) \quad \text{if} \quad V_d - V_{Q\,\text{min},d} > V_{\text{max},d} \quad (2c)$$

where V_{outfl} is the volume of the reservoir outflow, V_{Qmin} is the volume of required minimum outflow from the reservoir (in million m^3/d), V_{min} and V_{max} are the user defined minimum and maximum volumes of the reservoir, respectively (in million m^3). V_{min} can be set to zero. In this case, water is released as long as there is any water in the active storage. However, following the precautionary principle for prolonged droughts it might be useful to set minimum volumes for the reservoir. Then the current release might be curtailed to a certain degree, but some water is left to be released in subsequent periods. V_{max} can be set to the value of the active storage. If the flood protection storage is changed in the course of the year, also the active storage will change. In that case, V_{max} must be set to the corresponding values of the active storage. Furthermore, in very wet years the reservoir can release more water then the required minimum outflow from the reservoir. Then, V_{max} can be set with different values for all months, depending on the annual cycle of inflow. The idea behind this management is that in extremely wet years even during the usual low flow periods the reservoir has a high water level. Approaching the rainy season, the reservoir is filled to the maximum water level rapidly and large volumes of water are flowing out of the reservoir. If the reservoir is equipped with a hydropower plant the capacity of this plant might rapidly be exceeded. Lowering the water level, i.e. releasing more water, before the onset of the rainy season can increase the quantity of energy produced and the utilization ratio of the hydropower plant.

Reservoir release option ii

The reservoir release for option (ii) is calculated under the following assumptions. The reservoir is managed to produce a certain quantity of electricity in a hydropower plant. The electricity generated is calculated:

$$P_{\text{el}} = Q \times h \times k \qquad \text{with} \quad Q = \text{MIN}(V_{\text{outfl}}, \text{Cap}_{\text{HPP}}) \quad (3)$$

where P_{el} is the electricity produced (kW), Q is the flow through the turbine (m^3/s), h is the water head (m), k is the efficiency factor (kN/m^3), and Cap_{HPP} is the maximum turbine flow capacity (m^3/s). Rearranging Equation (3) and assuming a certain quantity of electricity, the required outflow depending on the water level in the reservoir can be calculated. The water level–volume relationship is used to calculate the required outflow depending on the actual volume.

Under option (ii), maximum release rates as part of the current volume of the reservoir can be set. These restrictions can be used to keep some water in the reservoir for prolonged drought periods. This maximum release is calculated:

$$V_{\mathrm{MAXoutfl},d} = V_d \times \mathrm{Part}_V (\text{for } \mathrm{Part}_V > 0) \tag{4}$$

where V_{MAXoutfl} is maximum outflow from the reservoir (in million m^3/d), Part_V is the part of the current volume that can be released, with a minimum value of zero and a maximum value of unity. Since the available quantity is calculated using the current volume V_d, no lower boundary must be set. The reservoir release is calculated:

$$V_{\mathrm{outfl},d} = V_{\mathrm{Qreq},d} \quad \text{if} \quad V_{\mathrm{Qreq},d} < V_{\mathrm{MAXoutfl},d} \tag{5a}$$

$$V_{\mathrm{outfl},d} = V_{\mathrm{Qreq},d} + \left(V_d - V_{\mathrm{Qreq},d} - V_{\max,d}\right) \quad \text{if} \quad V_d - V_{\mathrm{Qreq},d} > V_{\max,d} \tag{5b}$$

where V_{Qreq} is the required outflow from the reservoir (in million m^3/d) to produce a certain quantity of electricity. $V_{\max}$ here can be set to the value of the active storage. Otherwise the setting as stated for option (i) can be used.

Reservoir release option iii

Using option (iii) the reservoir is managed in such a way that the natural annual cycle of flow for an unregulated reservoir can be represented. Volume–discharge relationships must be provided.

$$V_{\mathrm{outfl},d} = V_d \times x \quad \text{with} \quad x = f(V) \tag{6}$$

where x is the parameter for the calculation of outflow depending on the volume. In the simplest form, this parameter is the same for all volume stages. However, the parameter also might change with changing volumes. To enable the simulation of a wide range of possible reservoir management variants, the relationships can be increasing or decreasing with volume. An example of this kind of operation is the management of Lake Victoria in the Nile Basin, where more water is released with increasing water level (see Sutcliffe & Parks 1999).

Furthermore, withdrawals from the reservoir can be included, e.g. for agricultural irrigation or drinking water. Depending on parameter settings, these withdrawals are considered with higher or lower priority compared to the reservoir release. Although only option (ii) refers directly to hydropower production, the electricity produced can be calculated for all three options.

The inclusion of new reservoirs can have significant effects on discharges, water availability and hydropower production. Therefore, within the analysed time period new reservoirs can be included. Then the reservoir parameters, start of filling, active and dead storage, gross capacity, and minimum discharges during the filling process, must be defined by the user. When the reservoir is full, its operation mode switches into an active state.

Model integration

In order to integrate reservoir management as a module into SWIM, the outlet of the reservoir must be located at the same position as one outlet of a SWIM sub-basin. This has to be safeguarded in the pre-processing. The maximum reservoir water surface areas are considered as hydrotopes in the SWIM model. These hydrotopes are deactivated in order to prevent a doubling of the calculation of precipitation and evaporation over the reservoir surface (see Equation (1)). The reservoir model is called by the SWIM model during the routing procedure. If the routing routine reaches a reservoir-sub-basin outlet the reservoir routine is called and the simulation is carried out according to the management options set. After the simulation of the reservoir the outflow is routed through the next downstream sub-basin. To include possible changes in the water infrastructure, new reservoirs can be included. In this case the reservoir is empty at the start of the simulation.

RESULTS

In the first step, the SWIM model without the reservoir module was calibrated and validated on measured discharges for time-periods before the Sélingué dam was put into operation. Then the reservoir model as described above was integrated into the SWIM model and tested for its applicability to the Upper Niger Basin.

Natural discharges simulated by SWIM

The SWIM model for the Upper Niger Basin was calibrated on measured discharges at several gauges. The Sélingué dam was put into operation in 1982. Daily discharge measurements were available for the years 1970 to 1981, i.e. before the dam was operating. The gauge Guelelinkoro (see Figure 1) was used to calibrate the inflows into the reservoir for the period 1970 to 1975. The years 1976 to 1981 were used for validation. For the years 1987, 1990 and 1992, daily discharges downstream of the dam (gauge Sélingué) were available. The available discharge series show significant data gaps. Therefore, results of the combined model's quality are only shown for years where at least 340 measurements were available. It has to be pointed out that a special calibration effort was put on the simulation of low flows, since these are the phases where the management of reservoirs is of utmost importance for low flow augmentation. In a first step, the SWIM model was manually calibrated using the six most sensitive parameters related to evapotranspiration, Muskingum routing, soil properties, and groundwater. In a second step, the optimal parameter settings of these parameters where estimated using PEST (Model-Independent Parameter Estimation and Uncertainty Analysis, www.pesthomepage.org).

As shown in Table 1, the performance of the model for the gauge Guelelinkoro are good to excellent. Especially for the years 1971 and 1972 with low discharges, the performance is very high, while the performance for 1975 with rather high discharges is lower, but still sufficient.

Reservoir model

Main reservoir characteristics

The Sélingué dam at the river Niger tributary Sankarani (see Figure 1) was put into operation in 1982. The main characteristics of the dam as used in this study are presented in Table 2. As the efficiency factor for the hydropower plant, 7.7 kN/m^3 is used. This value is calculated using the values for maximum fall height and capacity of the hydropower plant as given in Table 2, and a maximum electricity production of 47.6 MW (see Zwarts *et al.* 2005).

Zwarts *et al.* (2005) assume mean annual seepage losses of 8 m^3/s for the Sélingué reservoir. Using this value for seepage at mean filling and assuming zero seepage if the reservoir is empty, a linear relationship between filling and seepage was derived, with maximum losses of 13 m^3/s if the reservoir is full. According to these values, approximately 520 m^3/d (or 0.052%) are lost for each million m^3 of filling.

Effects of Sélingué dam on discharge at the dam location

The management options (i) and (ii) available in the reservoir model were applied to the Sélingué dam. The results were compared with measured daily discharges downstream of the dam for the year 1992. For option (i) it was tried to resemble the real management by setting minimum discharges, minimum volumes and maximum volumes. All the given values are changing over the year. Applying option (ii), a firm energy yield of the hydropower plant at Sélingué dam of 18 MW as given by Zwarts *et al.* (2005) was the main objective of reservoir management. The results are shown in Figure 2. Since no measured inflow data were available, only the inflow as simulated by SWIM is displayed. Using options (i) and (ii) the management of the reservoir was reproduced well. It has to be noticed that according to data given in Zwarts *et al.* (2005) in the months of June and July 1992, the water level in the reservoir dropped to 341.5 and 341.2 m above mean sea level (a.m.s.l.), respectively, which is markedly below the minimum water level of the active storage (see Table 2). Because water cannot be released from the permanent storage in the model, simulated outflow for these months is lower than the measured outflow. According to

Table 1 | Performance results for the SWIM model, gauge Guelelinkoro

	1971		1972		1974		1975		1976	
Year	obs	sim	obs	sim	obs	sim	obs	sim	obs	sim
Mean [m^3/s]	300	302	252	292	347	337	439	322	349	325
RMSE [m^3/s]		101		70		149		247		124
BIAS [m^3/s]		2.0		8.1		−9.8		−116		−23
PBIAS [%]		1%		3%		−3%		−26%		−7%
NSE		0.94		0.95		0.89		0.69		0.89

Table 2 | Main characteristics of Sélingué dam

	Dead storage	Active storage
Volume [million m^3]	238.0	1,928.7
Max. water level [m a.s.l.]	342.2	349.2
Min. water level [m a.s.l.]	338.5	342.2
Max. water surface [km^2]	50	
Hydropower Plant (HPP)		
Max. fall height of HPP [m]	17.2	
Base of HPP [m a.s.l.]	332.0	
Capacity of HPP [m^3/s]	360.0	

data given in Zwarts *et al.* (2005), the total electricity generated by the hydropower plant at Sélingué dam in 1992 was 171,205 MWh. Simulated total electricity generation is 164,935 MWh using option (i), and 163,275 MWh using option (ii). The simulated electricity generation is somewhat lower than the value given by Zwarts *et al.* (2005), but this is caused by the restricted reservoir release in the months of June and July, as explained above. Comparing the total electricity generated in 1992 shows a somewhat higher production for option (i) despite the fact that option (ii) is to optimize hydropower production. Looking at the daily electricity production (not displayed) shows that the electricity production using option (ii) is much more equally distributed over the year, i.e. the lowest mean daily value for option (ii) is 3.9 MW, while it is 2.8 MW using option (i).

Effect of Sélingué dam on discharge at gauge Koulikoro

The effects of the Sélingué dam on the discharges at gauge Koulikoro (see Figure 1) and the electricity production at the hydropower plant are displayed in Figure 3. The reservoir was managed using option (ii). The simulation covers the period 1971 to 2001 to include a wide range of dry and wet years; the reservoir was assumed to be in operation in 1971 already. Hence 31 years could be analysed. Flows during the dry season were increased, while flows during the wet season were lowered. Although the differences were not huge, the discharge with reservoir Sélingué was more than double the natural discharge during the dry season in the first half of the year. After the start of the rainy season, the discharge with reservoir Sélingué was lower than the natural discharge, while they were corresponding for some time after the filling of the reservoir. These effects can be seen much better in Figure 2, where only data for 1 year and the location of the dam are displayed.

The electricity production at the hydropower plant was usually higher than 10 MW. However, as the mean values show, in years with a late start of the rainy season, the electricity production was below this value.

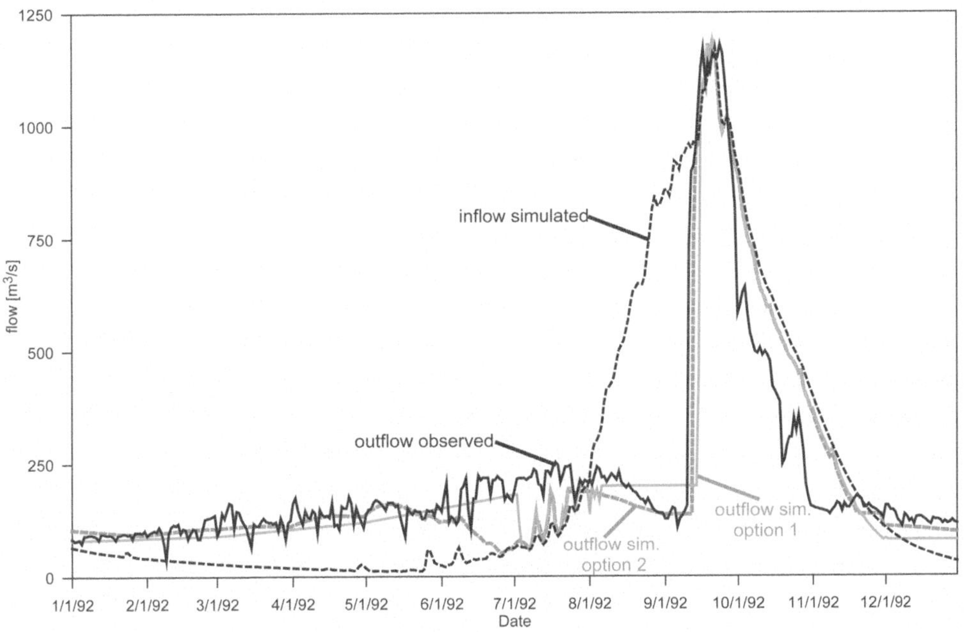

Figure 2 | Measured outflow, simulated inflow and outflow for Sélingué dam, year 1992; for management options see text.

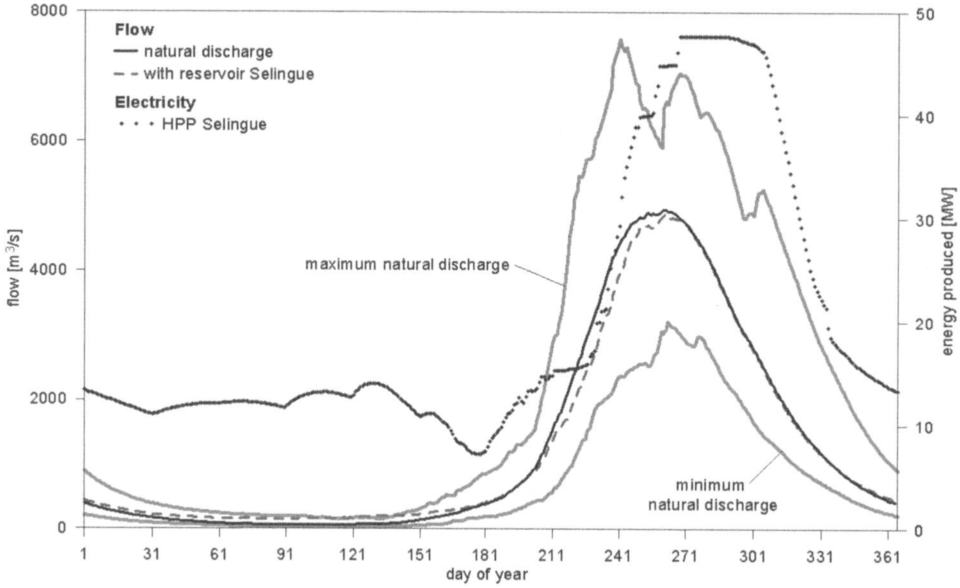

Figure 3 | Effect of Sélingué dam on the discharges at gauge Koulikoro and electricity production at the hydropower plant, mean values for years 1971 to 2001.

CONCLUSION

The SWIM model including the reservoir model is capable of simulating the management of existing and planned reservoirs. It is highly flexible to enable the simulation of different reservoir regulation schemes using a daily time step. Therefore it can be used to analyse the effects of reservoir management options, the construction of new reservoirs, or the effects of climate change on reservoir filling, hydropower production and discharges downstream of reservoirs. In the current version, reservoirs are simulated as a separate management facility. Future work will include the possibility of simulating reservoirs with joined management strategies.

As in the presented case of the Niger River Basin, water resources management has important implications for downstream regions. Reservoirs alter the natural flow regime usually by increasing low flows and decreasing flood peaks. Thus, the developed model supports the assessment of related consequences, e.g. in the IND. The agricultural production, e.g. of rice, is strongly influenced by the inflow and therefore depends on upstream reservoir management as well as future climate changes (see Liersch *et al.* 2012).

ACKNOWLEDGEMENT

The research leading to these results has received funding from the European Union Seventh Framework Programme (FP7/2007 - 2013) under grant agreement n°212300. It was carried out within the frame of the WETwin project.

REFERENCES

Arnold, T. 2006 Integrating a reservoir structure into the IMS framework, Documentation of Research 8/2006, Integrating governance and modelling project (CPFW).

Arnold, J. G., Allen, P. M. & Bernhardt, G. 1993 A comprehensive surface groundwater flow model. *Journal of Hydrology* **142**, 47–69.

FAO: Digital Soil Map of the World; http://www.fao.org/nr/land/soils/digital-soil-map-of-the-world/en/ (last accessed 31 Aug. 2011).

GLC2000: Global Land Cover; http://bioval.jrc.ec.europa.eu/products/glc2000/glc2000.php (last accessed 31 Aug. 2011).

GRDC: Global Runoff Data Centre; http://grdc.bafg.de (last accessed 31 Aug. 2011).

Krysanova, V., Meiner, A., Roosaare, J. & Vasilyev, A. 1989 Simulation modelling of the coastal waters pollution from agricultural watershed. *Ecological Modelling* **49**, 7–29.

Krysanova, V., Müller-Wohlfeil, D. I. & Becker, A. 1998 Development and test of a spatially distributed hydrological/water quality model for mesoscale watersheds. *Ecological Modelling* **106**, 261–289.

Krysanova, V., Wechsung, F., Arnold, J., Srinivasan, R. & Williams, J. 2000 SWIM (Soil and Water Integrated Model) User Manual. PIK Report 69. Potsdam Institute for Climate Impact Research, Potsdam, Germany.

Liersch, S., Cools, J., Kone, B., Koch, H., Diallo, M., Reinhardt, J., Fournet, S., Aich, V. & Hattermann, F. F. 2012

Vulnerability of rice production in the Inner Niger Delta to water resources management under climate variability and change. *Environmental Science and Policy*.

Ludwig, K. & Bremicker, M. (eds) 2006 The Water Balance Model LARSIM – design, content and applications. *Freiburger Schriften zur Hydrologie*, **22**, 141 pp.

Nash, J. E. & Sutcliffe, J. V. 1970 River flow forecasting through conceptual models, I-A, discussion of principles. *Journal of Hydrology* **10**, 282–290.

Neitsch, S. L., Arnold, J. G., Kiniry, J. R., Williams, J. R. & King, K. W. 2002 *Soil and Water Assessment Tool Theoretical Documentation*. Version 2000. Texas Water Resources Institute, College Station.

SRTM: Shuttle Radar Topography Mission; http://srtm.csi.cgiar.org/ (last accessed 31 Aug. 2011).

Sutcliffe, J. V. & Parks, Y. P. 1999 *The Hydrology of the Nile*. IAHS Special Publication (5). IAHS Press, Wallingford, Oxfordshire, UK.

US ACE 2010 *The Hydrologic Modeling System HEC-HMS Users Manual*. US Army Corps of Engineers, Hydrologic Engineering Center, Davis, CA, Version 3.5, August 2010.

Zwarts, L., van Beukering, P., Kone, B. & Wymenga, E. (eds) 2005 *The Niger, A Lifeline. Effective Water Management in the Upper Niger Basin*. RIZA, Lelystad/Wetlands International, Sévaré/Institute for Environmental studies (IVM), Amsterdam/A&W ecological consultants, Veenwouden, Mali/The Netherlands.

First received 9 January 2012; accepted in revised form 16 November 2012

Methodological challenges in evaluating performance, impact and ranking of IWRM strategies in the Jordan Valley

H. P. Wolff, L. Wolf, A. Subah, J. Guttman, A. Tamimi, A. Jarrar, A. Salman and E. Karablieh

ABSTRACT

The philosophy of integrated water resource management (IWRM), as formulated in several international summits, yielded numerous interpretations and extensions over the last decade but always focused on the overall objective of maximizing the welfare and livelihood of the people concerned. One of the major constraints of this concept is the gap between the well-defined philosophy and the fuzzy definition of operational and testable indicators for the achievement of its goals. This leads to difficulties in the evaluation of potential contributions from technological and managerial improvements. The experience of the multi-lateral IWRM research initiative SMART in the lower Jordan Valley shows that the evaluation and ranking of alternative IWRM strategies and their elements relies simultaneously on the identification of local goals and their interfaces with the superordinate national water sector policies. The documentation of the, still ongoing, development process of suitable assessment procedures describes their methodological embedding and conclusions drawn for the heterogeneous situation of water-related settings in this transboundary watershed.

Key words | IWRM, lower Jordan Valley, ranking, scenario impact analysis, strategy performance assessment

H. P. Wolff (corresponding author)
QUASIR Office for Quantitative Analyses, Karl-Pfaff-Straße 24a, D-70597 Stuttgart, Germany
E-mail: *hpwolff@quasir.de*

L. Wolf
CSIRO Land & Water, Dutton Park, Queensland, Australia

A. Subah
Ministry of Water and Irrigation, Jordan

J. Guttman
MEKOROT Water Company Ltd, 9 Lincoln St., 67134, P.O. Box 20128, Tel Aviv 61201, Israel

A. Tamimi
Palestinian Hydrological Group, Water Research Unit, P.O. Box 565, Al-Ma'ahed Street 4, Ramallah, Palestine

A. Jarrar
Palestinian Water Authority, Al-Balou - Baghdad St 2174, Ramallah, Palestine

A. Salman
E. Karablieh
University of Jordan, Faculty of Agricultural Economics and Agribusiness Management, Faculty of Agriculture, University of Jordan, P.O. Box 13204/13899, Amman 11942, Jordan

INTRODUCTION

The Jordan Valley is presumably one of the most reviewed watersheds with more than 30 books on documentation, overviews, analyses and knowledge construction about its water resources, water balance and related socio-political aspects (Allan 2002, p. 73). Its major natural water resources are nowadays strongly integrated into the man-made water infrastructure of surrounding areas and water imports from such areas represent an increasingly important additional resource for water users in the Valley. Rising use of low quality water, such as treated wastewater and saline water, as well as increasing concern about environmental consequences, e. g. the development of the Dead Sea and its surroundings, add continuously new elements to the objectives and requirements of sustainable water resource management.

The philosophy of integrated water resource management (IWRM) provides a promising paradigm for combining multiple objectives of water users, including nature, and complex interaction of water flows in the Jordan Valley. However, there is no unambiguous, generally accepted concept for its translation into commensurable objectives and criteria for decision making. This finds its reflection in the water sector strategies of Israel, Palestine and Jordan, which acknowledge the principles and need of IWRM, but provide different and not sufficiently specified

doi: 10.2166/wst.2013.310

benchmarks for the evaluation of technological implementations. Endeavors, which focus on the improvement of local technological set-ups, are thus forced to develop their own IWRM rating matrix for alternative combinations of solutions.

This paper does not intend to enrich the debate by another proposal of definitions for IWRM, but focuses on the question of operational approaches for assessing potential water management strategies, which endorse the IWRM paradigm. The approaches and experiences are the result of the first 6 years of the multi-lateral SMART research initiative in the Jordan Valley, which stands as an acronym for 'Sustainable Management of Available Water Resources with Innovative Technologies'.

THE METHODOLOGICAL CHALLENGE

The philosophy of IWRM, as formulated by the Dublin Conference on Water and the Environment (ICWE 1992) and specified by the Second World Water Forum and Ministerial Conference (2000), the International Conference on Freshwater (BMU 2001) and the United Nations World Summit On Sustainable Development (UN 2002), yielded numerous interpretations and extensions by different conferences and stakeholders over the last decade but always focused on the overall objective of improving the welfare and livelihood of the people concerned.

However, the definition of IWRM continues to be a matter of discussion despite the multitude of statements on its necessity (*cf.* van der Zaag 2005). Probably the most quoted definition comes from the Global Water Partnership (GWP) and reads '*IWRM is a process that promotes the coordinated development and management of water, land and related resources, in order to maximize the resultant economic and social welfare in an equitable manner without compromising the sustainability of vital ecosystems*' (GWP-TAC 2000). But even the authors of this definition point out in the same paper that IWRM is just an emerging collaborative framework that requires the development of local practices.

A number of criteria and components were added to this definition over the last decade, but a generally agreed list of parameters for the measurement of performance and success of IWRM strategies does not exist (*cf.* Medema *et al.* 2008). The 'three pillars' of IWRM, i.e. economic efficiency, equity and environmental sustainability (GWP-TAC 2004), are by themselves no hard criteria in terms of unambiguously interpretable parameters. The difficulties with a compilation of suitable evaluation criteria intensify in transboundary water systems, which connect countries and areas with significant differences in economic and societal developments, but common interests in the sustainability of water resources management (*cf.* GWP-TAC 2004).

Several authors and water professionals already formulated concerns about the fuzzy characterization of IWRM. Such concerns comprise the apprehension that IWRM is a mere reminder for the necessity of a holistic approach to natural resource management, but requests an unrealistic kind of institutional and organizational integration (*cf.* Biswas 2004), or even to be under a cloud of being not much more than a buzzword (van der Zaag 2005). Thus, the relevance of IWRM depends crucially on actual applications and the demonstrated improvement of existing water management practices, but this proof is still largely missing (Biswas 2004).

The development of appropriate, measurable indicators is a basic demand for the assessment of each management approach, but is characterized by iterative procedures in the narrow range between disciplinary limitations and functional arbitrariness in the case of IWRM (cf. Guenther 2007). Developed conceptual IWRM assessment methodologies, amongst which the output of the EU-funded project STRIVER ('*Strategy and methodology for improved IWRM – An integrated interdisciplinary assessment in four twinning river basins*', a project supported by the European Commission Sixth Framework Programme (FP6) from 2006 to 2009) may be one of the most recent ones (*cf.* Nesheim *et al.* 2008), provide valuable guidelines, but still lack convincing empirical validation of their transferability to different locations.

WATER RESOURCES MANAGEMENT IN THE LOWER JORDAN VALLEY

Economic opportunities, social necessities and aspirations in the Lower Jordan Valley entailed a significant growth in water demand over recent decades and will continue to do so in the future. This coincides with a situation where the exploitation of renewable natural water resources already exceeds a sustainable level and the reclamation of non-conventional water resources requires considerable investments. The resulting increase in water costs and values amplifies the role of socio-economic reflections in decision making on water resource management as well as on the allocation and use of water in the different sectors of water consumption.

Current water management approaches in Israel, Jordan and the Palestinian West Bank endorse elements of integration, but their levels, degrees and priorities differ significantly (*cf.* Froukh 2007; Fischhendler 2008; MWI 2009). A common denominator of the concepts of all parties is the desire to mobilize all potential water resources through advanced technologies.

The multi-lateral IWRM research initiative SMART focuses on the adaptation and implementation of improved approaches in technology and natural resource management for selected communities in Jordanian, Israeli and Palestinian areas of the Lower Jordan Basin (Wolf & Hoetzl 2011; Wolf *et al.* 2011) (Wolf *et al.* (2011) describes the full set of assessments, models and demonstration sites currently deployed in SMART). This corresponds to the second of the five main statements of the Bonn Recommendations for Action from the International Conference on Freshwater ('Bonn Keys', BMU 2001), which reads *'decentralization is key; national policy meets community needs'*. The overall objective of SMART is to contribute to the improved formulation of local IWRM strategies within the national water sector policies. Focal points are sub-strategies in technology implementation and sub-basin water management as well as the development of components for regional strategies. The strategic components are based on IWRM analyses at sub-basin level with details on implementation projects, which then enter a reporting framework to support larger scale IWRM planning. Components comprise improved techniques and management of decentralized water recycling, artificial aquifer recharge and desalination of brackish groundwater (Dropedia 2012).

These technical improvements deal with comparable hydro-geological and climatic conditions, but face different constraints in implementation and still require additional empirical knowledge on suitability and efficiency in the specific locations. A particular challenge in this context is the heterogeneity of social and economic situations, which suggests the hypothesis that suitable IWRM strategies for each location may also demand different combinations of the technical and managerial solutions under research, which corresponds to the statements of the GWP-TAC (2000, 2004) as well as to the concerns of those water professionals, who formulated critical thoughts on the applicability of IWRM as an operational concept (Biswas 2004).

A valuation and comparison of different strategies would not only help in the selection of best strategies for the selected focus areas (*cf.* Figure 1), but also allow for statements on the degree of direct transferability of developed technical solutions to other cases. This leads back to the methodological challenge and advocates the linking of IWRM to established concepts from general management sciences.

PERFORMANCE OF IWRM STRATEGIES

The performance of strategies and their measurement depends in general on the strategic objectives (Gamble *et al.* 2007). The definition of these objectives in IWRM strategies is a complex challenge due to the linkages of water resources management. Comparatively easy to grasp performance indicators, such as e.g. the provision of water quantities of a given quality at a certain point in time with minimal costs, are typical project and operational process measurements. Such indicators help in the isolated judgement on technical and managerial components, but fall short in the evaluation of strategies.

IWRM strategies are always just a part of superordinate water sector strategies and are simultaneously linked to strategies in other sectors, such as agriculture, industry, human settlement and nature. A generally valid definition of *per se* strategic objectives of IWRM is thus not possible which finds its reflection in the different national water sector strategies of the local partners. IWRM strategies on the level of basins, sub-basins or administrative entities, such as communities, are embedded in the respective national strategy, but may still deviate from the national mainstream due to the specific local characteristics. Table 1 gives an overview of the characteristics of major research locations of SMART and the preliminary identification of the local central objectives with regard to improved water management.

Prevailing performance measurement frameworks emphasize the central role of stakeholder satisfaction in any successful strategy and process development (Kennerly & Neely 2002). This finds its approximate analogy in the concurrent demand for stakeholder involvement by all definitions of IWRM (*cf.* Medema *et al.* 2008). Figure 2 displays an adaptation of the 'performance prism', which is an up-to-date performance framework of the Cranfield School of Management (*cf.* Neely *et al.* 2002), for the purposes of IWRM.

The prism approach to performance measurement understands 'strategy' as one out of five perspectives – or facets, in the terminology of the prism – which are logically interlinked and raise one key question each for measurement design (Neely *et al.* 2002). A predominantly technically oriented research project like SMART contributes by its nature not directly to the formulation of strategies, but in

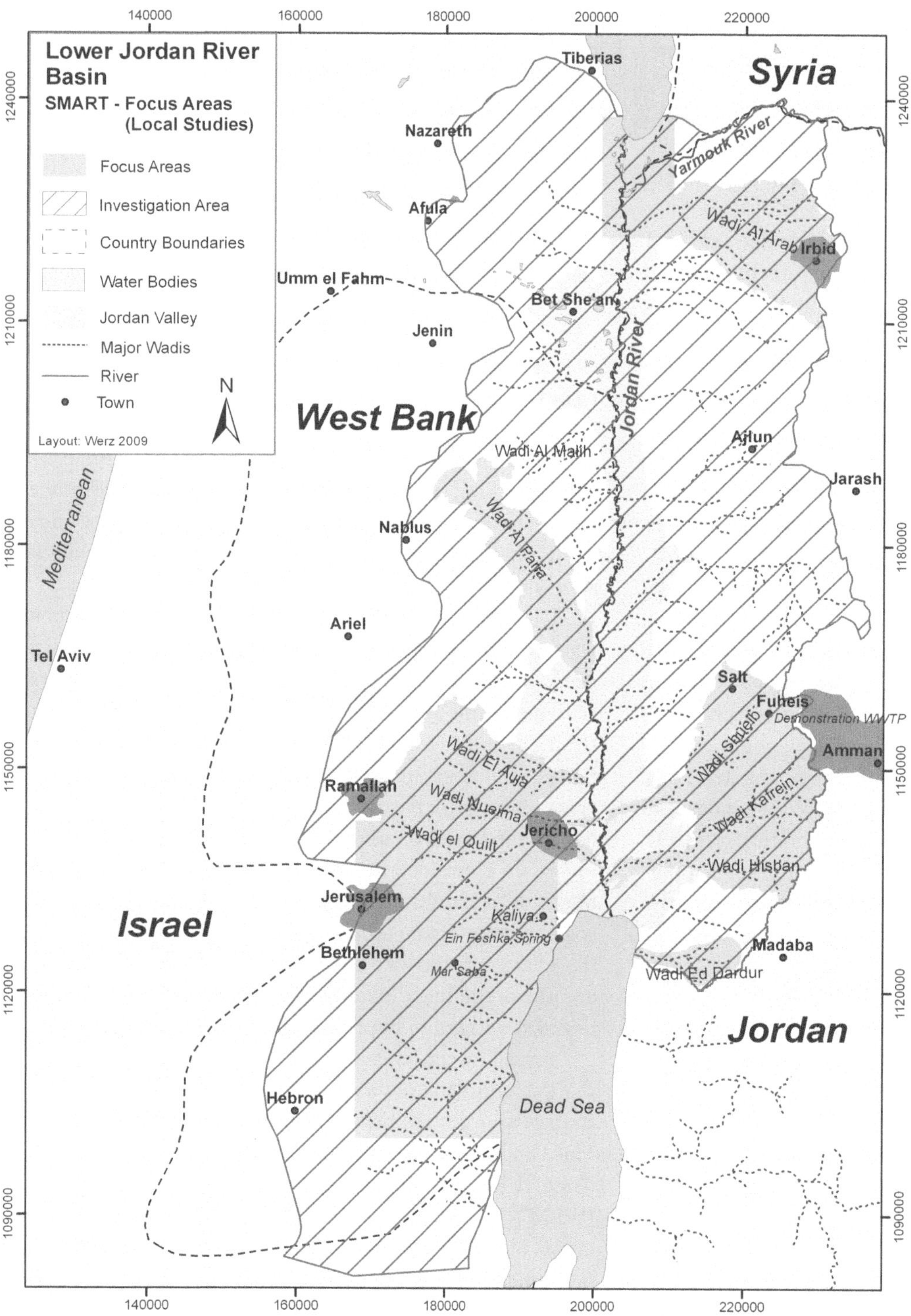

Figure 1 | SMART focus areas in the Jordan Valley. *source*: Wolf *et al.* (2011).

the first place to the interface 'solution development' between the facets 'Implementation of IWRM measures' (processes in the terminology of the prism) and 'IWRM tools and components' (capacities in the terminology of the prism). The key questions for both facets read:

1. For Implementation of IWRM measures – what critical processes do we require if we are to execute the currently agreed strategies?
2. For IWRM tools and components – what capabilities do we need to operate and enhance these processes?

Table 1 | Characteristics of major research locations of SMART in the lower Jordan Valley 2010

	Jericho (Palestinian Community)	Kaliya (Israeli Settlement)	Wadi Shueib (Jordanian side-valley)	Wadi Arab (Jordanian side-valley)
Population (capita)	19,783	1,150	121,709	449,735
Irrigated agricultural area (ha)	16	393	42	150
Water consumption (10^6 m³/year) 2010				
Municipal/domestic	1.48	0.23	3.67	13.28
Agriculture	0.14	5.85	2.83	10.12
Other (ex. nature)	0.46	0.31	1.12	2.53
Total	2.08	6.39	7.62	25.93
Primary objective of improved water management				
	Water for municipal use	Water for agriculture	Water for agriculture	Water for municipal use

Source: Compiled by the authors.

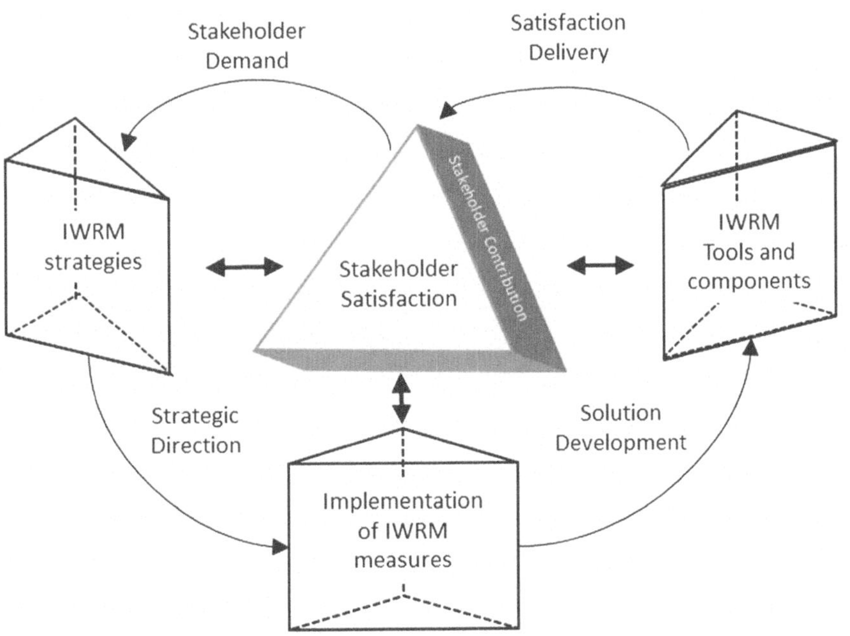

Figure 2 | Interpretation of the performance prism in the case of IWRM. *Source:* Based on Kennerley & Neely (2002), p. 153, adapted by the authors.

Both facets as well as their interface are linked to the core perspective of stakeholder satisfaction and, via iterative coupling, to the adaptation and improvement of the IWRM strategy. This allows for two conclusions with regard to the provision of performance indicators by researchers in natural and technical sciences to the assessment of IWRM strategies. The first is that such indicators must be at least functionally related to stakeholder satisfaction, whereby stakeholders are defined as those responsible for and affected by management intervention (Medema *et al.* 2008). The second is that technical indicators contribute to the development and assessment of IWRM strategy via their effects on stakeholder satisfaction, but are no direct determinants of those strategies.

SMART meets this insight with a set of internal and external stakeholder relations, which allow for a continuous communication and adjustment of solution development. Internal stakeholder relations rely on the direct participation of representative persons from national decision-making bodies as members of the project. External relations comprise workshops with representative persons from local interest groups in the concerned communities and – at least in Jordan – information exchange with international donor initiatives.

Table 2 | Relation between selected technical indicators and stakeholder satisfaction

Indicator	Relation to stakeholder satisfaction	
	Stakeholder (all study areas, if not mentioned otherwise)	Relation to satisfaction (m^3 = m^3 of water, ha = hectare)
Potential additional water supply (distinguished by water qualities) from improved managed aquifer recharge wastewater treatment and desalinization facilities in m^3/year	Water users	• Higher and more reliable water availability in liter/capita/day and m^3/production period
	Nation/society	• Impacts on water transfers from/to other regions in m^3/production period • Contribution to close demand-supply gap in m^3/year • Lower environmental costs in €/environmental objective • Environmental sustainability in saved costs or added value in €/ecosystem
	International donors (all, except Kaliya) & investors	• Contribution to achievement of Millennium Development Goals • More reliable water availability
Estimated costs of additional water through improved technologies in €/m^3	Water users & Nation/society	• Lower costs of water (Kaliya, Jericho) in €/m^3 • Lower costs of water reclamation and treatment) in €/m^3 • Higher return to investments in €/m^3
	International donors (all, except Kaliya) & investors	• Higher returns to investments in €/m^3
Controlled mass flows (e.g.) pollutants in water and land in ppt/ha, ppt/m^3 etc.	Water users	• Improved security in health and productivity of soils and water, €/household of saved health costs, €/ha and €/m^3 in production processes
	Nation/society	• Lower environmental costs in €/environmental objective • Higher returns to investments €/m^3 • Environmental sustainability in saved costs or added value in €/ecosystem
	International donors (all, except Kaliya) & investors	• Higher returns to investments in €/m^3

Source: Compiled by the authors.

Table 2 displays the potential contribution of some selected technical indicators to the formulation of IWRM strategies in the Jordan Valley. The decisive element in their usefulness is the possibility to determine and – if possible – to quantify the functional linkages to stakeholder satisfaction. The second essential component is the clear specification of units for the indicators as well as for the contributions to stakeholder satisfaction, even if the capture of some values may have to rely on indirect methods, such as e.g. in the case of environmental costs.

The list of considered stakeholders indicates that the definition of a stakeholder is not equivalent to the definition of a target group in development cooperation. Donors and investors are definitely a stakeholder in water resource management, but certainly no target group for development support.

IMPACT OF IWRM STRATEGIES

The preceding reflections on strategies already highlight the fact that measures of IWRM performance must go beyond partial, disciplinary assessment of success, be it from natural sciences or economics. Management experts also caution against the derivation of measures from strategies themselves, since strategy is not about destination but about the route to reach the desired destination (Neely & Adams 2005). Impact assessments have to start from overall goals and their indicators of success and work the way down to the individual parameters of technical and managerial innovations.

The various definitions of IWRM tend to formulate a set of normative goals, which Medema *et al.* (2008) summarized as: (1) the constitution of a sustainable approach to managing water resources, (2) an enhancement in water resource sustainability, (3) the production of a better understanding of human–ecosystem interactions, (4) the maximization of social and economic welfare generated from water and land resources, and (5) the reduction of disruption to the water cycle and to aquatically dependent ecosystems.

However, these goals may reflect the aspirations of the international community of water experts for the best rather than practical objectives of national water administrations, which will have to compromise with constraints, as e.g. power structures, that are beyond the exclusive scope of water management. In addition, they leave a wide field of interpretation with regard to the selection of relevant criteria and the question of how to measure them. The most relevant objection against the use of universal or even local normative goals as parameters for impact analyses of innovations is the dynamic character of interactions between nature, society and within the elements of society. IWRM is a continuous process rather than an end state.

An alternative approach to the evaluation of impacts is the measurement by scenario analyses, which may evade or at least alleviate the concerns with regard to normative goals (*cf.* Swart *et al.* 2004). An additional advantage of an evaluation within the overall picture of a scenario is the possibility to consider changes in values added by alternative strategies, which are a function of the surrounding conditions. Theoretical foundations and applicable methodologies for scenario development and analysis in environmental and natural resource assessments received a substantial boost through their increasingly important role as the tool of choice by the International Panel on Climate Change (IPCC, *cf.* Nakicenovic *et al.* 2000) and the Millennium Ecosystem Assessment (*cf.* Carpenter *et al.* 2005).

A notable difficulty in the application of scenarios as a tool for impact assessment in a transboundary IWRM environment arises from the selection of drivers for change, i.e. externalities, which shape future developments in the water sector and determine its potential states under different scenario assumptions. Experiences from the research area show that opinions and perceptions on major drivers differ between key experts from the participating nations. A recent regional scenario exercise on the areas around the Jordan River built its regional development scenarios on variations in both the economic and political development drivers for all three nations (*cf.* GLOWA Jordan River project, Anon 2009), while the scenario development team of the Jordanian Ministry of Water and Irrigation saw the combination of economic and demographic development as a more suitable set of drivers for developments in the water sector (*cf.* MWI-AFD 2011).

The probability of different relevant drivers increases with the number of research locations. The resulting problem is the required aggregation of local scenarios towards the establishment of regional scenarios for the evaluation of impacts from innovations on a larger scale. This would lead in the worst case – with the four research locations of SMART and under the assumption of just two drivers, which may adopt two potential states for each location – to the formulation of $4^4 = 256$ different regional scenarios. It is most unlikely that the capacities of any evaluation process will allow for the subsequent step of quantitative assessments for each of these scenarios. Apart from that, comparative interpretations and ranking may suffer from blurring and not very distinctive results between several of these scenarios. The problem is that such overlapping may become evident only after the results of the quantitative assessments are available.

Potential solutions to the dilemma that are under research by SMART include the identification of: (1) super-drivers, which describe higher-level regional development and may allow for a pre-selection of possible combinations of local scenarios, and (2) driver links, which capture relationships between the development of the specific drivers in different areas and may allow for a selection of regional scenario combinations based on their probability. Research on both concepts rely strongly on general econometric models and is still in its initial phase.

RANKING OF IWRM STRATEGIES

Primal comparisons and assessments of methods for the ranking of IWRM strategies were conducted by the EU-funded STRIVER project, which worked in an Asian-European context. Publications of this project list the four most frequently applied quantitative methodologies for ranking of alternative strategies and processes and provide a comparison of advantages and disadvantages (Nesheim *et al.* 2008, p. 129). The discussed approaches comprise multi-criteria analysis (MCA), cost–efficiency analysis, benefit–cost analyses and Bayesian networks. The evaluation of pros and cons of these approaches focused on suitable stakeholder layers, the flexibility of the methodology with regard to the incorporation of information, data demands, software requirements and the transparency of the ranking process for decision makers.

The STRIVER project opted for an MCA framework for ranking of IWRM approaches, despite its high process intensity, which requires a number of iterations with experts and stakeholders. However, published results of the method's application by this project are restricted to policy briefs on some experiences in limited pilot regions in South-East Asia (*cf.* Dang & Phi 2008; Nesheim *et al.* 2009). This implies that a comprehensive application of this generic methodology with the required adaptations to the purposes of IWRM is still outstanding.

The particular constellation of water related questions in the Jordan Valley and the subordinate role of IWRM with regard to the existing national water sector strategies of the countries concerned may require a different angle in the perspective of ranking. The chances for implementation will depend strongly on the conformity of proposed IWRM strategies and innovative elements with the formulated objectives and procedures in the national strategies. It would already be a valuable contribution of the SMART initiative to the improvement of local water resource management, if its research efforts would yield a basic, but tested methodological approach for the ranking of system conformity of suggested innovations and changes.

CONCLUSION

The wealth of definitions, philosophies, backgrounds and curricula of IWRM stands against a comparatively small number of reports on evaluations of applied IWRM strategies. The actual experiences of SMART in the Jordan Valley support the assumption that indicators and approaches for the performance measurement of IWRM strategies are, unlike the performance of its individual elements, highly case specific. This does not exclude the transferability of generic parts of applied methodologies, but advocates a very flexible course of action with intensive stakeholder participation not only in the procedural area of IWRM but already in the setting of local and regional IWRM objectives. The pursued solution to the deplored current restriction of IWRM strategy applications to the micro-level (Biswas 2004, p. 255) is the identification and – if possible – definition of interfaces in national macro and meso-scale water policies, which allow for the integration of complementary local IWRM strategies.

ACKNOWLEDGEMENT

The authors express their gratitude to the German Ministry of Education & Research (BMBF) for supporting this study through the SMART Project, funding no.:02WM0801.

REFERENCES

Allan, J. A. 2002 *The Middle East Water Question*. I.B. Tauris & Co. Ltd, London, paperback edition.

Anon 2009 *GLOWA Jordan River Scenarios of Regional Development under Global Change*. Center for Environmental Systems Research, University of Kassel, Germany.

Biswas, A. K. 2004 Integrated water resources management: A reassessment. A water forum contribution. *Water International* **29** (2), 248–256.

BMU – German Federal Ministry for the Environment, Nature Conservation and Nuclear Safety 2001 International Conference on Freshwater – Final Report. Bonn, 3–7 December 2001. Available from: http://www.bmu.de/english/water_management/doc/3468.php (accessed 22 July 2011).

Carpenter, S. R., Bennett, E. M., Zurek, M. & Pingali, P. (ed.) 2005 *Ecosystems and Human Well-Being: Millennium Ecosystem Assessment Scenarios for the Future of Ecosystem Services*. Island Press, Washington, D. C.

Dang, K. N. & Phi, T. T. H. 2008 Pressure Impact Multi-Criteria Environmental Flow Analysis in the Sesan River. STRIVER Technical Brief no. 7. Available from: http://kvina.niva.no/striver/Portals/0/documents/STRIVER_TB7_EFSesan.pdf (accessed 31 August 2011).

Dropedia 2012 Knowledge management platform about Integrated Water Resources Management in the Lower Jordan Rift Valley. Available from: http://129.13.109.100/~dropedia/index.php/Main_Page (accessed 08 April 2012).

Fischhendler, I. 2008 Institutional conditions for IWRM: the Israeli case. *Ground Water* Jan-Feb; **46** (1), 91–102.

Froukh, L. J. 2007 Water Demand Management of the West Bank. Conference paper, Water demand management in the Mediterranean, progress and policies, Zaragoza, Spain 19–21.03.2007. Available from: http://www.planbleu.org/publications/atelier_eau_saragosse/Poleau_2_PS_38_Froukh_final_EN.pdf (accessed 23 August 2011).

Gamble, J., Strickland, A. & Thompson, A. 2007 *Crafting & Executing Strategy*. 15th edition, McGraw-Hill, New York.

Global Water Partnership – Technical Advisory Committee (GWP-TAC) 2000 Integrated water resources management. TAC Background Paper No. 4. GWP, Stockholm, Sweden.

Global Water Partnership – Technical Advisory Committee (GWP-TAC) 2004 Integrated water resources management (IWRM) and water efficiency plans by 2005. Why, what and how? TAC Background Papers No. 10. GWP, Stockholm, Sweden.

Günther, D. 2007 Success factors for and impacts of participatory approaches on development of management indicators in IWRM. In: *Paper presented at 'Earth System Governance: Theories and Strategies for Sustainability', Amsterdam Conference on the Human Dimensions of Global Environmental Change*, Vrije Universiteit, Amsterdam, 24–26 May 2007.

ICWE – International Conference on Water and the Environment 1992 *The Dublin Statement on Water and Sustainable Development*. Available from: http://www.wmo.int/pages/prog/hwrp/documents/english/icwedece.html (accessed 2 February 2012).

Kennerly, M. & Neely, A. 2002 Performance measurement frameworks: a review. In: *Business Performance Measurement: Theory and Practice* (A. D. Neely, ed.) Cambridge University Press, Cambridge, UK, pp. 145–155.

Medema, W., McIntosh, B. S. & Jeffrey, P. J. 2008 From premise to practice: a critical assessment of integrated water resources management and adaptive management approaches in the water sector. *Ecology and Society* **13** (2), 29. Available from: http://www.ecologyandsociety.org/vol13/iss2/art29/ (accessed 31 August 2011).

MWI-AFD 2011 *Jordan Water Demand Management Study*. Jordanian Ministry of Water and Irrigation in cooperation with the Agence Française de Développement. Amman, Jordan.

MWI 2009 *Water for Life. Jordan's Water Strategy 2008–2022*. Ministry of Water and Irrigation, Jordan.

Nakicenovic, N., Alcamo, J., Davis, G., de Vries, B., Fenhann, J., Gaffin, S., Gregory, K., Grübler, A., Jung, T. Y., Kram, T., La Rovere, E. L., Michaelis, L., Mori, S., Morita, T., Pepper, W., Pitcher, H., Price, L., Riahi, K., Roehrl, A., Rogner, H.-H., Sankovski, A., Schlesinger, M., Shukla, P., Smith, S., Swart, R., van Rooijen, S., Victor, N. & Dadi, Z. 2000 *IPCC Special Report on Emissions Scenarios*. Cambridge University Press, Cambridge, NY, USA.

Neely, A., Adams, C. & Kennerley, M. 2002 *The Performance Prism: The Scorecard for Measuring and Managing Business Success*. Financial Times, Prentice Hall, London, UK.

Neely, A. D. & Adams, C. 2005 Perspectives on performance: the performance prism. In: *Business Performance Measurement: An Introduction* (S. S. Kambhammettu, ed.) Le Magnus University Press, India, pp. 229–248.

Nesheim, I., McNeill, D., Campbell, D., Barton, D., Stålnacke, P., Gooch, G. D., Rieu-Clarke, A., Saravanan, V. S., Berge, D., Beguería-Portugés, S., Bouraoui, F., Grizzetti, B., Joy, K. J., Machado, M., Manasi, S., Nhung, D. K., Paranjape, S., Portela, M. M., Raju, K. V. & Taron, A. 2008 Conceptual IWRM assessment methodology. STRIVER Report No. D5.2, The Centre for Development and The Environment, University of Oslo, Norway.

Nesheim, I., McNeill, D., Campbell, D., Barton, D., Stålnacke, P., Gooch, G. D., Rieu-Clarke, A., Saravanan, V. S., Berge, D., Beguería-Portugés, S., Bouraoui, F., Grizzetti, B., Joy, K. J., Machado, M., Manasi, S., Nhung, D. K., Paranjape, S., Portela, M. M., Raju, K. V. & Taron, A. 2009 Comparative assessment of IWRM methods. STRIVER Technical Brief no. 13. Available from: http://kvina.niva.no/striver/Portals/0/documents/STRIVER_TB13_IWRM_methods.pdf (accessed 31 August 2011).

Second World Water Forum and Ministerial Conference 2000 *Ministerial Declaration of The Hague on Water Security in the 21st Century*. Forum Press Release, 22nd March 2000, The Hague, Netherlands. Available from: http://www.waternunc.com/gb/secwwf12.htm (accessed 22 July 2011).

Swart, R. J., Raskin, P. & Robinson, J. 2004 The problem of the future: sustainability science and scenario analysis. *Global Environmental Change*, Elsevier **14**, 137–146.

UN – United Nations 2002 World Summit On Sustainable Development Reports, 26 August – 4 September 2002, Johannesburg, South Africa. Available from: http://www.un.org/jsummit/html/documents/summit_docs.html (last visited 25 July 2011).

Wolf, L. & Hoetzl, H. 2011 *SMART – IWRM: Integrated Water Resource Management at the Lower Jordan Valley, Project Report Phase 1*. KIT Scientific Publishing, Karlsruhe.

Wolf, L., Subah, A., Tamimi, A., Guttman, J., Bensabat, J., Mueller, R., Geyer, S., Sauter, M., Wolff, H. P., Tiehm, A., Riepl, D., Ali, W. & Hoetzl, H. 2011 SMART IWRM at the Lower Jordan River Basin – Reviewing models, results and uptake from large scale integrated water resources research. In: *Proceedings of the International Conference on Integrated Water Resources Management (IWRM)*, October 12–13, 2011, Dresden, Germany.

van der Zaag, P. 2005 Integrated water resources management: relevant concept or irrelevant buzzword? A capacity building and research agenda for Southern Africa. *Physics and Chemistry of the Earth* **30**, 867–871.

First received 6 January 2013; accepted in revised form 26 April 2013

Theme II

Groundwater management

Irrigated agriculture and groundwater resources – towards an integrated vision and sustainable relationship

Stephen Foster and Héctor Garduño

ABSTRACT

Globally, irrigated agriculture is the largest abstractor, and predominant consumer, of groundwater resources, with large groundwater-dependent agro-economies now having widely evolved especially in Asia. Such use is also causing resource depletion and degradation in more arid and drought-prone regions. In addition crop cultivation practices on irrigated land exert a major influence on groundwater recharge. The interrelationship is such that cross-sector action is required to agree more sustainable land and water management policies, and this paper presents an integrated vision of the challenges in this regard. It is recognised that 'institutional arrangements' are critical to the local implementation of management policies, although the focus here is limited to the conceptual understanding needed for formulation of an integrated policy and some practical interventions required to promote more sustainable groundwater irrigation.

Key words | groundwater, irrigated agriculture, irrigation efficiency, water resources management

Stephen Foster (corresponding author)
E-mail: *GWMATEfoster@aol.com* or *IAHfoster@aol.com*

Héctor Garduño
E-mail: *hector.garduno@live.com.mx*

BACKGROUND AND SCOPE OF PAPER

This paper is based upon the worldwide experience of the World Bank Groundwater Management Advisory Team (GW-MATE) in supporting 'public administrations' in their efforts to confront excessive groundwater resource use for agricultural irrigation during the period 2001–11. It focuses on using scientific understanding to achieve an integrated cross-sector vision of groundwater use in irrigated agriculture (the drivers, its benefits and risks, and possible management interventions) as a basis for the formulation of policy to promote a more sustainable relationship between irrigated agriculture and groundwater resources, rather than new scientific research.

The paper is based on detailed field work in a series of World Bank-supported 'pilot groundwater management projects' in South & East Asia and Latin America, whose setting and scope is summarised in Table 1, and which will be referenced where appropriate in this paper. These pilot projects covered a wide range of hydrogeological conditions (from some of the world's largest aquifers to much more localised groundwater bodies of limited potential) and agricultural production (from major irrigation of export crops to small-scale irrigation in subsistence farming). In all the aquifers of the pilot project areas there were growing concerns about groundwater depletion and/or incipient salinisation. Although potentially important, the leaching of agrochemicals to groundwater from irrigated agricultural soils was not generally considered by the pilot projects and is not discussed here.

doi: 10.2166/wst.2013.654

CONTEXT OF MANAGEMENT CHALLENGE

The 'global boom' in groundwater irrigation

The last 20–40 years have witnessed massive increases in groundwater use for irrigation in the world's more arid areas, which are subject to extended dry seasons and/or regular droughts (except as yet in Sub-Saharan Africa). Satisfactory statistics are now available through a UN-FAO initiative (Siebert *et al.* 2010), and indicate that 38% of the total cultivated land under irrigation (301 million ha) is equipped by water wells, with consumptive groundwater irrigation use being estimated at 545 km^3/a (43% of the total). The nations with the largest groundwater-equipped areas are India (39 million ha) and China (19 million ha).

From the outset it must be emphasised that access to groundwater has contributed greatly to increasing food security – principally by ensuring water availability at critical times in the crop-growth cycle and by mitigating the devastating effects of surface-water drought on crop yields.

Table 1 | Summary of groundwater problems and technical management interventions in selected GW-MATE pilot areas (after Garduño & Foster 2010)

Location of pilot areas/ experiences	Pre-existing groundwater resource status	Nature & outcome of technical intervention(s)
Guantao County, North China Plain	Very large dual-layered alluvial aquifer system in semi-arid area (500 mm/a); water-table decline of >0.5 m/a over 20 years (more in deeper aquifer)	General improvement in irrigation-water management on winter wheat has made >40 mm/a real water-saving. reducing water-table decline to 0.2 m/a
Central Punjab State, India	Thick alluvial peneplain aquifer in semi-arid region (500–800 mm/a); water-table decline of 0.5 m/a over 20 years (greater in more arid area and more generally in recent years)	Statewide statutory delay in transplanting paddy-rice (by 35–40 days to specified date) from 2008 has reduced consumptive groundwater use by >90 mm/a without impacting crop yields
Hivre Bazaar Micro-Watershed, Maharashtra, India	Shallow low-storage weathered hard-rock (basaltic) aquifer in semi-arid area (450 mm/a); almost completely depleted with widespread well-yield failures in dry season	Community ban on sugarcane/banana cultivation; major effort on recharge enhancement and dry-season crop planning based on antecedent water-table have recuperated aquifer system
Jaunpur Canal Command, Uttar Pradesh, India	Thick alluvial multi-layered aquifer in humid area (900 mm/a); soil waterlogging and salinisation in head-water zones but water-table falling to >10 m below ground level in tail-end zones 20 km distant	Farmer awareness plus investment in reducing canal seepage and excessive land application in head-water zones and diversification to higher-value crops in tail-end micro-management zones
Carrizal Aquifer, Mendoza, Argentina	Alluvial outwash peneplain in hyper-arid area (150 mm/a); rising salinity in shallow groundwater from irrigation returns and reduced aquifer discharge	Ban on relocation/deepening of saline water wells has constrained abstraction, and modifications to Mendoza River appear to have augmented recharge
Lower Ica Valley, Peru	Thick alluvial valley-fill aquifer in hyper-arid area (<50 mm/a); water-table decline of >0.5 m/a over past 10 years due to replacement of spate irrigation by pressurised systems	Efforts underway to restore recharge from flood-flows in Ica River by diversion to recharge canals/lagoons and to constrain irrigation abstraction in line with reduced estimates of recharge
Pampa Villacuri, Ica, Peru	Sedimentary aquifer of moderate thickness in hyper-arid area (50 mm/a); >2.0 m/a water-table decline with significant saline up-coning since introduction of irrigated commercial agriculture in late 1990s	Export asparagus production already has high irrigation efficiency and water-productivity, but efforts underway to reconcile with very weakly recharged aquifer and salinisation threat through surface-water transfer/artificial recharge
Silao-Romita Aquifer, Guanajuato, Mexico	Upland basin with layered volcanic and lacustrine deposits in semi-arid area (500 mm/a); shallow aquifer exhausted and deep aquifer depleting mainly due to intensive abstraction for alfalfa and maize production as livestock feed	Aquifer user management association formed and 'stabilisation plan' elaborated 10 years ago, but this has not conceptually or administratively confronted harsh reality of limited renewable groundwater resources

Groundwater is a 'very popular commodity' with most farmers since it is:

- Under their direct control for crop needs (given a reliable energy source for pumping)
- Usually found close to the point-of-use (often only a well's depth away)
- Well-suited to pressurised irrigation (and high-productivity precision agriculture).

A large proportion of the global investment in irrigation water wells has been on a private basis by individual farmers, albeit this has often been facilitated and stimulated by government through grants and low-cost loan finance, together in some cases with the provision of highly subsidised rural electrical energy for pumping.

General concerns about resource sustainability

In most regions that experience an extended dry season, consumptive water use by agriculture (if unconstrained) usually generates a demand for crop irrigation in excess of the availability of renewable groundwater resources, given that extensive areas of cultivatable land usually occur above aquifers. This situation has led to widespread depletion of groundwater resources with a number of collateral effects (Garduño & Foster 2010), which vary considerably in occurrence and intensity with hydrogeological setting:

- Counterproductive competition between irrigation users
- Conflicts with rural and/or urban drinking-water provision
- Incipient and progressive aquifer salinisation (occurring by a variety of different mechanisms), with serious long-term implications for agricultural productivity
- Degradation of important groundwater-dependent aquatic ecosystems.

Some discussion of the concepts of resource 'sustainability' and 'overexploitation' is relevant here, whilst not getting hung-up over semantics. Clearly all groundwater abstraction has an 'impact' – since it diverts flow from elsewhere in an aquifer system and reduces natural discharge. The real question is, when do such impacts become cumulatively significant? On economic criteria this would be when 'the sum cost of long-term third-party effects, environmental impacts and lost opportunity exceeds the short-term use benefits', but in practice these costs can be difficult to assess. Such an approach, however, does not address the 'efficiency-versus-equity issue' – given that less-depleted aquifers favour more equitable access and better protect ecological interests.

Accepting the harsh reality of weakly recharged aquifers

In areas where current average annual rainfall is less than 500 mm/a or so, the associated rate of diffuse groundwater recharge to shallow aquifers is sensitive to soil type and vegetation cover, and can fall off markedly. Moreover, even in more humid areas, deeper aquifers may also be weakly recharged due to physical isolation from the land surface. In all such cases groundwater-irrigated agriculture will have developed under conditions of limited contemporary aquifer recharge (less than 50 mm/a), or even 'non-renewable groundwater resources' (Foster & Loucks 2006), and there is need for public administrations and private groundwater users to come to terms with this reality and to plan accordingly by:

- Making every effort to ensure high efficiency and productivity of resource use
- Undertaking careful metering of groundwater use, with continuous monitoring and periodic evaluation of aquifer response
- considering the issue of intergenerational equity by investment in implementable 'exit-solutions', such as surface-water transfer and/or low water-use activities.

The nexus with rural electricity-supply policy

The major cost component of groundwater production (once water wells are constructed) is the energy required to lift water, which will depend on water-table depth, aquifer characteristics, well efficiency and unit energy price. Rural electricity pricing can thus be a useful tool to constrain groundwater abstraction, especially in the absence of volumetric measurement and/or administration capacity – although in cases where energy costs are only a small proportion of total crop production costs the scope may be restricted.

Paradoxically energy pricing is often used in the opposite way, with major subsidies of rural electricity in place to decrease farming costs (and reduce the price differential between groundwater and highly subsidised canal water) (Shah 2009). Although rural energy subsidies can be politically justified it has to be recognised that:

- The adoption of flat-rate rural electricity tariffs is perverse, since it results in farmers becoming completely insulated from groundwater resource status and water well inefficiencies, and thus from the energy consumption of crop production
- While it is legitimate to support poor farmers to improve their livelihoods, better targeted subsidies to cover part of their estimated energy bill are preferable since they incorporate an incentive to use water more efficiently.

An alternative approach, applicable in some situations, is regulating (and in effect rationing) the provision of electrical energy for groundwater pumping, which is showing much promise in combating excessive abstraction in the Gujarat Jyotigram Scheme, India (Verma & Shah 2008). This could be especially appropriate for weathered hard-rock aquifers, whose shallow groundwater production is characterised by rapidly escalating energy consumption with excessive drawdown – but parallel action has to be taken to deter corrupt practices, protect poor farmers and constrain use of alternative energy sources.

Need for action by public administrations

Where groundwater use is unsustainable it is appropriate to ask whether it is necessary for the public administration to intervene or better to allow nature to take its course through steadily rising groundwater production costs that will eventually act as a disincentive for excessive abstraction. However, the latter approach is usually unacceptable where:

- An aquifer system is susceptible to irreversible degradation from the intrusion or invasion of saline water, or other effects
- Village and small-town groundwater sources are negatively impacted, making it more difficult to achieve the Millennium Development Goals
- There are serious reductions of natural aquifer discharge, which impact unacceptably on 'downstream' water availability and groundwater-dependent ecosystems
- The user community is highly heterogeneous and water-table depletion would eliminate groundwater access for its poorer members, aggravating social inequality.

In promoting a more balanced approach to groundwater use in irrigated agriculture, which also values its other roles, public administrations also have to bear in mind that:

- Undesirable side-effects from resource exploitation can sometime commence well before groundwater abstraction exceeds average replenishment – and both these effects, and natural susceptibility to irreversible degradation, will vary considerably with hydrogeological setting
- Maintaining groundwater stocks against all depletion is rarely appropriate, especially in more arid regions where (given the long periodicity of major recharge events) groundwater storage is very important for mitigating the impacts of surface-water drought and for providing time to allow transition to lower water-use economies.

Pragmatic approach to management interventions

A fundamental paradigm that emerges from the GW-MATE experience is that 'the hydrogeologic setting of a given aquifer supporting irrigated agriculture both defines the groundwater resource problem itself and constrains potential management solutions' (Garduño & Foster 2010). Thus a 'one-size-fits-all' approach to groundwater resource management is inadequate – it being necessary to tailor the suite of instruments and measures deployed to local hydrogeologic setting and socioeconomic circumstance. Moreover adaptive management is advocated, with periodic assessment of progress and adjustment of approach, guided by monitoring and modelling of aquifer system behaviour.

GW-MATE has evolved a 'pragmatic framework' to guide the selection of the preferred management approach in any given situation to achieve an appropriate balance between:

- Groundwater resource administration, through use regulation and charging
- Community awareness raising, participation and self-regulation
- Investments in demand-side measures to reduce consumptive use of groundwater and (where feasible) supply-side techniques for enhancing groundwater recharge
- Command-and-control measures (such as water well drilling bans, prohibiting certain cropping practices and electrical-energy rationing)
- Groundwater demand constraints through macro-policy interventions (such as agricultural crop guarantee pricing and rural electrical-energy subsidies).

The pilot projects conducted and initiatives evaluated by GW-MATE (Table 1) have achieved useful (and in some cases inspiring) progress towards sustainable groundwater irrigation. But major reductions of consumptive water use have not been easy to achieve and some pilots have also revealed vulnerabilities of a socioeconomic and institutional character – although the main focus of this paper is to review the applicability and limitations of the various technical management interventions employed.

APPROACH IN 'GROUNDWATER-ONLY' IRRIGATION AREAS

Demand-side versus supply-side management

Mobilisation on rainwater harvesting and recharge enhancement measures will provide a useful initial focus for community participation in groundwater management. Thus some financial provision for this needs to be incorporated in management action plans. However, whilst 'managed aquifer recharge' should be encouraged, it is not usually the solution to groundwater resource imbalance and if pursued in isolation (rather than as part of a balanced suite of management measures) may merely result in increased demand. Moreover, volumetrically the effect of 'groundwater-friendly' agricultural land-use practices is generally more significant (because much larger land areas are involved).

The most direct approach to reducing groundwater irrigation demand (and consumptive use) is to constrain abstraction and effect a reduction in irrigated area.

However, without concomitant action to sustain farmer incomes, by increasing water-use productivity through improved husbandry to improve crop yields or cultivation of higher-value crops, this policy can prove very difficult to implement and sustain (Garduño & Foster 2010).

Potential for selected control of agricultural cropping practices

National food policies can be very important drivers of groundwater use, and improving their alignment with water resource objectives facilitates local management efforts – for example eliminating guarantee prices or subsidies for highly water-intensive crops (like sugarcane or paddy rice) in water scarce areas will greatly help groundwater management. Unilateral bans on such crops by community groundwater user associations may also form a critical component of local resource management. Another important intervention that can, in some cases, be taken at provincial government level is exercising control over the date of planting out paddy rice (Table 1). Moreover, any overview of groundwater use in irrigated agriculture has to challenge the wisdom of some long-standing agricultural practices, such as groundwater irrigation of animal feed (typically alfalfa and/or maize) in arid regions using (in some cases non-renewable) groundwater resources (Table 1).

Effectiveness of improving irrigation water-use efficiency

Mobilising finance for improving 'irrigation water efficiency' can be the key to increasing water-use productivity, reducing unit energy consumption, and a useful component of groundwater resource management action plans. But such improvements do not necessarily equate to 'real water resource savings', and without parallel investments in demand management the reverse often occurs (Foster & Perry 2010). This is because a substantial proportion of the so-called 'losses' associated with 'inefficient irrigation' are in fact returns to groundwater. Moreover, progressive changes from gravity (flood) irrigation to pressurised (drip) irrigation inevitably result in a substantial increase in groundwater consumptive use, even if actual abstraction is successfully capped.

An extreme example of the effect of land management change in irrigated agriculture on groundwater recharge (and thus on resource availability) is abandonment of the traditional practice of spate irrigation in mountain-front areas (in which land is deliberately flooded with wet-season run-off to encourage groundwater recharge and increase storage availability in the dry season) – very sound water resource conservation practice in mountainous arid regions. While there may be overriding agricultural reasons for abandoning such practice (since it is not compatible with investments in modern pressurised irrigation), if such decision is taken alternative methods of ensuring groundwater recharge from flood run-off will need to be introduced (Table 1).

OPPORTUNITIES FOR CONJUNCTIVE MANAGEMENT

Spontaneous conjunctive use by farmers

The spontaneous drilling of water wells by farmers, in and around major irrigation-canal commands on extensive alluvial plains, has occurred widely as a coping strategy (Shah 2009; Foster *et al.* 2010) in face of inadequate canal-water service levels associated with:

- Poor canal maintenance and inability to sustain design flows
- Poorly administered canal-water, allowing unauthorised or excessive off-takes
- Insufficient surface water availability for dry season diversion
- Rigid canal-water delivery schedules, unresponsive to crop needs.

These factors lead to high groundwater dependence, with excessive exploitation in tail-end sections.

In effect conjunctive use of groundwater and surface water, in some form or other and with varying degrees of effectiveness, is capable of achieving:

- Much greater water-supply security – by taking advantage of natural aquifer storage
- Larger net water-supply yield – than generally possible using only one source alone
- Better timing of irrigation-water delivery – since groundwater can be rapidly deployed to compensate for shortfalls in canal water at critical times in the crop-growth cycle
- Reduced environmental impact – by counteracting land waterlogging and salinisation.

It is noteworthy also that private groundwater use is often characterised by higher water productivity, despite (or perhaps because of) the fact that the unit cost of this water supply to the user is much higher (Foster *et al.* 2010).

Limits and threats to groundwater resource sustainability

It is very sound practice to use natural aquifer storage to buffer temporal and spatial variability in the availability of canal water for irrigation. But spontaneous (unplanned, unregulated and unmanaged) groundwater resource use sometimes results in aquifer depletion to water-table levels that complicate the deployment of low-cost ground-level lift pumps for irrigation and/or that induce saline groundwater encroachment.

Clearly there are upper limits on how much groundwater can be sustainably abstracted on a conjunctive basis with surface water for consumptive use in agricultural irrigation – which varies with hydrogeological setting and surface-water delivery scenario. Where groundwater storage is large (which is normally the case), it will be 'long-term recharge rates' (averages or trends) over the entire area under consideration that will constrain conjunctive use development – and the key issue is to find a balance of groundwater use which overall avoids long-term water table decline whilst also countering the rising water table and the menace of land waterlogging and soil salinisation.

Spontaneous conjunctive use sometimes encounters increasing groundwater salinity, which if not adequately diagnosed and controlled will result in a serious subsequent decline in agricultural productivity and threat to drinking-water supply security. The salinisation threat varies significantly down the length of major river basins (Figure 1), as well as with climatic regime, and requires detailed understanding and proactive management (Foster *et al.* 2010). It arises through a number of distinct mechanisms:

- Rising water-table due to excessive canal seepage and/or field application in head-water areas leading to soil water-logging and phreatic salinisation (Table 1), or sometimes naturally saline shallow groundwater becoming mobilised
- Leaching of soil salinity across irrigation areas on first habilitation of arid soils and/or salt fractionation by 'efficient' irrigation, with accumulation in tail-end sections of canal commands if no groundwater discharge/drainage occurs (Table 1)
- More classical intrusion and encroachment of saline groundwater due to excessive abstraction of fresh groundwater, both in arid inland basins and coastal areas
- Additionally there are hyper-arid areas in which virtually all groundwater is naturally saline, except where some infiltration from surface watercourses and irrigation canals forms 'freshwater lenses', which require very careful management.

It follows that lining of primary and secondary irrigation canals will be a high priority:

- On arid alluvial plains where the phreatic aquifer is naturally saline (with fresh groundwater confined at greater

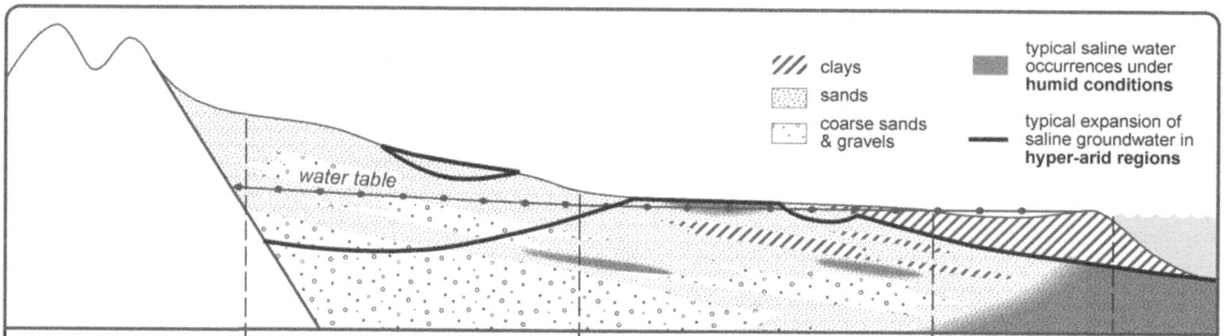

ALLUVIAL SETTING	OUTWASH & PENE-PLAIN	FLOOD PLAIN	COASTAL PLAIN
Hydrogeological Conditions	unconfined high-yielding aquifer with increasingly deep water table but rapid vertical recharge	multi-layered aquifer with shallow water table and moderate waterwell yields but with some 'rejected recharge'	confined aquifer with very restricted vertical recharge and reducing effective thickness
Water Resource Management Implications as in Humid Regions	much care needed to avoid 'double-accounting' due to infiltration of surface water via irrigation systems	poorly managed surface water irrigation results in soil waterlogging/salinisation; river base-flow used for irrigation can be reduced by 'upstream' consumptive use	aquifer highly vulnerable to saline water intrusion on intensive exploitation and reversal of natural hydraulic gradients
Water Resource Management Implications as in Hyper-Arid Regions	**high soil-profile salinity under natural vegetation mobilised when irrigated cultivation introduced**	**groundwater at depth naturally saline with freshwater restricted to lenses beneath major rivers/irrigation areas**	**surficial aquitard may also contain saline water which can seep downward**

Figure 1 | Schematic long-section of an alluvial groundwater system showing typical groundwater–surface water relationships and salinisation threats in a humid region, with comparative situation for a hyper-arid region.

depth), since canal seepage represents a 'non-recoverable resource loss' contributing to rising water table and soil salinisation

- On humid alluvial plains with rising water table in a shallow fresh groundwater system, since excessive canal seepage will be contributing to soil waterlogging and associated secondary salinisation.

In sharp contrast, on highly permeable alluvial terraces and peneplains (especially in more arid areas) the secondary and tertiary canal systems are often found to carry water for relatively few days per year, and the majority of irrigation users depend entirely on water wells, but with canal seepage being responsible for much aquifer recharge (Foster *et al.* 2010). An important corollary is that any attempt to line these canals to 'save water' for use in other areas can be very detrimental to existing users.

Advantages of planned conjunctive management

If conjunctive use can be more planned it offers a major opportunity of increasing agricultural production (through improvements in overall cropping intensity and irrigation water productivity) without compromising groundwater sustainability. Planned conjunctive use of groundwater and surface water for irrigated agriculture is also a preferred adaptation strategy for climate change. Serious impediments have to be overcome, however, for the implementation of such strategies (Foster *et al.* 2010), which are primarily institutional in character, given that provincial government organisations often simply mirror current water-use realities and tend to perpetuate the status quo, rather than offering an enabling structure for promotion of conjunctive management.

Integrated numerical modelling of irrigation canal flows, groundwater use and aquifer response, soil-water status and crop water-use are a great aid to evaluating the potential benefits of varying the spatial and temporal use of groundwater and distribution of surface water, and thus of improving conjunctive use efficiency and sustainability.

FORWARD LOOK

The greatly increased use of groundwater for irrigated agriculture in many developing nations over the past 15–25 years has resulted in widespread excessive exploitation, but does not yet represent a 'resource crisis' because the large volumes of groundwater in aquifer storage can generally buffer 'over-exploitation' for numerous years. But resource sustainability issues need to be confronted, especially where there is threat of insidious mobilisation or accumulation of groundwater salinity and/or where a significant component of the groundwater resources abstracted are non-renewable.

In most developing nations, groundwater resource accounting in areas of irrigated agriculture remains rather weak. This problem has a number of facets:

- Little momentum towards universal metering of larger abstractions and thus inevitable uncertainty over resource use (because of the limitations of indirect estimation)
- Restricted dialogue and mutual understanding between agronomists and hydrologists of soil-water balances for irrigated cropping on permeable soils and of seepage from irrigation-canal networks for aquifer recharge.

This inevitably often means that the only data to guide groundwater management are water-table trends, with the handicap that these are usually tardy indicators and often cannot be directly related to specific cropping and irrigation practices.

Integrated policy to reduce groundwater consumptive use and improve resource sustainability whilst maintaining or increasing farmer incomes should focus on:

- Cultivating higher-value crops on smaller irrigated areas – the rising demand for 'precision irrigation' with pressurised systems offers an adaptable platform for this, but whether it follows a 'sustainable path' will depend on the detail of irrigation-water management (and whether 'real water-resource savings' are pursued and groundwater use rights or allocations are capped in consumptive use terms), and there will also be market-related and risk-defined limits on the scope for such conversion
- For the critical staple-crop (wheat, maize, rice, etc) production the need is to increase yields from smaller irrigated areas – this can only be achieved through improving soil management, seed density and type, fertiliser and pesticide use to eliminate nutrient constraints or pest impacts on crop growth, but could impact on groundwater recharge (quantity and quality) through increasing both consumptive use per unit area and nutrient and/or pesticide leaching.

While the impact of climate change on groundwater replenishment (and on long-term sustainable resources) remains uncertain, and requires more detailed monitoring and analysis before reliable predictions can be made, it is clear that groundwater storage reserves will be a critical element in climate-change adaptation to confront more frequent and extended droughts, and through providing time for socioeconomic transformation. Other climate-change responses, however, such as the stimulation of biofuel (sugarcane, soya beans, maize, etc) cultivation, could imply more groundwater irrigation and also increase pressure to extend the 'irrigation frontier'.

ACKNOWLEDGEMENTS

The authors wish to express thanks to the World Bank GW•MATE Programme Managers (Karin Kemper, Catherine Tovey & Amal Talbi), the World Bank-Water Anchor (under Abel Mejia & Julia Bucknall) and the Global Water Partnership-Secretariat (Ania Grobicki & Aurelie Vitry) for facilitation of the work on which this paper is based. The valued support and encouragement of Jacob Burke (UN-FAO) and Mohamed Ait-Kadi (GWP-Technical Committee Chair) is also acknowledged. It should, however, be recorded that the opinions expressed are those of the authors alone and not necessarily of the World Bank Group or the Global Water Partnership.

REFERENCES

Foster, S. & Loucks, D. P. 2006 *Non-renewable Groundwater Resources–a Guidebook on Socially-sustainable Management for Water-policy Makers*. UNESCO IHP-VI Series on Groundwater 10, Paris, France.

Foster, S. S. D. & Perry, C. J. 2010 Improving groundwater resource accounting in irrigated areas: a prerequisite for promoting sustainable use. *Hydrogeology Journal* **18**, 291–294.

Foster, S., van Steenbergen, F., Zuleta, J. & Garduño, H. 2010 Conjunctive use of groundwater and surface water–from spontaneous coping strategy to adaptive resource management. *GW • MATE Strategic Overview Series* **2**. http:/www.worldbank.org/gwmate (accessed 15 June 2011).

Garduño, H. & Foster, S. 2010 Sustainable groundwater irrigation–approaches to reconciling demand with resources. *GW•MATE Strategic Overview Series* **6**. http:/www.worldbank.org/gwmate (accessed 15 June 2011).

Shah, T. 2009 *Taming the Anarchy: Groundwater Governance in South Asia*. Resources for Future Press, Washington DC, USA.

Siebert, S., Burke, J., Faures, J. M., Frenken, K., Hoogeveen, J., Doell, P. & Portman, F. T. 2010 Groundwater use for irrigation – a global inventory. *Hydrology and Earth System Science* **14**, 1863–1880.

Verma, S. & Shah, T. 2008 Co-management of electricity and groundwater: an assessment of Gujarat's Jyotigram Scheme. *Indian Economic and Political Weekly* **43** (7), 59–66.

First received 3 January 2013; accepted in revised form 15 October 2013

An expert system for real-time well field management

B. S. Marti, G. Bauser, F. Stauffer, U. Kuhlmann, H.-P. Kaiser and W. Kinzelbach

ABSTRACT

Well field management in urban areas faces challenges such as pollution from old waste deposits and former industrial sites, pollution from chemical accidents along transport lines or in industry, or diffuse pollution from leaking sewers. One possibility to protect the drinking water of a well field is the maintenance of a hydraulic barrier between the potentially polluted and the clean water. An example is the Hardhof well field in Zurich, Switzerland. This paper presents the methodology for a simple and fast expert system (ES), applies it to the Hardhof well field, and compares its performance to the historical management method of the Hardhof well field. Although the ES is quite simplistic it considerably improves the water quality in the drinking water wells. The ES knowledge base is crucial for successful management application. Therefore, a periodic update of the knowledge base is suggested for the real-time application of the ES.

Key words | decision support, expert system, groundwater, urban well field, well field management

B. S. Marti (corresponding author)
F. Stauffer
W. Kinzelbach
Institute of Environmental Engineering,
ETH Zurich, Wolfgang-Pauli-Strasse 15,
8093 Zurich,
Switzerland
E-mail: *marti@ifu.baug.ethz.ch*

G. Bauser
Former affiliation: Institute of Environmental Engineering,
ETH Zurich, Wolfgang-Pauli-Strasse 15,
8093 Zurich,
Switzerland
New affiliation: Camille Bauer AG,
Aargauerstrasse 7, 5610 Wohlen,
Switzerland

U. Kuhlmann
TK Consult AG, Seefeldstrasse 287,
8008 Zurich,
Switzerland

H.-P. Kaiser
Stadt Zurich Wasserversorgung,
Hardhof 9, 8021 Zurich,
Switzerland

INTRODUCTION

Groundwater is the most important drinking water resource in Switzerland. The natural reservoirs typically lie in valleys and are mainly fed by infiltrating water from rivers. However, the majority of the Swiss population live in valleys close to rivers. Thus, pressure on the quality of the drinking water resource is high. Industrial sites, transportation lines for hazardous materials, sewage lines and old waste disposal sites result in a high potential for groundwater contamination and constitute a considerable challenge for well field operation. In this paper we present the methodology for the design of an expert system (ES) for real-time well field management and apply it to a case study: Hardhof well field in Zurich, Switzerland.

Well field management became popular in the 1980s with the optimization of pump and treat schemes in groundwater remediation (e.g., Gorelick 1983; Gorelick & Voss 1984). The solutions to the optimization problems were obtained analytically for simple groundwater flow configurations. However, these analytical solutions are not applicable to general groundwater flow problems with complex geometry. Alternative management strategies for reservoirs have been proposed for synthetic problems (e.g., Gordon *et al.* 2000). The first successful applications of real-time management of underground reservoirs were implemented in the oil industry (e.g., Saputelli *et al.* 2006). With the increasing availability of online measurements of hydraulic head and water quality indicators, real-time management of groundwater becomes relevant (e.g., Bauser *et al.* 2010; Cheng *et al.* 2011).

Site description

The Hardhof well field is fed directly by the river Limmat in the north and west. In the south, it is artificially

doi: 10.2166/ws.2013.021

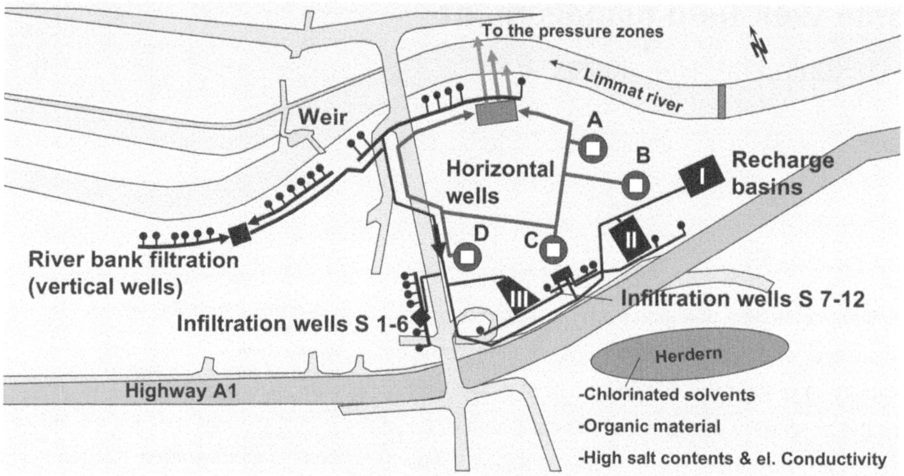

Figure 1 | Overview of the Hardhof well field. River bank filtrate is drawn in 19 bank filtration pumps and recharged to the aquifer through three infiltration basins (I, II, and III) and 12 infiltration wells (S1 to S12). Drinking water is drawn in the four horizontal wells A, B, C, and D and pumped to the different pressure zones of the drinking water distribution network of the city of Zurich.

recharged with river bank filtrate in three infiltration basins and 12 infiltration wells (Figure 1). Through this recharge, the capacity of the well field is increased and a hydraulic barrier against potentially contaminated water from the former waste disposal site, Herdern, is formed. The dimension of the solute plume spreading from the Herdern site westwards was determined by Jäckli (1992). In 2001, Kaiser analyzed the origin of the water drawn in the four horizontal wells (Kaiser 2001). Based on these studies the water in the four drinking water wells can be assigned either to river bank filtrate (low electrical conductivity (EC) close to that of the river Limmat) or to city water (groundwater coming from below the city area and possibly passing through the Herdern site, characterized by elevated EC). A detailed description of the Hardhof well field including technical data can be found in Bauser *et al.* (2010).

Central to this well field is the fact that the drinking water quality is influenced not only by pumping rates in the production wells but also to a considerable part by the artificial recharge. Although ES may be applied for the management of any well field this work focuses on well fields heavily influenced by artificial recharge.

For this study, daily measurements of hydraulic head in 85 locations were available for the years 1992 to 2011. The locations of the piezometers are given in Huber *et al.* (2011). Furthermore, the daily pumping rates of all wells and basins as well as daily discharge data of the rivers Limmat and Sihl, precipitation and climate data were available.

METHODOLOGY

An ES is a collection of logic rules mimicking expert knowledge. It is a widely applied decision support system (e.g., Shu-Hsien 2005). The principal layout of an ES follows a tree of if-then relationships, leading from a given situation to a decision. The algorithm follows human reasoning and therefore it is intuitive and easy to understand. In order to establish the if-then relationships, a knowledge base has to be built. In our case, this was done by analyzing the daily time series of model input and output data.

For the design of the ES we proceed in four steps:

- Phase 1: Modeling the well field.
- Phase 2: Building the knowledge base; expert knowledge is gathered, and the control and actuating variables that influence the control variable are determined.
- Phase 3: Setting up the rule base of the ES; the knowledge is organized in a set of if-then relationships leading from a given situation to a decision.

- Phase 4: Identifying parameters and adjusting the rule base; the ES is iteratively calibrated on the design period and validated on a new set of data.

Phase 1: Modeling of the well field

The design of a new management strategy often cannot be undertaken in the real well field because flawless operation of the well field cannot be guaranteed during the design phase. Therefore, a model of the well field is needed which reproduces the relevant processes of the real well field with an appropriate accuracy over the management horizon.

In the present case, drinking water quality has to be maintained at an appropriate level. The relevant information for management is the distribution of hydraulic head in the aquifer and the quality of the water in the four drinking water wells. The management horizon in this case study is 1 day. The well field was modeled with a transient three dimensional finite element groundwater flow model (SPRING® 3.4, Delta 2006) and calibrated with a modified pilot point method similar to Alcolea *et al.* (2008). Because of computational constraints (large number of nodes) and the large uncertainty of the input data for solute transport (i.e., the initial and boundary conditions), fully coupled flow and solute transport has not been feasible for the Limmat valley aquifer up to now. Bauser *et al.* (2010) therefore implemented a particle tracking tool which they coupled to the groundwater flow model to verify the origin of the water in the drinking water wells. Particles are tracked back from the four horizontal wells along a quasi-stationary flow field until they either pass the virtual boundary between Hardhof and city area (along Highway A1 in Figure 1) or it is clear that the particle path lines do not lead to the city area (i.e., when they reach a control boundary, e.g., the river bank). Thus Bauser *et al.* (2010) were able to compute the fraction of water in each well coming from the city area, hereafter called fraction of city water. They further showed that the fraction of city water can be used as an indicator for water quality in the drinking water wells. A detailed description of the particle tracking tool is given in Bauser *et al.* (2012). More details on the model can be found in Hendricks Franssen *et al.* (2011). We have used their deterministic central model in the present study.

Phase 2: Building of the knowledge base

A throughout process oriented understanding of the system was necessary for the modeling of the system. In this phase, knowledge about the system is organized in a control oriented manner: the control variable as well as the relevant disturbances are determined. In the case of well field management, the control variable can be the water quality produced or the pumping rates in the wells. In this work we focus on the drinking water quality. Relevant disturbances with regard to drinking water quality in a well field are processes which dominate the flow field (i.e., recharge and abstraction of groundwater) or sources of pollution (e.g., periodicity or threshold behavior). In order to be controllable, the disturbances have to be known. Unknown or minor disturbances are treated as uncertainties of the model.

Figure 2 shows the concept of a control oriented system description where Σ denotes the model of the system to be managed. The simulated water quality (the output of the model Σ) is fed back to the input of the ES. We will refer to it as the feedback variable.

Already, the major sources and sinks of the system are known from phase 1, the modeling of the system. Delay times between the disturbances and the control variable may be found through an analysis of Pearson's cross-correlation between each major disturbance and the control variable. The delay times of the system should be verified for example with an estimation of travel times in the aquifer and taken into account in the input of the ES.

Phase 3: Setting up of the rule base

In order to cope with multiple inputs to the ES we propose combining the impacts of the disturbances and the feedback variable on the simulated water quality in one artificial variable. This variable contains the

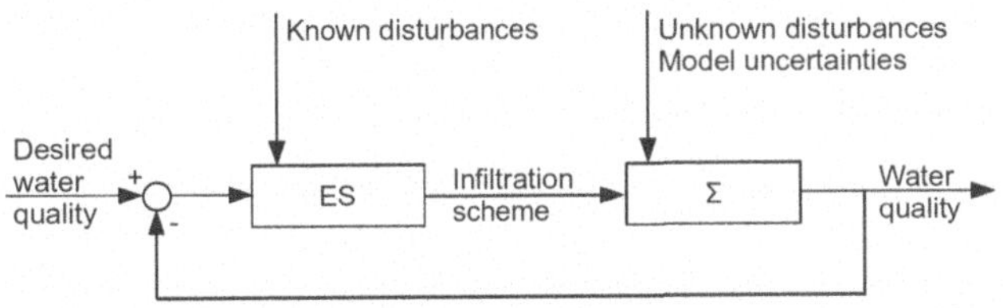

Figure 2 | Concept of the control loop. Σ denotes the model of the system to be managed. Note that the simulated water quality is fed back to the input of the ES.

information about how probable a reduced water quality in the current management horizon is. First, each of the $i = 1, \ldots, N$ disturbances and the feedback variable are attributed a weight w_i. As a first guess this weight corresponds to the maximum of Pearson's cross-correlation. The weight w_i accounts for the relative impact of the input variable i to the ES on the output of the model. The following steps are accomplished for each management horizon:

- For a first guess, the input variables are classified approximately according to their 30th, 60th and 90th percentile and attributed a value v_i. For example invalid measurements are classified as 'no measurement' and attributed the value 0. Values below the 30th percentile are classified as low and attributed the value 1. See Table 1 for an overview of the classes and the corresponding values.
- The values of the input variables are then multiplied with the weights of the variables and summed up over all input variables $\Sigma(w_i \cdot v_i)$. The resulting value is a measure for the probability of an impairment of the drinking water quality in the given time step.
- The resulting value is again classified into low, medium, high and extreme, and an infiltration scheme is attributed. The infiltration scheme determines the amount for artificial recharge and (if necessary) its distribution to the artificial recharge infrastructure(s). For the determination of the infiltration scheme, expert knowledge again plays a central role. The artificial recharge modifies the flow field of the groundwater and the expert has to understand in which way it has to be changed in order to maintain an appropriate water quality.

Phase 4: Identification of parameters

The well field management is now simulated with the ES management and the parameters of the ES (weights, values of the classes) are adapted manually where needed. If the calibration of the ES is satisfactory, a validation period is simulated.

RESULTS AND DISCUSSION

An ES was designed according to the methodology described above to manage the water quality in the Hardhof well field. The simulation of the well field management is based on historical data. The following paragraphs describe the design phases 2 to 4 (phase 1, the modeling of the well field, is described in detail in Bauser *et al.* (2010), and Hendricks Franssen *et al.* (2011)).

Phase 2: Building of the knowledge base

The goal of the ES is to maintain a high water quality in the Hardhof well field by adapting the amount and distribution of water infiltrated in basins I to III and in the injection wells S1 to S12. Due to its location close to the southern boundary of the well field, well C is the most vulnerable one from which to withdraw city water. This is confirmed by the measurement of EC in the

Table 1 | The values of the classes and the initial and adjusted parameters of the ES

		Initial guess			Adjusted values		
Class	**Value [-]**	**$Q_C(t)$ [m³/d]**	**$fr_C(t-1)$ [-]**	**$h_L(t-14)$ [m a.s.l.]**	**$Q_C(t)$ [m³/d]**	**$fr_C(t-1)$ [-]**	**$h_L(t-14)$ [m a.s.l.]**
No value	0	≤ 0	≤ 0	≤ 395.0	≤ 0	≤ 0	≤ 396.0
Low	1	$\leq 2'500$	≤ 0.05	≤ 398.9	$\leq 2'500$	≤ 0.05	≤ 398.9
Medium	10	$\leq 6'500$	≤ 0.10	≤ 398.4	$\leq 6'500$	≤ 0.10	≤ 398.5
High	100	$\leq 12'000$	≤ 0.20	≤ 398.2	$\leq 10'000$	≤ 0.20	≤ 398.3
Extreme	1,000	$> 12'000$	< 0.20	> 398.2	$> 10'000$	> 0.20	> 398.3
Weight w	–	0.5	0.8	0.2	1	0.8	0.8

drinking water wells, with highest values in well C. For the ES, we thus only use the fraction of city water in well C fr_c from the model output. For the decision about the infiltration scheme of the next time step the fraction of city water of time step t–1 is fed back to the input of the ES. The desired water quality is zero percent of city water in well C.

In this study we have assumed that the Herdern deposit constitutes a constant source of pollution. Therefore, we can concentrate on the flow field as the major influence on the quality of the drinking water. The factors influencing the flow field are recharge and abstraction from the aquifer. The major abstractions in the Hardhof area are effected by the Zurich water works. The rates are determined by the demand of the city of Zurich and regarded as known disturbances in this work. So is the distribution of the abstraction in the four drinking water wells. Recharge has two important components in the Hardhof area: infiltration from the river Limmat and artificial recharge. The rate of infiltration from the river depends highly on the water level in the river (h_L). The water level can be determined on a daily basis but it cannot be influenced by the Zurich water works, it is therefore a known disturbance. The artificial recharge in the Hardhof area is determined by the Zurich water works. It can be used by management to act on the system in order to maintain an appropriate drinking water quality. The degrees of freedom for management can be reduced from 12 + 3 (12 injection wells + 3 recharge basins) to 6 because the 12 injection wells cannot be operated individually but in the following groups: group 1 contains S1 to S6, group 2 corresponds to well S7, group 3 contains S8, S9, and S10, and group 4 contains S11 and S12. Group 1 lies downstream of the well field and is never operated.

The cross-correlation analysis between EC measurements in the river Limmat, the horizontal wells, the infiltration infrastructures, and a representative piezometer in the city revealed the typical lag times (in days) between the measurement locations.

The following disturbance variables were found to have the highest correlation with EC in well C on a given day t: Abstraction in well C on day t, $Q_c(t)$, fraction of city water in well C on day t–1, $fr_c(t-1)$, and water level of the river Limmat on day t–14, $h_L(t-14)$. The corresponding correlations between the input variables and the water quality measure are given in Table 1 in the form of the weights w of the variables (initial guess). The lag times correspond to the average travel times of the water found with tracer tests (e.g., Kaiser 2001).

Phase 3: Setting up of the rule base

In phase 2, two known disturbances and one feedback variable were identified and their influence on water quality was estimated (initial guess of the weights w_i in Table 1). The initial estimates of the boundaries between the classes for each ES input variable are determined based on the cumulative frequency distribution of daily values from 2004 for each variable (Table 1). The initial estimate for the infiltration scheme is given in Table 2. The infiltration focuses on injection well groups 2, 3 and 4 and basins I, II and III.

An exemplary path through the ES algorithm for the computation of the infiltration scheme of day t could look like this (the numbers refer to the initial guess of the parameters of the ES): the river water level on day t–14 was

Table 2 | Infiltration scheme for the initial and the adjusted parameters of the ES. The daily total abstraction rate is multiplied by the factor f and distributed between the basins and the injection wells with the ratios given for each class. Among the basins and the injection well groups, the infiltration rate is further distributed according to the ratios given here below

	Initial guess				Adjusted values			
	Low	Medium	High	Extreme	Low	Medium	High	Extreme
Factor f	1	1	1.3	1.5	1	1.2	1.2	1.5
Basins:wells	2:1	2:1	1:1	1:2	2:1	1:1	1:1	1:2
B I:II:III	1:2:2	1:6:4	1:6:4	1:6:5	1:2:2	1:6:4	1:6:4	1:6:5
Group 2:3:4	1:3:2	1:4:2	1:4:2	1:4:2	1:3:2	1:4:2	1:6:2	2:4:2

below 397 m a.s.l. It is therefore classified as low and attributed the value 1. This value is multiplied with the weight 0.2 yielding 0.2. Then, the abstraction rate of day t is classified as high, attributed the value 100, multiplied with the corresponding weight of 0.8 (yielding 80), and summed up with the result of the previous decision (yielding $80+0.2=80.2$). The same procedure is repeated for the fraction of city water in well C on day t–1. The resulting value is again classified. The class is usually higher or the same as the highest classified input variable. A classification of low means that the danger of attracting city water into the well is low and less infiltration is needed in order to maintain the hydraulic barrier.

In order not to allow a depletion of the aquifer, the minimum total infiltration rate is set equal to 1–1.5 times the total abstraction rate in the Hardhof well field on a given day, depending on the classification of the infiltration scheme ($Q_{\text{infiltration}} = f \cdot Q_{\text{abstraction}}$, with $f \in [1,1.5]$, Table 2).

Phase 4: Identification of parameters

The ES management is now simulated with historical data from 2004. Iteratively, the parameters of the ES are tuned in order to reduce the fraction of city water in the drinking water wells. Thereby, the visualizations of particle path lines for selected days are consulted. A significant reduction of the fraction of city water is achieved after three iterations. A first validation period in January 2005, however, exhibits a very low groundwater table which never occurred during 2004 and where the ES management does not perform satisfactorily.

Phases 2 to 4 are completed again with an extended knowledge base from January 1992 to January 2005. Although the statistics of the prolonged time series change, the delays between the input variables and the water quality remain the same. After three iterations of manual parameter tuning, the water quality computed with the ES management between January 2004 and February 2005 was again reduced considerably compared to the historical management scheme (Figure 3). The final adjusted parameters of the ES are given in Tables 1 and 2. The weights of the adjusted parameter set do not refer to the statistics of the knowledge base any more but have a very similar weight. As an alternative to the correlations, the input variables of the ES could be given the same weight as an initial guess.

Figure 4 shows the particle path lines before (on the left) and after (on the right) tuning of the parameters. On the right side of Figure 4 no particle path lines computed with the ES management cross the boundary line between Hardhof area and city area, whereas this is not the case in the left hand figure. Accordingly, the fraction of city water is also smaller after the tuning of the parameters than before tuning. The figure further shows the particle path lines of the historical management where the fraction of city water was elevated (also reflected in the elevated EC measurement). Compared to the historical management, the

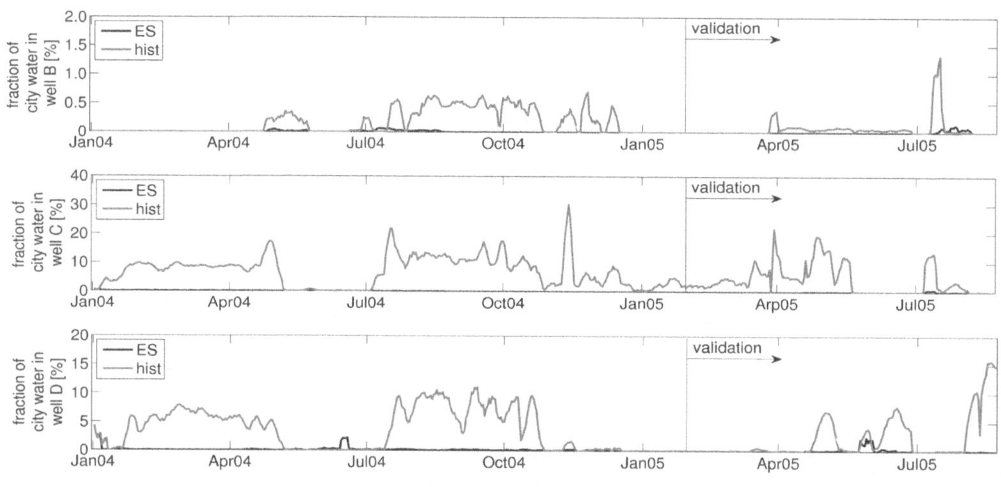

Figure 3 | Fraction of city water of the calibration and validation period of the ES with extended knowledge base. The validation period starts on 1 February 2005 and is marked with a vertical line. The fraction of city water in well A is zero at all times for the historical as well as for the ES management. The fraction of city water in the drinking water wells computed with the historical management (hist) are given in gray.

Figure 4 | Particle path lines on 4 May 2004 (simulation day number 124) for the initial parameters (dark gray in the left figure) and the iterated final parameters (dark gray in the right figure). The light gray lines were computed with the historical well field operation. The black line south of the well field describes the boundary between Hardhof and the city water area.

infiltration rate in basin III was increased by a factor of 2 using the adjusted ES. While with the historical management scheme, less water was infiltrated than abstracted for several days before 4 May 2004 (path lines for this day are depicted in Figure 4), the ES management infiltrates 50% more water than is abstracted during the same period and thus succeeds in maintaining the fraction of city water at a low level (see Figure 3).

The validation period was prolonged to 8 months from February to August 2005. The ES with an enlarged knowledge base and tuned parameters performed well during the validation phase (Figure 3). Except for two peaks of 3% of city water in well D in June 2004 and June 2005, all peaks of city water in the drinking water wells are reduced significantly with the ES management scheme. The two peaks are caused by particle path lines crossing the boundary line twice (see Figure 5). Even though the path lines pass through the potentially contaminated city area, the water quality in the well is not endangered since the water originates from one of the

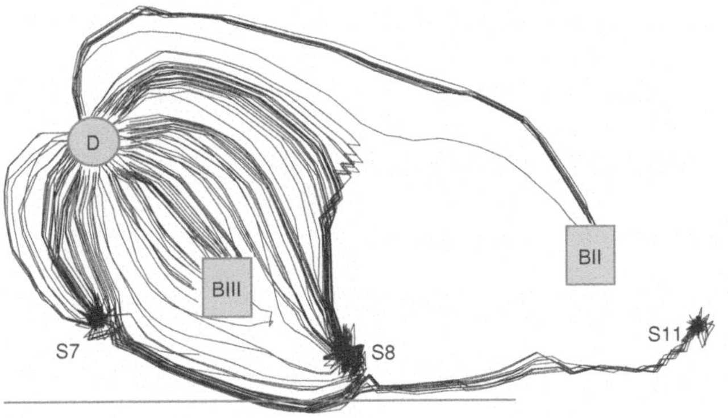

Figure 5 | Particle path lines from well D on simulated day 515 at the beginning of June 2005.

injection wells of the water works. The fraction of city water could be reduced by increasing the infiltration rates. This is not recommendable, however, because excessive infiltration could lead to water logging in the Hardhof area.

The ES management holds a comparison with the optimal control presented in Bauser *et al.* (2012). Even though the infiltration scheme found with the ES is not optimal, it nevertheless reduces the amount of city water in the drinking water wells to an acceptable level with a reasonable simulation time (because no iterative model runs are needed).

Applicability to other settings

The design procedure of the ES presented here may be applied to an arbitrary well field management problem, given that the problem can be solved. The essential limitation of the ES is the knowledge base. The management results can only get as good as the knowledge base is wide. If essential processes influencing the well field are unknown the application of the presented method yields unreasonable results.

A drawback of the ES is the need for manual adjustment of the parameters. The procedure is tedious and prone to conceptual errors (i.e., if the understanding of the system is not profound enough). A periodic updating of the knowledge base and the parameters of the ES with the newest data is recommendable for management application in a real well field. The implementation of the ES in a real well field is assumed to be straight forward because the rule base is intuitively understandable without background knowledge in control engineering, which enhances its acceptance among professionals operating the well fields.

Model uncertainty and unknown disturbances

In the presented methodology we do not consider model uncertainties and unknown disturbances. However, the model is assumed to yield a conservative estimate of the water quality (Bauser *et al.* 2010). So we base management decisions on a cautious estimate of the fraction of city water in the wells.

CONCLUSIONS

In this paper, the methodology for the design of an ES for well field management is presented. The applicability of the method was demonstrated in a study of the Hardhof well field.

It has been shown that, given that the main processes influencing the well field are known, the ES is an efficient alternative for well field management. We propose periodic updating of the knowledge base and the parameters of the ES to newly available data in order to maintain good performance.

Although the ES management is not optimal, it is close to it, as shown by a comparison with Bauser *et al.* (2010, 2012). Furthermore it features a small computation time and a simple, intuitive structure which are key advantages for implementation in a well field.

REFERENCES

Alcolea, A., Carrera, J. & Medina, A. 2008 Regularized pilot points method for reproducing the effect of small scale variability: application to simulations of contaminant transport. *Journal of Hydrology* **355**, 76–90.

Bauser, G., Hendricks Franssen, H.-J., Kaiser, H.-P., Kuhlmann, U., Stauffer, F. & Kinzelbach, W. 2010 Real-time management of an urban groundwater well field threatened by pollution. *Environmental Science and Technology* **44**, 6802–6807.

Bauser, G., Hendricks Franssen, H.-J., Stauffer, F., Kaiser, H.-P., Kuhlmann, U. & Kinzelbach, W. 2012 A comparison study of two different control criteria for the real-time management of urban groundwater works. *Journal of Environmental Management* **105**, 21–29.

Cheng, W.-C., Putti, M., Kendall, D. R. & Yeh, W.-G. 2011 A real-ime groundwater management model using data assimilation. *Water Resources Research* **47**, W06528.

Delta, H. 2006 Software SPRING® 3.4. Ingenieurgesellschaft GmbH, Parkweg 67, 58453 Witten, Germany.

Gordon, E., Shamir, U. & Bensabat, J. 2000 Optimal management of a regional aquifer under salinization conditions. *Water Resources Research* **36** (11), 3193–3203.

Gorelick, S. M. 1983 A review of distributed parameter groundwater management modeling methods. *Water Resources Research* **19** (2), 305–319.

Gorelick, S. M. & Voss, C. I. 1984 Aquifer reclamation design: the user of contaminant transport simulation combined with nonlinear programming. *Water Resources Research* **20** (4), 415–427.

Hendricks Franssen, H.-J., Kaiser, H.-P., Kuhlmann, U., Bauser, G., Stauffer, F., Müller, R. & Kinzelbach, W. 2011 Operational real-time modeling with EnKF of variably saturated subsurface flow including stream-aquifer interaction and parameter updating. *Water Resources Research* **47**, W02532.

Huber, E., Hendricks Franssen, H.-J., Kaiser, H.-P. & Stauffer, F. 2011 The role of prior model calibration on predictions with Ensemble Kalman Filter. *Groundwater* **49** (6), 845–858.

Jäckli 1992 *Abstell- und Unterhaltsanlage SBB Zürich-Herdern, Hauptuntersuchung Umweltverträglichkeit, Teilbericht Hydrogeologie (Environmental Impact Assessment Report for the Industrial Site Herdern in Zurich, Section on Hydrogeology)*. Geologisches Büro Dr. Heinrich Jäckli AG, Zurich, Switzerland.

Kaiser, H.-P. 2001 *Tracerversuche (Tracer Tests)*. Stadt Zürich Wasserversorgung, Zurich, Switzerland, unpublished.

Saputelli, L., Nikolaou, M. & Economides, M. 2006 Real-time reservoir management: A multiscale adaptive optimization and control approach. *Computers and Geosciences* **10** (1), 61–96.

Shu-Hsien, L. 2005 Expert system methodologies and applications – a decade review from 1995 to 2004. *Expert Systems with Applications* **28**, 93–103.

First received 3 January 2013; accepted in revised form 12 March 2013

Riverbank filtration in India – using ecosystem services to safeguard human health

C. Sandhu and T. Grischek

ABSTRACT

India has great potential to use riverbank filtration (RBF) for drinking water production as an ecosystem service for human health, principally through effective removal of common waterborne pathogens, even during monsoon. Water quality results from site investigations in North India have shown a removal of total and faecal coliform (indicator) bacteria in the range of 1.3 to >5.2 log for total coliforms and 2.3 to >4.2 log for faecal coliforms at the bank filtration schemes of Haridwar, Nainital, Patna, and Mathura. At rural RBF sites, where bank filtrate is collected and supplied by Koops ('well' in Hindi), a removal of 1.0–3.4 log and 0.3–2.8 log was observed for total and faecal coliforms respectively. At the RBF sites in Haridwar and Patna, there was only minimal breakthrough of coliforms during monsoon floods, for which disinfection using conventional chlorination was sufficient.

Key words | coliform removal, drinking water, ecosystem service, monsoon, riverbank filtration

C. Sandhu (corresponding author)
T. Grischek
University of Applied Sciences Dresden,
Faculty of Civil Engineering & Architecture,
Division of Water Sciences,
Friedrich-List-Platz 1,
D-01069 Dresden,
Germany
E-mail: *sandhu@htw-dresden.de*

INTRODUCTION

The resources and processes provided by natural ecosystems for the benefit of humans, including the provision of drinking water, are as a whole termed ecosystem services. High quality drinking water is paramount to human health. Aquifer recharge through induced infiltration of surface water from rivers and channels, termed riverbank filtration (RBF), is an important process which is used the world over to provide raw water for drinking water production and industrial water use. Large cities and industrial centres often developed at locations where surface water was available for water supply and transport, but the water quality has degraded as development progressed. Compared to direct abstraction of surface water for drinking water supply, RBF provides the following advantages:

- Protection against contamination by chemicals and pathogens.
- Sufficient water treatment to meet drinking water quality standards at some sites.
- Cost savings in water treatment if RBF is used as a pretreatment step.

These advantages are a direct result of the natural purification properties of the aquifer, an integral part of the ecosystem, and combine to yield (pre)treated water for drinking purposes.

RBF also provides sufficient water quality for irrigation even if the surface water is polluted by pathogens. Investigations at the Zarqa River, Jordan, demonstrated that faecal indicator bacteria and bacteriophages were removed from river water by RBF by 3.4–4.2 log and 2.7–3.3 log, respectively (Saadoun *et al.* 2008). In a well used for irrigation in Muzaffar Nagar, by the Kali River in the state of Uttar Pradesh in North India, total and faecal coliforms were removed by 1.7–1.9 log and >1.5 log respectively (Thakur *et al.* 2009). Furthermore, RBF is a type of naturally occurring and induced surface water – groundwater interaction and can be managed to enhance sanitation of surface water bodies due to very

doi: 10.2166/ws.2013.054

efficient sorption and biodegradation processes in the hyporheic zone. Most RBF sites are robust against predicted climate change, especially to a higher temperature and a higher frequency of extreme events (Eckert *et al.* 2008; Schoenheinz & Grischek 2011; Sprenger *et al.* 2011).

In some towns and cities in India, existing bank filtration schemes (mainly on rivers, but also at some lakes) currently serve as both sustainable alternatives and supplements to existing surface water and groundwater sources for the public water supply. Water diversions for irrigation, hydropower generation and discharge of partially treated and untreated wastewater to surface water bodies with extremely low flows have aggravated the water supply situation and increased the vulnerability to pathogens of many Indian cities using surface water. Fortunately, most existing bank filtration systems achieve very effective pathogen and turbidity (often associated with the presence of pathogens) removal and do not require much additional treatment or disinfection of the filtrate for their water supply. Hence the further development of bank filtration in India represents a potential ecosystem service that can provide pathogen-free drinking water to many cities and towns located near perennial surface water bodies and having suitable hydrogeologic conditions.

The goal of this paper is to provide an overview of the system design, capacity, and pathogen removal efficiency of selected bank filtration sites in India, aiming to illustrate the ecosystem service they represent.

BANK FILTRATION SCHEMES IN INDIA

Overview

The benefit of obtaining very low-turbidity water via natural bank filtration as a result of the percolation of surface water during and after the monsoon has been recognised in India for a very long time. In Nainital, bank filtrate has been abstracted from Nainital Lake by production wells adjacent to the lake since 1956. The first large-diameter (~10 m) caisson wells abstracting bank filtrate in Haridwar were constructed in the 1980s. In the cities of Ahmedabad on the Sabarmati River and Delhi and Mathura on the Yamuna, bank filtration supplements existing surface and groundwater abstraction for drinking water supply. In Haridwar and Patna on the Ganga River, and Medinipur and Kharagpur on the Kangsabati River, RBF is used as an alternative to surface water abstraction and to supplement groundwater abstraction.

System capacity and design parameters of urban bank filtration schemes

For an effective RBF scheme, the adjacent river should at a minimum be in hydraulic contact with the aquifer at the proposed site, but the location and design of a RBF scheme must also be based on the hydrology of the river basin, site hydrogeology and the specific water abstraction goals. A summary of system designs from selected RBF systems indicates that while in the United States a combination of vertical and horizontal or radial collector wells (RCW) is used, older RBF schemes in Germany mainly use a series of vertical wells (Grischek *et al.* 2002). Although a RCW is expensive to construct compared to a vertical well, the advantage of a RCW lies in its greater production capacity. This is beneficial for urban areas with a high water demand, where the production of a single RCW equals that of numerous vertical wells. This aspect is reflected in Table 1, where the RBF schemes of Ahmedabad and Mathura using RCWs have comparatively higher per-well production capacities. The RCWs in these cities are normally installed beneath the river bed in fine to medium alluvium. On the other hand, in Haridwar along the Ganga River, 22 large-diameter (~10 m) vertical caisson wells without laterals are used to generate bank filtrate (Table 1). Due to their typical close proximity to rivers, the operation of RCWs should be accompanied by intensive water quality monitoring with respect to pathogen removal, especially during floods. Price *et al.* (1999) report coliform breakthrough in a RCW at the Russian River during high river stage, which resulted in temporary well closure. Also, Mutiyar *et al.* (2011) found lower removal of organochlorine pesticides in a RCW compared to vertical wells at the Palla well field in Delhi.

Up to the end of 2009, bank filtrate was pumped from 16 large-diameter caisson wells whose total production was 33,000 m^3/day, comprising around 50% of the total

Table 1 | System capacity and design parameters of some bank filtration schemes in India

Location	Source water body	Well type (number of wells)	Production capacity in m^3/day	Depth in m	Lateral distance from source water in m	Travel-time of bank filtrate	References
Haridwar	Ganga	CW (22)	>43,000	7–10	15–110	2– > 100 days	Own data
Srinagar*	Alaknanda	VFW (6)	>4,000	18–20	3–165	n. d.	Own data
Patna	Ganga	VFW (6)	>3,500	150–300	9–236	n. d.	b
Karnaprayag*	Alaknanda	VFW (1)	>700	20	53	n. d.	Own data
Agastmuni*	Mandakini	VFW (1)	>280	30	33	n. d.	Own data
Satpuli*	East Nayar	VFW (1)	720	26	43–45	2 days (monsoon)– 2 weeks	Own data
Kesarwala	Song River	VFW (1)	900	48	40	n. d.	Own data
Nainital	Lake Nainital	VFW (9)	>24,100	22–37	4–94	8– > 30 days	Own data
Bhimtal	Lake Bhimtal	VFW (1)	>320	48	16	n. d.	Own data
Dehradun	Bandal	RCW(s)	140–430	1.5–2	Beneath river bed	2–4 min	a
Sahaspur (Dehradun)	Swarna	RCW(s)	210–570	3–4	Beneath river bed	>150 min	Own data
Medinipur	Kangsabati	RCW (1)	15,900	6–11	Beneath river bed	n. d.	a
Kharagpur	Kangsabati	RCW (1)	22,700	6–8	Beneath river bed	n. d.	a
Muzaffar Nagar	Kali	VFW	29–300	8–15	68	n. d.	c
Palla (Delhi)	Yamuna	VFW (~90)	~100,000 (in 2007)	45–54	Few meters to 600 m	Few weeks	e, f
Mathura	Yamuna	RCW (1)	2,400	15.5–18	Beneath river bed	1.5–3 days	d
Ahmedabad	Sabarmati	RCW (7)	110,000	10–11	Beneath river bed	n. d.	a

CW: large-diameter (10 m) caisson well; VFW: vertical filter well (production well); RCW: radial collector well; RCW(s): small-scale radial collector well; n. d.: not determined; a: Sandhu *et al.* (2011a); b: Sandhu *et al.* (2011b); c: Thakur *et al.* (2009); d: Singh *et al.* (2010); e: Rao *et al.* (2007); f: Lorenzen *et al.* (2010); *RBF site under development since construction in 2010.

supply, with the balance made up by groundwater from 50 tube wells (Sandhu *et al.* 2011a). The volume of bank filtrate abstracted as of January 2010 has increased following the construction of six new caisson bank filtrate wells of a similar design to the existing wells (Figure 1). Since then more than 43,000 m^3 of water are abstracted every day in Haridwar by a total of 22 large-diameter caisson wells, which draw a high proportion of bank filtrate (>70%) and provide 68% of the total supply. These new wells were installed to meet the large increase in drinking water demand that was anticipated to (and did) occur during the huge religious gathering of the Kumbh Mela in January–April 2010.

Development of Koop (wells) for small-scale bank filtration in rural Uttarakhand

Many rural drinking water production schemes in the steeply sloped areas of the mountainous North Indian state of Uttarakhand are typically supplied by gravity tapping surface water from springs, rivers and streams with a highly variable seasonal discharge. The surface water is normally collected in boulder-filled galleries (BFGs), and does not undergo substantial pre- or post-treatment other than rapid sand filtration, and in some instances chlorination. However, the removal of turbidity and microbial pathogens by the BFGs is insufficient, especially during the monsoon (June–September) when the settling basins fill up with silt, and coagulation and filtration are inadequate to remove the turbidity. Especially large monsoon flows can also physically damage or completely wash away BFGs. Consequently, the interruption of water supply is a frequently recurring problem during the monsoon.

As a substitute to the BFG, the Uttarakhand State Water Supply organisation (UJS), developed a so-called Koop (in Hindi: well) in 1997, with more than 800 Koops installed since then, primarily for rural drinking

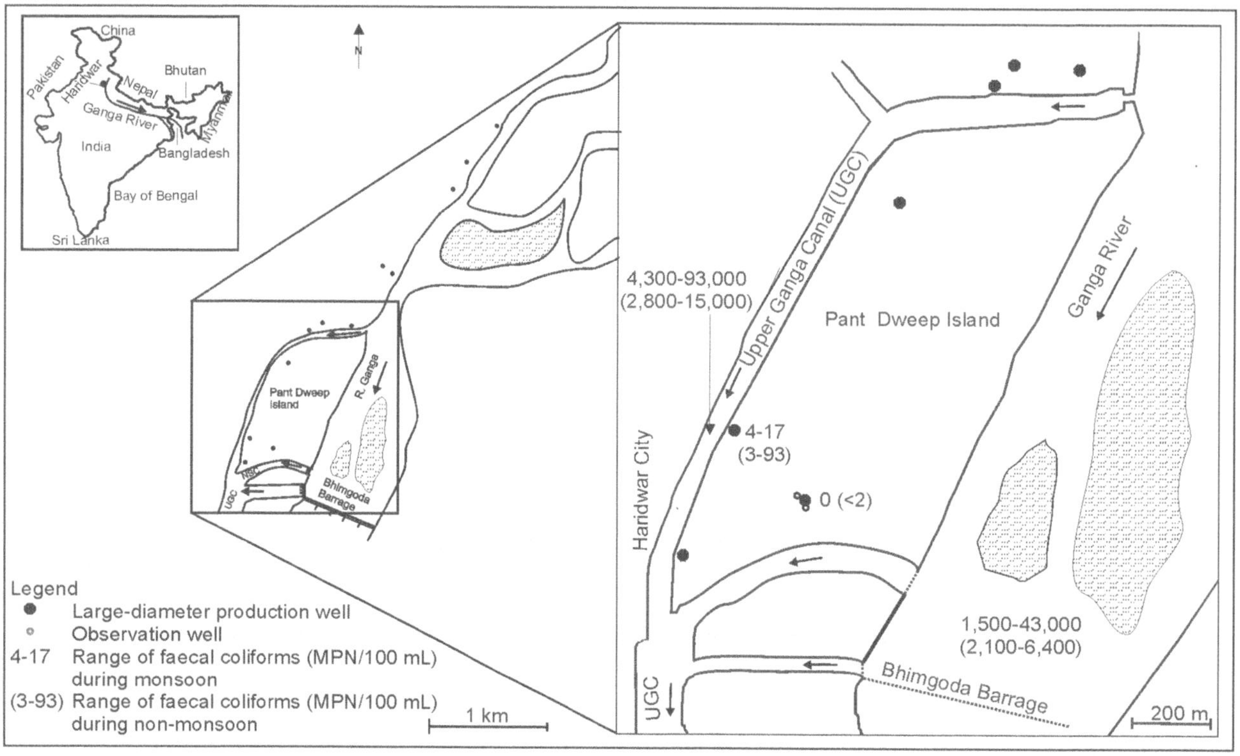

Figure 1 | Observed faecal coliform concentrations for some RBF wells in Haridwar compared to the faecal coliform concentrations in surface water.

water production. However, due to the short retention time of the water from the surface to the Koop (usually a few minutes only), the attenuation of pathogens is limited. Thus, in order to obtain improved water quality by increasing the retention time of the filtrate in the subsurface, a Koop was constructed with a geotextile and improved filter-media layering (Figure 2) at a site in Sahaspur by the Swarna river (30 km west of Dehradun in Uttarakhand).

The Koop abstracts filtrate from beneath the stream bed and thus is protected from floods and has low operational costs. The Koop assembly typically consists of one vertical collector cylinder (1–2 m tall) and four perforated radial pipes. Each radial pipe is 0.5–1.0 m long and has a diameter of 0.05 m. The assembly is made from mild steel and is painted to prevent corrosion. The top of the Koop has a waterproof rubber seal and a fixed steel plate. A welded outlet socket for attaching the supply pipe is located in the

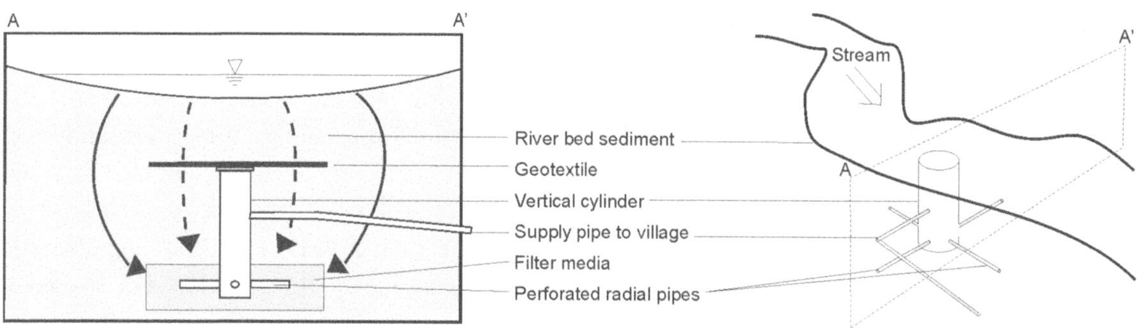

Figure 2 | Principle of a Koop with geotextile.

middle of the vertical cylinder. The assembled Koop is installed 3–4 m below (or beside) the bed (or bank) of the stream.

REMOVAL OF PATHOGENS AND HARMFUL SUBSTANCES DURING RBF

Urban bank filtration schemes

India is currently in a phase of rapid population growth and urbanisation that has resulted in sharply increasing quantities of domestic sewage being discharged (mostly untreated) into river systems. The efficiency of bank filtration in the pre-treatment of raw water for drinking and especially the removal of pathogens has been demonstrated for various schemes in India.

Studies on water quality changes during RBF in Haridwar commenced in 2005 on Pant Dweep Island (Figure 1). Sampling in December 2005, March 2006 and September 2006 revealed bank filtrate having very low (less than 1 mg/L) dissolved organic carbon content under aerobic conditions, an arsenic concentration of less than 0.01 mg/L and other trace metals below the Indian Standard IS 10500 (1991) limit. Dash *et al.* (2010) report that bank filtrate abstracted from production well 18 on Pant Dweep Island, when compared to raw Ganga River water, showed 2.5-log removal of total coliforms, 3.5-log removal of faecal coliforms, 0.7-log removal of turbidity in the non-monsoon period (November 2005–June 2006) for the shortest travel time of 84–126 days at a minimum distance of 115 m, and 4.7-log removal of total coliforms, 4.4-log removal of faecal coliforms (Table 2), 2.5-log removal of turbidity and 1.0-log removal for organics as measured by UV absorbance during the monsoon period (July–September 2006) for the shortest travel time of 77–126 days. All other water quality parameters were also within the limits prescribed by the Indian Standard IS 10500 (1991).

Lake Nainital is the primary source of drinking water for the town of Nainital. Nainital's population before 2006 was estimated at around 42,500 people. The seasonal tourist influx temporarily adds an average of 100,000 people to this area each year. From 1990–2007 seven RBF production wells were installed adjacent to Lake Nainital. In 2008,

Table 2 | Coliform counts in surface water at selected RBF sites in India, and coliform removal during RBF

		RBF site					
Season	**Parameter (coliform counts in MPN/100 ml)**	**Haridwar (b, own data)**	**Nainital (a)**	**Patna (d)**	**Mathura (c)**	**Satpuli**	**Delhi, Palla well field (e)**
Non-monsoon	TCC in SW	4,300–15,000	46,000–240,000	24,000–160,000	2,300–1,500,000	240	1,200 CFU/100 ml
	TCC in RBF well	<2–93	<2	8–170	43–75,000	<2	<1 CFU/100 ml
	Log removal of TC	2.5	>5.2	2.3–4.2	1.3–1.7	>2.1	>3.1
	FCC in SW	2,100–15,000	5,000–24,000	n. d.	150–230,00	n. d.	500–40,000
	FCC in RBF well	<2–93	<2	n. d.	43–9,300	<2	<1
	Log removal of FC	3.5	>4.2	n. d.	2.3–3.2	n. d.	2.7–4.6
Monsoon	TCC in SW	9,300–230,000	50–500,000	90,000–160,000	n. d.	920	Similar to non-monsoon
	TCC in RBF well	<2	<2	8–300	n. d.	49	As above
	Log removal of TC	4.7	>5.4	2.6–4.4	n. d.	1.3	n. d.
	FCC in SW	1,500–93,000	50–50,000	n. d.	n. d.	8	1,000–3,000
	FCC in RBF well	0–17	<2	n. d.	n. d.	<2	<1
	Log removal of FC	4.4	>4.4	n. d.	n. d.	>0.6	3.0–3.5

TCC: total coliform counts in abstracted water (before disinfection); FCC: faecal coliform counts in abstracted water (before disinfection); SW: surface water; n. d.: not determined. a: Dash *et al.* (2008); b: Dash *et al.* (2010); c: Singh *et al.* (2010); d: Sandhu *et al.* (2011b); e: Sprenger *et al.* (2008).

24,100 m^3/day of water was abstracted from the production wells. The water from the production wells has been found to be of better bacteriological quality than that taken directly from the lake and treated with sand filter units. According to Dash *et al.* (2008), only 1.9-log (>98.6%) reduction of total and faecal coliform counts was achieved by the sand filter units whereas 5.2-log (>99.999%) and 4.2-log (>99.99%) reductions respectively were measured in the non-monsoon period while using bank filtration (Table 2). In monsoon, the removal of coliforms was greater. The observed log removal rates for faecal coliforms are in agreement with findings of Hijnen *et al.* (2005) and Weiss *et al.* (2005) in laboratory and field studies in the USA.

In Patna, the removal of total coliforms was observed to be in the range of 2.3 to 4.2 log, with a higher log removal during monsoons, on account of the higher pathogen loading and turbidity in the Ganga River (Table 2). In Mathura, the maximum breakthrough of coliforms in the RCW beneath the river bed is considerably high (Table 2) similar to the RCW at the Russian River (Price *et al.* 1999), not only due to the extremely high coliform counts in the Yamuna River water on account of the large quantities of untreated sewage discharged into the river, but also due to the relatively short travel time of 1.5–3 days and short flow paths to the radials positioned at depths of 15.5 and 18 m directly beneath the river bed (Table 1). Yet, RBF in Mathura serves as a significant pre-treatment step by reducing organics, contaminants, colour, UV-absorbance and coliform bacteria by around 50% from river to filtrate (Singh *et al.* 2010).

Up to now there are only a few data available on the occurrence of viruses and protozoa in surface water and bank filtrate in India. But viruses are regarded as the most critical microorganisms where the effectiveness of soil passage for removing pathogens from source water is concerned. However, faecal indicator bacteria also seem to be able to penetrate into an aquifer as far as viruses, and may therefore be useful indicators of faecal contamination (Schijven 2002). Foppen (2007) reports that due to the presence of dissolved organic carbon (DOC) (1–100 mg/L) attachment of *E. coli* to soil grains was 2 to 80 times less than in cases without DOC. Except for the bank filtration sites at the Yamuna River in Delhi and Mathura and the Kali River at Muzaffar Nagar, all investigated sites showed low DOC concentrations (0.5 to 3 mg/L) in river/lake water. As the DOC in River Ganga water and many of its tributaries and mountainous streams in Northern India is only around 1 mg/L, no negative effect on pathogen removal by attachment is expected. Another advantage of low DOC is a low risk of disinfection by-product formation if chlorination is used as the only further treatment step as often found in Northern India.

Removal of coliforms by bank filtration through Koops in rural areas

Travel times for Koops were determined to be between 2 and 4 minutes only. For a new Koop installed using sorted filter media and a geotextile, the travel time was more than 150 minutes. An initial comparison of the removal efficiency of coliforms by conventionally built Koops for six different sites in Uttarakhand and the new Koop by the Swarna river in Sahaspur is shown in Table 3. Due to the

Table 3 | Coliform counts in surface water and in wells at Koop sites in Uttarakhand, and subsequent coliform removal during bank filtration

	Koop	
Parameter	**Coliform counts in MPN/100 ml without geotextile at 6 sites in Uttarakhand ($n = 6$)**	**Coliform colony forming units in CFU/100 ml with geotextile and improved filter-media at Sahaspur in Uttarakhand (own data; $n = 1$)**
TCC in surface water	2,200–46,000	244
TCC in Koop	19–1,950	Not detectable
Log removal of TC	1.0–3.4	>2.4
FCC in surface water	90–11,000	220
FCC in Koop	5–780	Not detectable
Log removal of FC	0.3–2.8	>2.3

TCC (total coliform counts) and FCC (faecal coliform counts) data provided by UJS for six conventionally constructed Koops without geotextile.

very short travel times and non-standardised design geometry of Koops, the removal rates are highly variable. Surprisingly, even a very short travel time and flowpath length in the stream bed result in a considerable removal of coliforms. Although the first data from the new Koop indicate a higher removal of coliforms, prolonged monitoring is necessary to obtain comparable results with Koops installed without geotextiles, especially for monsoon conditions.

CONCLUSIONS AND PROSPECTS OF RBF IN INDIA

Most of the cited examples of bank filtration schemes in India illustrate the ecosystem service for providing drinking water for improved human health. Bank filtrate from a few sites monitored in recent years has shown a significantly higher quality when compared to directly abstracted surface water. However, log removal rates for faecal coliforms presented in the paper for selected sites cannot simply be transferred or assumed for other pathogens and other sites. For example, a recent study by Metge *et al.* (2011) suggests that in evaluating the efficacy of RBF operations to remove the protozoan pathogen *Cryptosporidium parvum* oocysts, it may be necessary to consider not only the geochemical nature and size distribution of the sediment grains, but also the degrees of sediment sorting and the concentration, reactivity, and aquifer penetration of the source water DOC as in the case of *E. coli* described above.

At most sites in India where directly abstracted surface water is used for drinking, the removal of pathogens has the highest priority. At some sites, the contamination by organic substances and heavy metals is also considerable. Bank filtration is known to provide good removal of such contaminants (Stuyfzand 1998, Ray *et al.* 2002). For lesser known pollutants such as persistent trace organics, detailed investigations are rare. Sprenger (2011) reports that a series of organic trace compounds including polar to nonpolar substances from household, industrial and agricultural sources were considerably attenuated following RBF by the Yamuna River.

The drinking water production of many cities and towns in India will be or is already in the process of being expanded and optimised to meet the growing water demand, such as through the City Development Plans or various schemes such as the 'Jawaharlal Nehru National Urban Renewal Mission', the Asian Development Bank's current 'North Eastern Region Capital Cities Development Investment Programme' and the planned 'Uttarakhand Urban Sector Development Investment Programme' and the World Bank's 'National Ganga River Basin Project'. The technical and socio-economic feasibility of using RBF for urban and decentralised water supply schemes should be investigated for locations (covered by these programmes) having suitable hydrogeological conditions. This would also serve as a first step towards meeting the goals of the Government of India's 'National Action Plan on Climate Change' (NAPCC 2008).

ACKNOWLEDGEMENTS

The authors are grateful for the financial support for the project 'Indo-German Riverbank Filtration Network' (BMBF IND 08/156) funded by the BMBF within its programme 'India and Germany – Strategic Partners for Innovation'. The financial support of the project 'Saph Pani – Enhancement of natural water systems and treatment methods for safe and sustainable water supply in India' by the European Commission within the 7th Framework Programme (grant agreement no. 282911) and the logistical support of Mr P. C. Kimothi and his team from the Riverbank Filtration Cell of Uttarakhand Jal Sansthan is also kindly acknowledged. The authors also wish to thank Verena Müller, Silke Kraus, Anandi Iyer and Jan Palkoska of the BMBF for their administrative support, and Thomas J. Voltz and Rico Bartak for their edits.

REFERENCES

Dash, R. R., Mehrotra, I., Kumar, P. & Grischek, T. 2008 Lake bank filtration at Nainital, India: Water-quality evaluation. *Hydrogeology Journal* **16** (6), 1089–1099.

Dash, R. R., Bhanu Prakash, E. V. P., Kumar, P., Mehrotra, I., Sandhu, C. & Grischek, T. 2010 River bank filtration in Haridwar, India: Removal of turbidity, organics and bacteria. *Hydrogeology Journal* **18** (4), 973–983.

Eckert, P., Lamberts, R. & Wagner, C. 2008 The impact of climate change on drinking water supply by riverbank filtration. *Water Science and Technology: Water Supply* **8** (3), 319–324.

Foppen, J. W. 2007 *Transport of Escherichia Coli in Saturated Porous Media*. UNESCO-IHE, A.A. Balkema Publ., p. 263.

Grischek, T., Schoenheinz, D. & Ray, C. 2002 Siting and design issues for riverbank filtration schemes. In: *Riverbank Filtration – Improving source-Water Quality* (C. Ray, G. Melin & R. B. Linsky, eds). Kluwer Academic Publishers, Boston, Water Science and Technology Library, 43, pp. 291–302.

Hijnen, W. A. M., Brouwer-Hanzens, A. J., Charles, K. J. & Medema, G. J. 2005 Transport of MS2 phage, *Escherichia coli, Clostridium parvum*, and *Giardia intestinalis* in a gravel and a sandy soil. *Environmental Science and Technology* **39** (20), 7860–7868.

Indian Standard IS 10500 1991 *Indian Standard Specifications for Drinking Water*. Bureau of Indian Standards, New Delhi, India.

Lorenzen, G., Sprenger, C., Taute, T., Pekdeger, A., Mittal, A. & Massmann, G. 2010 Assessment of the potential for bank filtration in a water-stressed megacity (Delhi, India). *Environmental Earth Sciences* **61** (7), 1419–1434.

Metge, D. W., Harvey, R. W., Aiken, G. R., Anders, R., Lincoln, G., Jasperse, J. & Hill, M. C. 2011 Effects of sediment-associated extractable metals, degree of sediment grain sorting, and dissolved organic carbon upon *Cryptosporidium parvum* removal and transport within riverbank filtration sediments, Sonoma County, California. *Environmental Science and Technology* **45** (13), 5587–5595.

Mutiyar, P. K., Mittal, A. K. & Pekdeger, A. 2011 Status of organochlorine pesticides in the drinking water well-field located in the Delhi region of the flood plains of river Yamuna. *Drinking Water Engineering Science* **4**, 51–60.

NAPCC 2008 *National Action Plan on Climate Change – Technical Document*. Prime Minister's Office, Government of India, New Delhi. pp. 52. Available from: http://pmindia.nic.in/Pg01-52.pdf (accessed 27 July 2011).

Price, M. L., Flugum, J., Jeane, P. & Tribbett-Peelen, L. 1999 Sonoma county finds groundwater under the direct influence of surface water depends on river conditions. In *Proc. Int. Riverbank Filtration Conference*, 4.–6.11.1999, Louisville, Kentucky, 25–27.

Rao, S. V. N., Kumar, S., Shekhar, S., Sinha, S. K. & Manju, S. 2007 Optimal pumping from skimming wells from the Yamuna River flood plain in north India. *Hydrogeology Journal* **15** (6), 1157–1167.

Ray, C., Melin, G. & Linsky, R. B. 2002 *Riverbank Filtration – Improving Source Water Quality*. Water Science Technol. Library, 43, Kluwer Academic Publishers, Boston.

Saadoun, I., Schijven, J., Shawaqfah, M., Al-Ghazawi, Z., Al-Rashdan, J., Blandford, W., Boving, T., Ababneh, Q. & Berg, H. V. D. 2008 Removal of fecal indicator coliforms and bacteriophages by riverbank filtration (RBF) in Jordan. In *Proc. Int. Conf. on Construction and Building Technologies*, 16.–20.06.08, Kuala Lumpur, pp. 191–198.

Sandhu, C., Grischek, T., Kumar, P. & Ray, C. 2011a Potential for riverbank filtration in India. *Clean Technologies and Environmental Policy* **13** (2), 295–316.

Sandhu, C., Grischek, T., Schoenheinz, D., Prasad, T. & Thakur, A. K. 2011b Evaluation of bank filtration for drinking water supply in Patna by the Ganga River, India. In: *Riverbank Filtration for Water Security in Desert Countries* (C. Ray & M. Shamrukh, eds). NATO Science for Peace and Security Series, Springer, Dordrecht, pp. 203–222.

Schijven, J. F. 2002 Virus removal by soil passage at field scale and groundwater protection. In: *Riverbank Filtration: Understanding Contaminant Biogeochemistry and Pathogen Removal* (C. Ray, ed.). Kluwer Academic Publishers, Boston, pp. 55–84.

Schoenheinz, D. & Grischek, T. 2011 Behavior of dissolved organic carbon during bank filtration under extreme climate conditions. In: *Riverbank Filtration for Water Security in Desert Countries* (C. Ray & M. Shamrukh, eds). NATO Science for Peace and Security Series, Springer, Dordrecht, pp. 51–67.

Singh, P., Kumar, P., Mehrotra, I. & Grischek, T. 2010 Impact of riverbank filtration on treatment of polluted river water. *Journal of Environmental Management* **91** (5), 1055–1062.

Sprenger, C. 2011 Surface/groundwater interactions associated with river bank filtration in Delhi (India) – Investigation and modelling of hydraulic and hydrochemical processes. PhD Thesis, Freie Universität Berlin, Germany.

Sprenger, C., Lorenzen, G. & Pekdeger, A. 2008 Occurrence and fate of microbial pathogens and organic trace compounds at riverbank filtration sites in Delhi, India. TECHNEAU integrated project: deliverable D 5.2.6. Available from: http://www.techneau.org (accessed April 2012).

Sprenger, C., Lorenzen, G., Hülshoff, I., Grützmacher, G., Ronghang, M. & Pekdeger, A. 2011 Vulnerability of bank filtration systems to climate change. *Science of the Total Environment* **409** (4), 655–663.

Stuyfzand, P. J. 1998 Fate of pollutants during artificial recharge and bank filtration in the Netherlands. In: *Artificial Recharge of Groundwater* (J. H. Peters, K. van den Akker, H. Bouwer, J. Carrera, P. Dillon, K. H. Jensen, U. Schöttler, P. Stuyfzand, H. G. de Jonge, E. Roosma, T. Olsthoorn & I. Johnson, eds). Balkema, Rotterdam, pp. 119–125.

Thakur, A. K., Ojha, C. S. P., Grischek, T., Ray, C. & Jha, R. 2009 River bank filtration in extreme environment conditions. In *Proc. Int. Conf. Water, Environment, Energy and Society (WEES-2009)*, 12.–16.01.09, New Delhi, 3, pp. 1340–1347.

Weiss, W. J., Bouwer, E. J., Aboytes, R., LeChevallier, M. W., ÓMelia, C. R., Le, B. T. & Schwab, K. J. 2005 Riverbank filtration for control of microorganisms: results from field monitoring. *Water Research* **39** (10), 1990–2001.

First received 17 February 2013; accepted in revised form 14 May 2013

A groundwater perspective on the river basin management plan for central Portugal – developing a methodology to assess the potential impact of N fertilizers on groundwater bodies

M. P. Mendes, L. Ribeiro, J. Nascimento, T. Condesso de Melo, T. Y. Stigter and A. Buxo

ABSTRACT

The Water Framework Directive establishes that the river basin management plans must have a summary of the pressures and impacts of human activities, such as agriculture, on the chemical and quantitative status of groundwater bodies. In order to identify those areas where a potential impact from agricultural activities on groundwater bodies exists, but currently lacking groundwater monitoring data, a methodology was developed that combines the use of gross nitrogen balance values with the results of a specific vulnerability assessment index. A farm management efficiency parameter is added, to identify the factors that contribute to nitrogen use efficiency and to assess the near-future scenarios. This methodology allows the identification of significant pressures that may be responsible for a groundwater body failing good status where there is no representative monitoring network.

Key words | nitrates, river basin management plan, significant pressures, susceptibility index, Water Framework Directive

M. P. Mendes (corresponding author)
L. Ribeiro
J. Nascimento
T. Condesso de Melo
T. Y. Stigter
A. Buxo
CVRM – Centre of Geo-Systems,
Instituto Superior Técnico,
Av. Rovisco Pais,
1049-001 Lisbon,
Portugal
E-mail: *mpaulamendes@ist.utl.pt*

INTRODUCTION

Groundwater is the largest body of freshwater in the European Union and is also a main source of public water supply in many regions (EC 2006). In order to ensure the implementation of water policy in compliance with the Water Framework Directive (WFD), Directive 2000/60/EC (EC 2000), it is necessary to implement river basin management plans (RBMPs). With respect to groundwater, the RBMP must include, for each river basin district, a summary of pressures and impacts of human activities on groundwater quantitative and qualitative status.

The WFD principal objective is the achievement of good groundwater status by the end of 2015. The assessment of the impacts and pressures on groundwater status can help identify the groundwater bodies (GWBs) at risk of failing to achieve good status at the present and, considering development scenarios, in the near future.

The assessment of significant pressures on GWBs due to nitrogen (N) fertilizer management requires an understanding of the magnitude and cumulative effects of these pressures and the properties and behaviour of groundwater systems (EC 2003). It should also require the availability of monitoring data provided by representative networks, which however are often not available.

A methodology was developed with the objective of identifying pressures and assessing impacts of significant diffuse source pollution from agricultural activities for a certain area, based on: (i) a N loss indicator (NLI) integrating gross N balance calculations and groundwater vulnerability assessment; (ii) subsequent identification of the areas with significant pressures; (iii) integration of the factors that contribute to N use efficiency into a farm management efficiency parameter.

NLIs are environmental management tools for assessing the agricultural pollution potential at the ground surface (EC 2003; CEC 2006; Brentrup & Palliere 2010; Buczko *et al.* 2010). According to Buczko *et al.* (2010) NLIs can be classified based on: (1) source terms, (2) transport terms, and (3) composite approaches. N balances are (among the

doi: 10.2166/wst.2013.427

most commonly used) NLIs of type 1, whereas groundwater vulnerability indexes are of type 2. Among the latter, DRASTIC (Aller *et al.* 1987), EPIK (Doerfliger & Zwahlen 1997) and VULFRAC (Fernandes & Rudolph 2001) are often used to assess the groundwater vulnerability to diffuse nitrate pollution from agricultural areas. In this study the Susceptibility Index (SI) developed by Ribeiro (2005) is applied. Adapted from DRASTIC, the main differences include the elimination of the parameters representing soil, vadose zone and hydraulic conductivity, and the addition of a parameter defining land use. Intrinsic vulnerability, i. e. determined by the natural hydrological and geological conditions, is thus transformed into specific vulnerability, including potential impacts of specific land uses or contaminant type (Vrba & Zaporožec 1994).

STUDY AREA

The Central River Basin District of Portugal (CRBD) includes three major river basins: Mondego, Vouga, and Lis, with a total area of 11,477.50 km^2.

On the Portuguese mainland four hydrogeological units can be identified: the Tagus–Sado Tertiary Basins, the Meridional and Western Meso-Cenozoic Borders and the Hercynian Massif. In the Western Meso-Cenozoic Border there is a wide variety of geological features, including detrital formations of Tertiary and Quaternary age, Cretaceous sandstones and Jurassic marls and limestones, supporting 27 groundwater bodies that were defined according to Article 5 and Annex VII of the WFD (EC 2000). This geological diversity is responsible for the GWBs' high heterogeneity and distinctive hydrodynamical behaviour (Ribeiro 2002). Of the 27 GWBs, 17 are located in the CRBD (Figure 1).

Ten GWBs of this region are dominated by intergranular porosity, three are characterized as double porosity systems and one by mixed porosity type (intergranular and karst); three GWBs consist of small aquifers, installed in fractured and fissured hard rock, and three are undifferentiated bodies. The Undifferentiated Old Massifs of the Vouga and Mondego River Basins are the two largest GWBs of the region and consist of small aquifers with poor groundwater potential (Figure 1).

Land use in the CRBD is dominated by forest (almost 50% of the total area), with 18.6% representing coniferous forest, 18.4% transitional woodland-shrub and 12.7% mixed forest. Permanently irrigated land represents 1.7% of the total area and is located in the western part, whereas 1.1% is occupied by rice fields, located in the alluvial plain of the Mondego River. Among the temporary crops (27.4% of land use), cereals for the grain production and temporary grass are the most important (40 and 51% of temporary crops, respectively). Permanent crops account for 15.5% of total land use, with vineyards and olive plantations among the most representative (87% of permanent crop land). Close to the coast the majority of farms are small (≅2.5 ha) and the utilized agricultural area (UAA), i.e. total

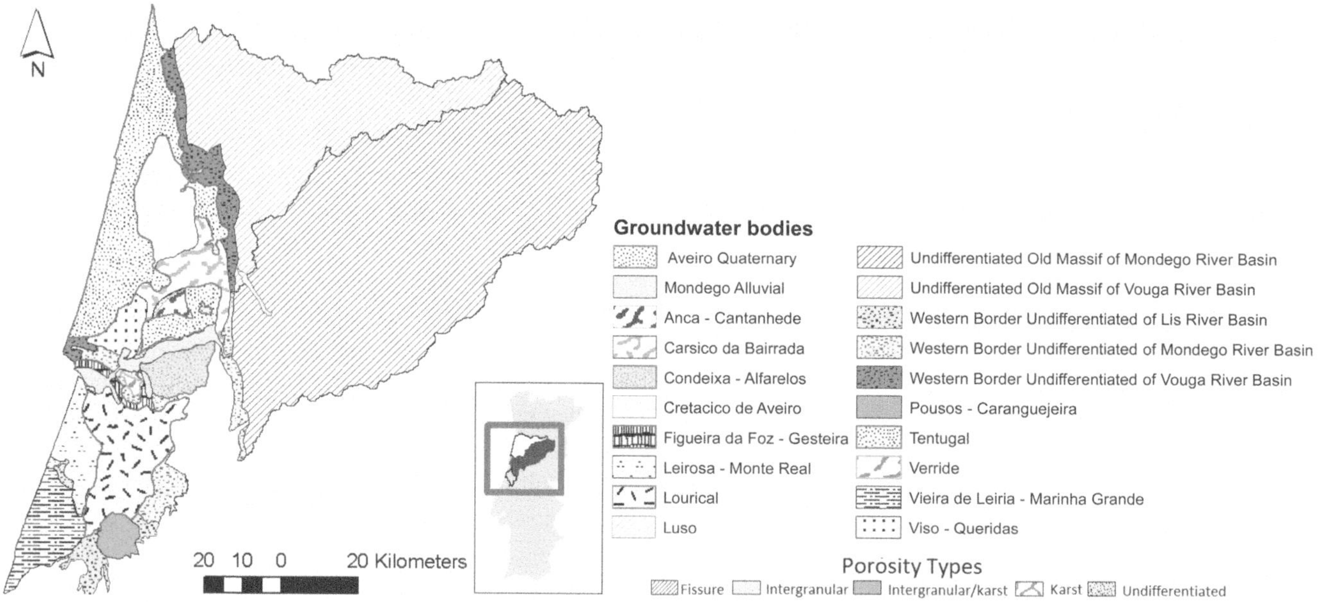

Figure 1 | Location of the study area, its three main drainage basins and its GWBs on the Portuguese mainland, indicating their main porosity types.

land used by a farmer for agriculture, increases towards the interior (≅10 ha) (INE 2011). Discontinuous fabric corresponds to 3.8% of the area, whereas the continuous urban fabric merely represents 0.1% (Figure 2).

According to a regional rank, the western part of the CRBD showed the highest growth rate of ageing rural population during the period of 1999 to 2009 in Portugal, with only 13% of the producer's household income originating from agricultural activities (INE 2011).

METHODS

The potential impact of nitrate contamination on groundwater is assessed here by considering the pollution pressure (where this occurs at the ground surface) through a N balance approach, in combination with a measure of vulnerability of the groundwater body to pollution. The concept of 'potential impact' describes the effects that a pressure is likely to have on a GWB, and is used in the evaluation of whether the body is 'at risk' of failing the good status by the end of 2015 (EC 2003).

The gross N balance for Portugal was estimated by the Instituto Nacional de Estatistica (INE) according to OECD/EUROSTAT (2003). For N balance calculations, INE (2009) used agricultural statistics data, i.e. amounts of inorganic N fertilizers sold, livestock units, crop production and several coefficients to convert the statistical data into N tonnages. The N output on the farm land is calculated considering the harvest of marketed crops, including marketed fodder crops and non-marketed fodder crops and grass. Based on the ratio of N balance over UAA, three levels were established according to the 40th and 80th percentiles for the entire country (INE 2009): high

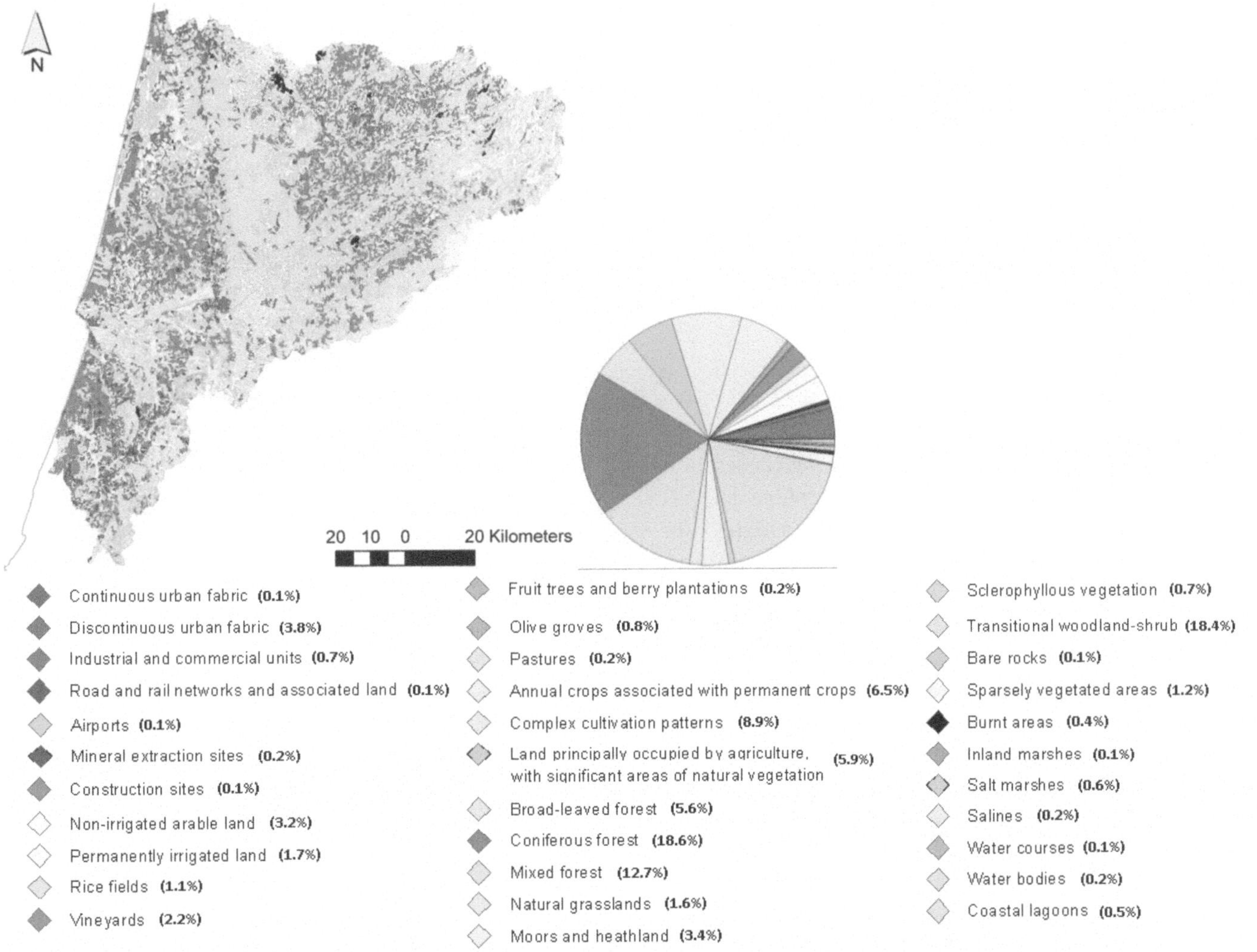

Figure 2 | Map of land use (level III) for the studied region (adapted from Corine Land Cover 2006) and the corresponding land use percentages plotted in a pie-chart.

(>83 kg/ha), medium (−15 to 83 kg/ha) and low (<− 15 kg/ha). A negative N balance occurs where agriculture removes more N from the environment than it adds.

To assess the vulnerability of the GWBs to pollution from agricultural activities the susceptibility SI was used. The SI method (Ribeiro 2005) is an adaptation of the DRASTIC method (Aller *et al.* 1987). The main differences include the addition of a parameter defining land use (*LU*) and the elimination of three DRASTIC parameters: Soil (*S*), Impact of vadose zone (*I*) and hydraulic Conductivity (*C*). This was done to simplify the index, based on the criteria of parameter redundancy, as these parameters are qualitatively represented by the other parameters: Depth to water (*D*), Recharge (*R*), Aquifer media (*A*) and Topography (*T*). Though soil type can largely influence the attenuation potential for certain contaminants, it is indirectly represented by land use, particularly because natural soils are frequently altered during cultivation of the land (Stigter *et al.* 2006).

The five parameters of SI are subdivided into ranges or media types (following the DRASTIC method), which are rated between 10 and 100 according to their relative impact on the pollution potential. The rating of *LU* is established according to legend level III of Corine Land Cover, with, for example, permanently irrigated lands rated 90, permanent crops 70, pastures 50, forests 0 (see Ribeiro (2005) and Stigter *et al.* (2006) for full classification of parameters and assigned ratings). Following an Analytic Hierarchy Process the weight string for the SI is calculated:

$$SI = 0.186D + 0.212R + 0.259A + 0.121T + 0.222LU$$

The obtained values are distributed among classes, and as the SI is a relative measure of how susceptible a GWB is to pollution, a qualitative degree is assigned to each class, ranging from 'very low' to 'very high' (Ribeiro 2005).

For the estimation of the *D* and *R* parameters in the study area an extensive bibliographic research was carried out and piezometric maps were elaborated. The average values for *R* and *D* are respectively 157 mm/year and −4.8 m for intergranular GWBs, 250 mm/year and −6.0 m for karstic GWBs and 60 mm/year and −29.5 m for undifferentiated and fissured GWBs. The spatial distribution of *T* was obtained using a digital elevation model.

From a groundwater perspective, combining N balance calculations (NLI based on source terms) with the vulnerability assessment (NLI based on transport terms) will establish not only the N amount potentially available to diffuse N losses, but also the areas where leaching of N surplus can affect important groundwater resources (Figure 3).

It is also important to assess the farm management efficiency, which can be done by calculating correction factors. Leão & Morais (2011) used correction factors to assess the loss of irrigation efficiency related to: (a) the characteristics of used irrigation systems and (b) the conditions of irrigation. These factors were adjusted according to the user profiles: legal personality of the agricultural holding (*LPAH*), age (*AG*), level of education (*LE*) and degree of agricultural training of the farm manager (*AT*) (Table 1). Here, the correction factors for loss of efficiency concerning farm management were applied to

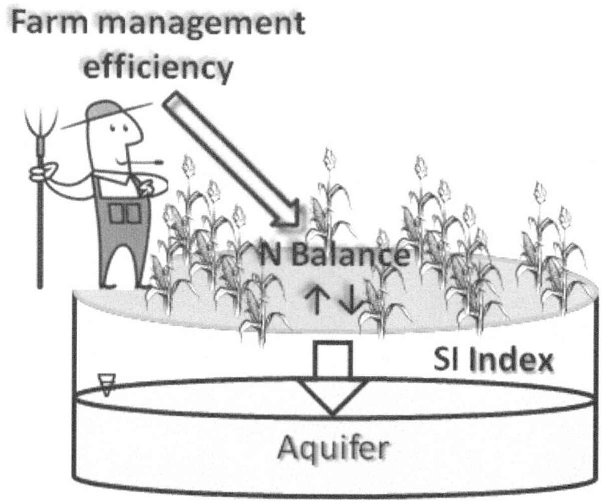

Figure 3 | Methodology applied to assess the N fertilizers potential impact on GWBs.

Table 1 | Classes of the indicators and respective rank of efficiency

Indicators	Classes	Rank
LPAH	Singular	1
	Others	2
	Holding (agricultural enterprise)	3
AG	>65 years	1
	45 to 64 years	2
	<45 years	3
LE	First cycle of basic education or less	1
	High school degree or superior without agricultural training	2
	High school degree or superior with agricultural training	3
AT	Practical agricultural training	1
	Elementary (training courses)	2
	Complete	3

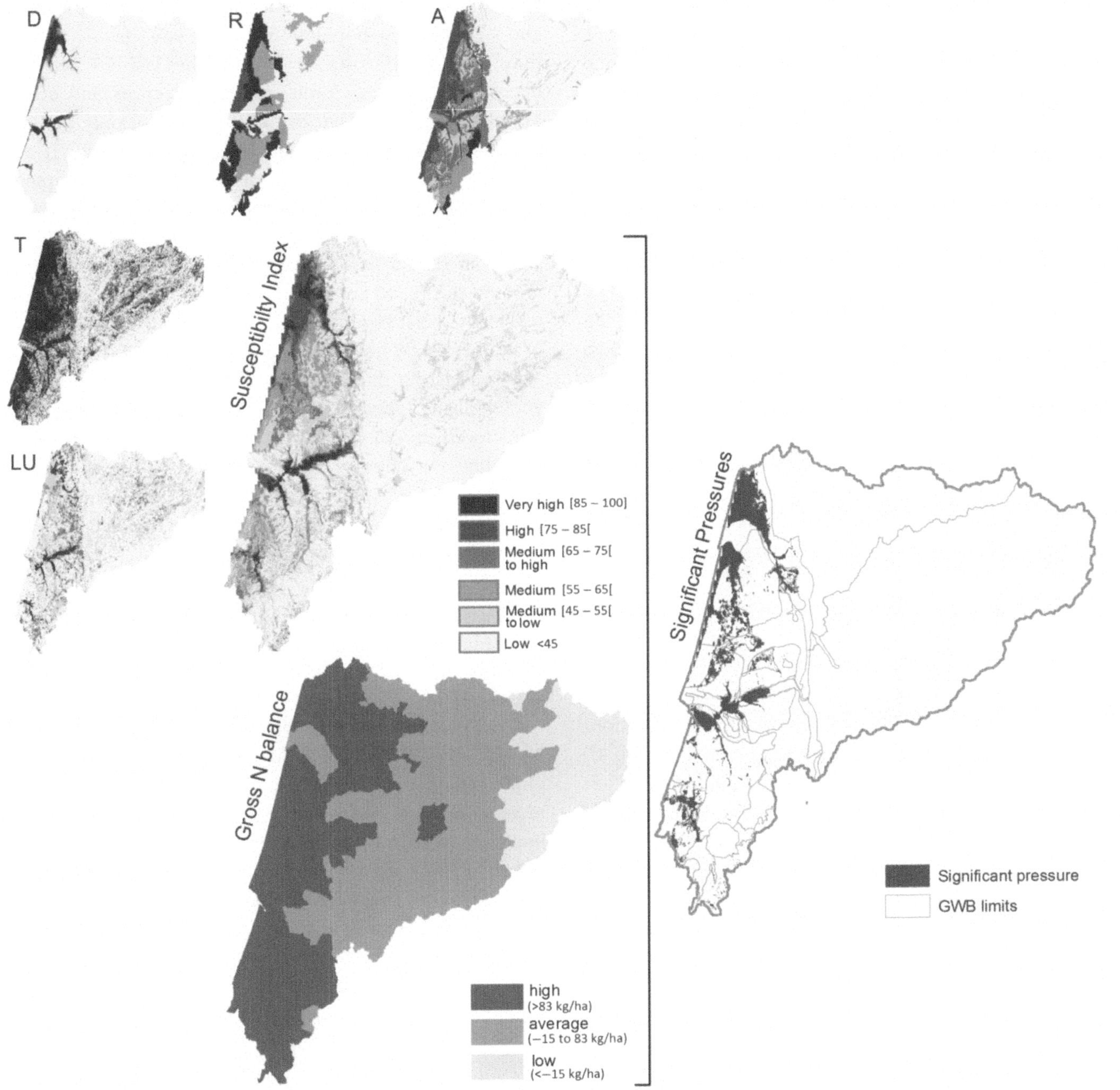

Figure 4 | Gross N balance and SI map (on the left); delimitation of the GWBs areas with significant pressures (on the right).

establish the underlying factors of N use efficiency. The efficiency (EFI) was calculated according to the following expression:

$$\mathrm{EFI} = (10LPAH + 15AG + 35LE + 40AT)/100$$

The level of EFI ranges from 1 to 3 and is established according to three classes where values below 1.5 (first quartile) are considered low and values above 2.5 (third quartile) are high.

In the current case study, EFI was also estimated to assist in the conceptualization of scenarios for the assessment of near-future development in agriculture. It was possible to assess the response indicators *LPAH*, *AG*, *LE* and *AT* using Agricultural Census data (INE 2011).

RESULTS AND DISCUSSION

The results for the gross N balance applied to the region are shown in Figure 4. They clearly indicate where agricultural pressures can cause impact on groundwater chemical status. The N surpluses are located near the coast in areas occupied by non-irrigated arable lands, permanently irrigated lands and rice fields (Figure 2). These western areas have the highest N loads from inorganic fertilizers with annual values exceeding 50 kg/ha (INE 2009).

According to the SI map also presented in Figure 4, the GWBs most vulnerable to contamination are the Aveiro Quaternary aquifer system and the Mondego Alluvial aquifer system. The medium high to very high SI values are related to flat topography (slopes less than 2%), high recharge rates and permeable aquifer material. On the other hand, the referred conditions also represent a high dilution potential, shown to be an important attenuation mechanism and sometimes a controlling factor on nitrate levels (Stigter *et al.* 2006, 2011).

The areas with the highest potential impact on groundwater quality, designated as areas with significant pressures, were delimited by combining the following characteristics: (i) SI index above 65, i.e. with medium high or higher vulnerability to groundwater contamination, and (ii) the highest N balance level (>83 kg/ha) (Figure 4). The Mondego alluvial, Aveiro Quaternary and Ança-Cantanhede aquifer systems have the highest proportions of areas with significant pressures, respectively 60, 45 and 28%.

EFI was used to identify the factors: *LPAH*, *AG*, *LE* and *AT* (Table 1), that affect N use efficiency in the administrative areas with significant pressures, i.e. the western part of the CRBD. This region has a low EFI value of 1.3 due to a traditional family farming activity and a rural ageing population without any specific agricultural training (Figure 5). During the last 20 years, a decrease in livestock units, permanent crops and arable land has been observed. In the same period only the pastures have slightly increased (INE 2009; Table 2). It can be assumed that this increase of pastures is directly related to the decrease in livestock numbers and a rural ageing of the population.

From 1995 to 2007 there was a negative evolution of N input in the inland areas, whereas the western coastal areas revealed a positive tendency, because of higher amounts of N fertilizers and low level of EFI.

Considering the facts presented, two scenarios can be contemplated (optimistic versus pessimistic). The first scenario foresees that legal obligations will be respected regarding the Nitrates Directive (EEC 1991) and Sewage Sludge Directive (EEC 1986) and respective amending acts, where sewage sludge is used in agriculture. The EFI will increase due to:

- professionalization of agriculture;
- the implementation of action programs on a compulsory basis in the GWBs with significant pressures by nitrates.

The second scenario, which would result in high rates of N leaching, is that of a large resistance towards the implementation of the measures required by the environmental protection legislation, due to several factors, including a lack of acceptance and credibility as to the potential for water contamination and the need for good agricultural practices, as well as lack of communication and cooperation with the farmers, and technical support and training.

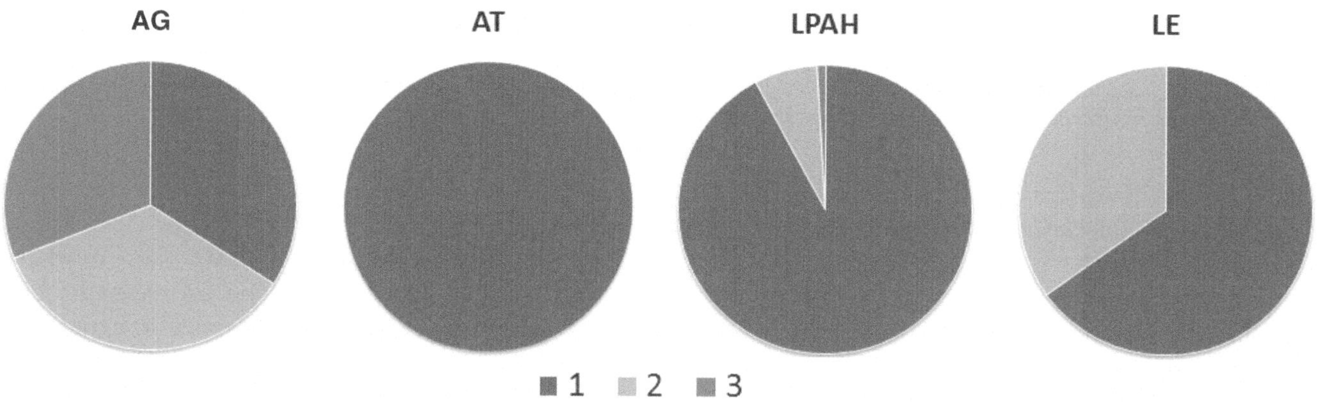

Figure 5 | Response indicator classes for EFI calculation.

Table 2 | Temporal tendency of UAA and number of animals (INE 2009)

Date	UAA per land use (hectares) Permanent crops	Arable land	Pastures	Livestock Number of animals (1000 livestock units)
1989	56460	153148	15027	315
1999	48300	103645	17864	260
2007	35211	78839	18603	186

CONCLUSIONS

A RBMP must express the driving forces that are responsible for significant pressures that can cause potential impact in the qualitative status of GWBs, but it must also reflect the ongoing changes in the farmed environment.

The proposed methodology can help establish areas with highest potential impact on groundwater quality, due to agricultural activities, where there is a lack of groundwater monitoring data, in compliance with the WFD.

As shown in this study, the farm management efficiency is relevant to explain the high rates of N leaching, and therefore the need for decision-makers to contemplate such socioeconomic constraints must be emphasized. The extrapolation of farm development scenarios can address issues related to short-term socioeconomic changes, which undoubtedly will influence the efficiency and performance of agro-environmental measures and consequently contribute to obtaining a good chemical status for the GWBs.

REFERENCES

Aller, L., Bennet, T., Lehr, J. H. & Petty, R. J. 1987 DRASTIC: a standardized system for evaluating groundwater pollution potential using hydrogeologic settings, U.S. EPA Report 600/2-85/018, 1987.

Brentrup, F. & Palliere, C. 2010 Nitrogen Use Efficiency as an Agro-Environmental Indicator. OECD Workshop: Agri-Environmental Indicators: Lessons Learned and Future Directions, 23–26 March 2010, Leysin, Switzerland. Available from: http://www.oecd.org/dataoecd/40/49/44810433.pdf (accessed 29 August 2011).

Buczko, U., Kuchenbuch, R. O. & Lennartz, B. 2010 Environmental indicators to assess the risk of diffuse nitrogen losses from agriculture. *Journal of Environmental Management* **91** (6), 1305–1315.

CEC – Commission of the European Communities 2006 Development of agri-environmental indicators for monitoring the integration of environmental concerns into the common agricultural policy, SEC(2006) 1136, Brussels, Belgium. Available from: http://eur-lex.europa.eu/LexUriServ/LexUriServ.do?uri=COM:2006:0508:FIN:EN:PDF (accessed 29 August 2011).

Corine Land Cover 2006 *Global Monitoring for Environment and Security Land Fast Track Service Precursor for continental Portugal*. European Environment Agency and European Commission Copenhagen, Denmark.

Doerfliger, N. & Zwahlen, F. 1997 EPIK–A new method for outlining of protection areas in karstic environment. In: *International Symposium and Field seminar on Karst Waters and Environmental Impacts* (G. Gunnay & A. I. Johnson, eds). Antalya, Turkey. Balkema, Rotterdam, pp. 117–123.

EC – European Commission 2003 *Common implementation strategy for the Water Framework Directive (2000/60/EC). Guidance Document No 3. Analysis of Pressures and Impacts. Produced by Working Group 2.1 - IMPRESS*. Office for Official Publications of the European Communities, Luxembourg, 153 pp.

EEC 1986 Directive 86/278/EEC of the Council of the European Communities on the protection of the environment, and in particular of the soil, when sewage sludge is used in agriculture. OJ L181, 04.07.86 pp. 6–12.

EEC 1991 Council Directive of 12 December 1991 concerning the protection of waters against pollution caused by nitrates from agricultural sources. OJ L375, 31.12.1991 pp. 1–8.

EC 2000 Directive 2000/60/EC of the European Parliament and of the Council establishing a framework for Community action in the field of water policy. OJ L 327, 22.12.2000, pp. 1–72.

EC 2006 Directive 2006/118/EC of the European Parliament and of the Council on the protection of groundwater against pollution and deterioration. OJ L 372, 27.12.2006, pp. 19–31.

Fernandes, A. J. & Rudolph, D. L. 2001 The influence of Cenozoic tectonics on the groundwater production capacity and vulnerability of fractured rocks: a case study in São Paulo, Brazil. *Hydrogeology Journal* **9** (2), pp. 151–167.

INE 2009 Agri-environmental indicators 1989–2007, Lisbon, Portugal. Available from: http://www.ine.pt/xportal/xmain?xpid=INEandxpgid=ine_publicacoesandPUBLICACOES pub_boui=74873737and PUBLICACOESmodo=2 (accessed 29 August 2011).

INE 2011 Recenseamento agrícola – análise dos principais resultados: 2009, (Census of agriculture - main results : 2009) Lisbon, Portugal. Available from: http://ra09.ine.pt/xportal/xmain?xpid=RA2009andxpgid=ine_ra2009_publicacao_detand contexto=puandPUBLICACOESpub_boui=119564579and PUBLICACOESmodo=2andselTab=tab1and pra2009=70305248 (accessed 29 August 2011).

Leão, P. & Morais, A. 2011 MECAR – Methodology to Estimate the Irrigation Water Consumption in Portugal. In: *O uso da água na agricultura - 2011* (Leão & Ribeiro, eds), INE, I.P., Lisbon, Portugal.

OECD/EUROSTAT 2003 OECD/EUROSTAT gross nitrogen balances handbook.

Ribeiro, L. (ed.) 2002 *Recursos Hídricos Subterrâneos de Portugal Continental*. (Groundwater Resources of Mainland Portugal) INAG, Lisbon, p. 94.

Ribeiro, L. 2005 Desenvolvimento e aplicação de um novo índice de susceptibilidade dos aquíferos à contaminação de origem agrícola. (Development and application of a new index of susceptibility of aquifers to contamination from agricultural sources.) 7° Simpósio de Hidráulica e Recursos Hídricos dos Países de Língua Oficial Portuguesa, APRH, 29 May to 2 June.

Stigter, T. Y., Ribeiro, L. & Carvalho Dill, A. 2006 Evaluation of an intrinsic and a specific vulnerability assessment method in comparison with groundwater salinisation and nitrate contamination levels in two agricultural regions in the south of Portugal. *Journal of Hydrology* **14**, 79–99.

Stigter, T. Y., Carvalho Dill, A. & Ribeiro, L. 2011 Major issues regarding the efficiency of monitoring programs for nitrate contaminated groundwater. *Environmental Science and Technology* **45**, 8674–8682.

Vrba, J. & Zaporožec, A. 1994 Guidebook on mapping groundwater vulnerability. *IAH International Contributions to Hydrogeology* **16**, 156.

First received 14 February 2013; accepted in revised form 14 June 2013

Consideration of emerging pollutants in groundwater-based reuse concepts

A. Tiehm, N. Schmidt, P. Lipp, C. Zawadsky, A. Marei, N. Seder, M. Ghanem, S. Paris, M. Zemann and L. Wolf

ABSTRACT

Elimination of pathogens and emerging pollutants represents a key factor in integrated water resources management in arid regions. Within the SMART Jordan Valley project it is the objective of this study to assess the occurrence and examine the elimination of selected emerging pollutants and pathogens in waste water treatment and aquifer recharge. In batch and soil column studies non-chlorinated organophosphorous compounds (tri-*n*-butylphosphate, triphenylphosphate) and endocrine disruptors (e.g. 17-ß-estradiol, bisphenol A) proved to be biodegradable, while the X-ray contrast agents iomeprol and iopromide were eliminated in the soil columns only, and the chlorinated trialkylphosphates showed persistency. Treating waste water in a membrane bioreactor (MBR) in combination with powdered activated carbon (PAC) resulted in considerable removal rates also for the more persistent compounds such as the antiepileptic carbamazepine. Viruses were shown to be present in most of the Jordan Valley surface water samples. MBR treatment resulted in a decrease of MS2 bacteriophages used as model viruses.

Key words | biodegradation, endocrine disruptors, membrane bioreactor, pharmaceuticals, trialkylphosphates, viruses

A. Tiehm (corresponding author)
N. Schmidt
P. Lipp
C. Zawadsky
Water Technology Center (TZW),
Karlsruher Str. 84, 76139 Karlsruhe, Germany
E-mail: *andreas.tiehm@tzw.de*

A. Marei
Department of Earth and Environmental Sciences,
P.O. Box 20002, Al-Quds University of Jerusalem,
Palestine

N. Seder
Jordan Valley Authority,
P.O. Box 2412, 11183 Amman, Jordan

M. Ghanem
Palestinian Hydrology Group,
P.O.Box 323, Ramallah, West Bank/Palestine

S. Paris
Huber SE, 92335 Berching, Germany

M. Zemann
Department of Applied Geology (AGK),
Karlsruhe Institute of Technology (KIT),
Kaiserstraße 12, 76128 Karlsruhe, Germany

L. Wolf
CSIRO, Queensland Biosciences Precinct,
306 Carmody Rd St Lucia Queensland,
Australia 4067

INTRODUCTION

In arid regions the use of non-traditional water resources such as treated waste water is an important issue. However, pathogens and emerging pollutants such as pharmaceutical residues, trialkylphosphates used as flame retardants (e.g. tri-*n*-butylphosphate, tris-(2-chloropropyl)-phosphate) and endocrine disruptors such as natural hormones and chemicals used as plasticizers (e.g. bisphenol A) can become potential health risks if adequate treatment is missing. The release of emerging pollutants into the aquatic environment has become an issue of increasing concern over recent years. Pharmaceuticals of all fields of prescription (e.g. betablockers, antiepileptics, antiphlogistics, X-ray contrast agents) have been detected in influents and effluents of waste water treatment plants (WWTPs) all over the world, in the range of ng/L up to µg/L (Sacher *et al.* 2008; Ying *et al.* 2009; Jelic *et al.* 2011). Removal rates in WWTPs and in the environment vary significantly and there is a lack of understanding with respect to the elimination processes. As part of the SMART Jordan Valley project, new integrated approaches for water management, aquifer recharge and waste water reuse are developed. The Jordan Valley is one of the driest areas in the world and groundwater is being exploited at about twice its recharge rate to meet the increasing water demand (MWI 2009). A groundwater-based reuse concept for the region would increase the water availability via subsurface storage and could provide an additional low cost treatment option making use of the natural degradation processes within the vadose zone and the aquifer itself (Dillon *et al.* 2010; Page *et al.* 2010).

In several sampling campaigns in 2007 and 2008 waste water, groundwater and surface water samples have been

doi: 10.2166/wst.2013.290

taken in Jordan and Palestine. Among the most frequently detected compounds in the study area were pharmaceutical residues such as lipid regulators (e.g. gemfibrozil, bezafibrate), antiepileptics (e.g. carbamazepine) and antiphlogistics (e.g. naproxen, ibuprofen, diclofenac) (Tiehm *et al.* 2011), which also have been previously reported in European and United States surface waters. Another aspect of artificial groundwater recharge is the unintentional introduction of pathogenic microorganisms contained in waste water. Therefore water samples were also analysed with respect to viruses. The objectives of our studies are: (i) to assess the occurrence of emerging pollutants and selected pathogens in water samples of the Jordan valley, (ii) to obtain more insight into biodegradation processes, and (iii) to monitor and improve the removal of key compounds in treatment processes such as membrane bioreactors (MBR) and soil-aquifer-treatment.

METHODS

Elimination of emerging pollutants during soil passage

Unsaturated soil columns were irrigated with treated waste water from a full-scale WWTP (Figure 1). Hydraulic retention time (HRT) in the columns was 6–7 days. By comparing leachate concentrations of the biologically inhibited (operated at 2 °C) and the column that was operated at 20 °C, biodegradation of spiked trace organics (5 μg/L) during soil passage was assessed. Samples were taken after 2, 3 and 5 weeks as well as after 7 and 13 months of operation time.

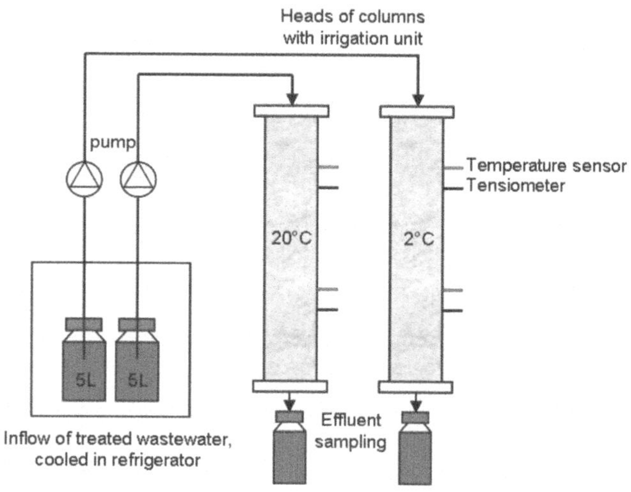

Figure 1 | Schematic illustration of the column experiment.

Elimination of emerging pollutants in batch assays

Using filtered raw waste water and treated waste water, batch tests were set up to examine biodegradation in the presence of two different background concentrations of organic load. Two batch assays were set up in parallel for each condition. A mixture of pharmaceutical residues (5 μg/L of each substance) was added. The batches were inoculated with activated sludge from a full-scale WWTP and incubated at room temperature under aerobic conditions during 12 weeks. Controls were autoclaved and incubated at 4 °C. The remaining concentration of each substance in the active assays was calculated against the concentration in the autoclaved control batch ($C/C_{control}$). Arithmetic means and standard deviations of the parallel assays were calculated.

MBR pilot treatment plant

Huber SE Germany provided an MBR pilot plant consisting of a process tank (800 L) in which two membrane modules (2 m) were submerged (Figure 2). The average filtrate flux was adjusted to 15 L/m^2/h. For mechanical cleaning of the membrane surface, after 2 min of filtration the pumping process was interrupted regularly for 1 min of relaxation time. Membranes were aerated continuously (0.25–1.5 Nm^3/m^2/h) in order to support removal of deposits on the membrane surface. For nitrification/denitrification, the process tank was aerated intermittently with 6 Nm^3/h. Sludge retention time was 25 d, HRT was 24 h and total suspended solids (TSS) content in the process tank was adjusted to 8–12 g/L by weekly removal of excess sludge. The MBR was continuously operated with the influent of a full-scale WWTP after primary treatment (Lipp *et al.* 2009). To improve the elimination of persistent pollutants, the MBR filtrate was further treated with 10 mg/L powdered activated carbon (PAC).

Chemical analysis

Analysis of pharmaceutical residues was done by high-performance liquid chromatography–electrospray ionization tandem mass spectrometry (HPLC-ESI-MS/MS) following a previously described procedure (Sacher *et al.* 2008). Also the analysis of X-ray contrast agents has been described previously (Sacher *et al.* 2005). Analysis of the trialkylphosphates was done by a gas chromatography–mass spectrometry (GC-MS) method using a GC-MSD Agilent 5973 (Agilent Technologies, Santa Clara, California) after solid-phase extraction on plastic cartridges filled with 200 mg of Bakerbond SDB 1 material (Mallinckrodt Baker, Deventer, The Netherlands). For GC

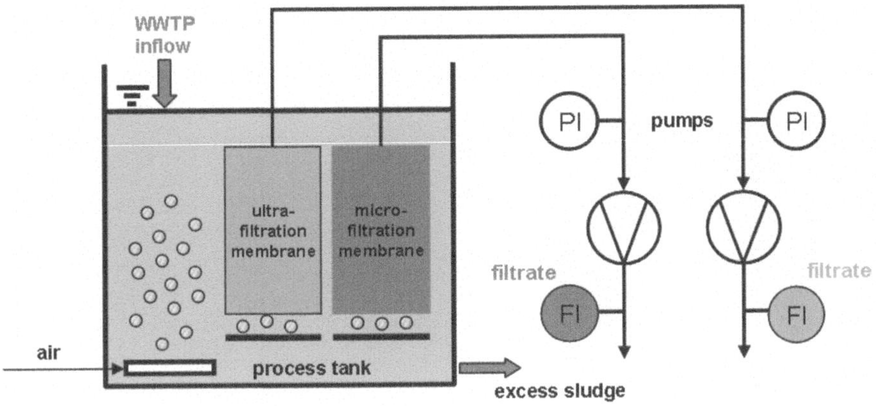

Figure 2 | Schematic illustration of the MBR pilot plant.

separation, a DB-5 (60 m) (Agilent Technologies, Santa Clara, California) was used. Injection temperature was 250 °C. The injection volume was 2 μL and total flow was 7.4 mL/min with helium as carrier gas. The temperature program started at 60 °C (held for 2 min), set at 20 °C/min to 130 °C (held for 5 min), set at 1 °C/min to 205 °C, set at 5 °C/min to 300 °C (held for 10 min). Detection with the MSD device was done in EI mode (SIM). Calibration was done between 20 and 500 ng/l. Detection of endocrine disruptors (steroid hormones, bisphenol A) was carried out by a GC-MS method using an Agilent 7890 A with an MSD 5975 C (Agilent Technologies, Santa Clara, California, USA) after solid-phase extraction on plastic cartridges filled with 100 mg of STRATA X material (Phenomenex, Aschaffenburg, Germany). For GC separation, a Rtx®-5Sil MS (Restek Corporation, Bellefonte, Pennsylvania) was used. Injection temperature was 280 °C. The injection volume was 1 μL and total flow was 1.2 mL/min with helium as carrier gas. The temperature program started at 120 °C (held for 1 min), set at 15 °C/min to 180 °C, set at 5 °C/min to 290 °C (held for 10 min). Detection with the MSD device was done in EI mode (SIM). Calibration was done between 1 and 500 ng/l. Regardless of the analytical methods applied, all water samples were put onto solid phase extraction cartridges after the main particles had settled, considering only dissolved compounds for the analysis. Analytical errors of measurement were between 25 and 36%.

Detection of microorganims

Field samples were examined for the occurrence of adenoviruses, rotaviruses, noroviruses and MS2 bacteriophages. For virus enrichment, the cation coated filter was used as described previously (Haramoto *et al.* 2005; Zawadsky & Tiehm 2009). Viral RNA und DNA were extracted simultaneously and samples were analysed by polymerase chain reaction (PCR). During 15 months MBR operation, two and three grab samples were taken from the MBR inflow and ultrafiltration filtrate, and the process tank and microfiltration filtrate, respectively.

RESULTS AND DISCUSSION

To gain an overview of relevant trace organics and pathogenic microorganisms along the Lower Jordan Valley, 45 samples from 38 different locations were taken in the West Bank and east of the Jordan Valley, between Wadi Kufrinja in the north and Wadi Kafrein in the south. Surface water samples were taken at the King Abdullah Canal, the Wadi Shueib and the Wadi Kafrein. Waste water samples originated from effluents of the WWTPs in Wadi Shueib and Wadi Kafrein. Surface water samples consisted either of surface runoff or of a mixture with effluent from the WWTPs. Additionally, groundwater samples from springs and wells were analysed. Key substances were subsequently chosen for biodegradation studies with batch assays and soil columns. Furthermore, their degradation potential in an MBR was tested, and elimination of viruses was assessed exemplarily for MS2 bacteriophages.

Screening and elimination of viruses and the MS2 bacteriophage

Screening of field samples

In total, 23 Jordan valley water samples were examined for viruses and MS2 bacteriophages (Figure 3). In 69% of the samples at least one group of viruses was detected. Rotavirus group A was dominating (72%). The results are consistent

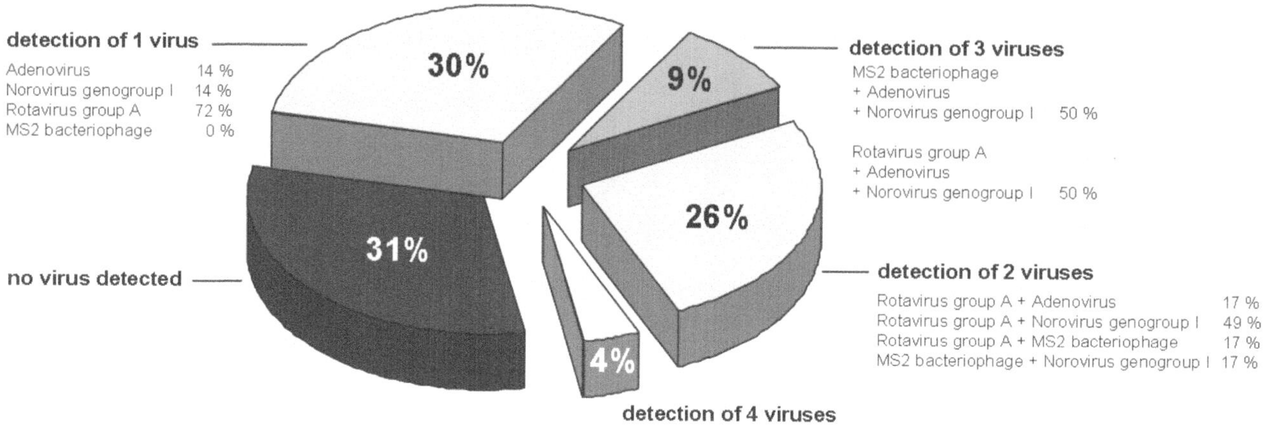

Figure 3 | Detection of viruses and the MS2 bacteriophage in Jordan valley water samples.

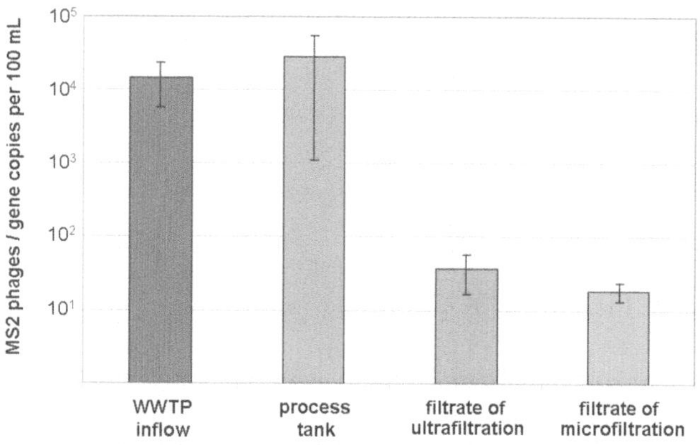

Figure 4 | MS2 bacteriophage removal in the MBR during ultra and microfiltration; error bars indicate standard deviations.

with previous reports as rotaviruses seem to be ubiquitous: 50–60% of cases of acute gastroenteritis of hospitalized children throughout the world are caused by human rotaviruses (WHO 2008).

MBR treatment

In order to include the elimination of viruses in subsequent studies focusing on new decentralized waste water treatment and reuse schemes, removal of MS2 bacteriophages as model organisms was investigated. A considerable removal was demonstrated during treatment of waste water in the MBR. After membrane filtration 2- to 3-log reduction of MS2 phages was observed in the filtrates (Figure 4).

Screening and elimination of emerging pollutants

Screening of field samples

The concentrations of some of the most frequently detected substances among pharmaceuticals in the Jordan Valley are presented in Table 1. The detected values for the

Table 1 | Minimum, maximum and median concentrations of selected pharmaceuticals in the Lower Jordan Valley

		Min/Max concentration (Median) [ng/L]		
Group	**Substance**	**GW**	**SW**	**WW**
Antiphlogistics	Diclofenac*	n.d.[a]	<20/160 (<20)	<100/390 (<20)
	Ibuprofen*	n.d.	<20/1,400 (<20)	<50/750 (140)
Lipid regulator	Gemfibrozil*	<20/70 (<20)	<20/2,100 (150)	<50/4,800 (1,400)
Antiepileptic	Carbamazepine*	<20/180 (<20)	<20/1,600 (240)	<50/3,600 (1,800)
X-ray contrast agents	Iopromide*	<20/1,600 (<20)	<100/3,000 (660)	<100/2,200 (190)
	Iomeprol	<20/48 (<20)	<100/5,300 (<100)	<100/360,000 (2,000)
Endocrine disruptor	Bisphenol A	<10/140 (15)	<25/53 (<25)	<25/340 (86)
Trialkylphosphate	Tris-(2-chloroethyl)-phosphate	<75/84 (<75)	<75/1,200 (<75)	260/780 (320)
	Total number of samples	14	19	5

(GW = groundwater, SW = surface water, WW = waste water), *adapted from Tiehm *et al.* (2011).
[a]n.d. = not detected.

Table 2 | Elimination of endocrine disruptors, trialkylphosphates and X-ray contrast agents during the batch and column experiments

		Batch studies[a]		Column studies[a]	
		FRWW	TWW	2 °C	20 °C
Endocrine disruptors	Estrone	+++	+++	+++	+++
	17-*β*-Estradiol	+++	+++	+++	+++
	Estriol	+++	+++	+++	+++
	Bisphenol A	+++	+++	+++	+++
Trialkylphosphates	Triethylphosphate	+	–	–	+
	Tri-*n*-butylphosphate	+	++	+++	+++
	Triphenylphosphate	++	++	+++	+++
	Tris-(2-chloroethyl)-phosphate	–	–	–	–
	Tris-(2-chloropropyl)-phosphate	–	–	–	–
X-ray contrast agents	Iomeprol	–	–	++	+++
	Iopromide	–	–	++	+++

(FRWW = filtered raw waste water, TWW = treated waste water.)
[a] – = less than 10%, + = 10–50%, ++ = 50–90% and +++ = more than 90% elimination.

emerging pollutants were in the same range as in Europe and the USA.

All compounds were detected in surface and waste water samples, and some trace organics have been found in groundwater as well. For instance, carbamazepine, a very persistent compound in the environment, was found in groundwater samples. Among the compounds with the highest concentrations was the X-ray contrast agent iomeprol, with 360 µg/L found in one waste water sample. The antiphlogistics ibuprofen and diclofenac were not detected in groundwater samples but with up to 1,400 and 160 ng/L, respectively, in surface water samples.

MBR treatment

Analyses of samples from the influent and effluent of the MBR showed that some substances (ibuprofen and naproxen) are removed, whereas others (carbamazepine and diclofenac) passed the MBR. However, addition of PAC to the MBR filtrate also improved elimination of the most persistent pollutants, i.e. carbamazepine and diclofenac (86 and 69% removal, respectively) (Lipp *et al.* in press).

Batch and column studies

Table 2 gives an overview of the compounds tested in this study and their behaviour during the batch and column studies.

Two of the trialkylphosphates (tri-*n*-butylphosphate, triphenylphosphate) showed 50–90% elimination during the batch test with treated waste water and complete elimination in the column studies. Endocrine disruptors were always completely eliminated. Triethylphosphate and the two chlorinated organophosphorous compounds showed only little to no removal under the conditions tested. This is in accordance with other findings in the literature, where chlorinated trialkylphosphates tend to pass through the WWTPs unaffected, while non-chlorinated species showed high elimination rates under aerobic conditions (Meyer & Bester 2004; Marklund *et al.* 2005). The X-ray contrast agents iomeprol and iopromide showed a recalcitrant behaviour in the batch assays but were eliminated during soil passage in the columns, both at 2 and 20 °C. Different microbial communities in the batch assays and the soil columns most probably are the reason for the differing results. During 1 year of operation, leaching of the substances to the column effluents would have been expected, if mainly sorption was responsible for elimination. But as none of the endocrine disruptors or the two trialkylphosphates were found in the effluent at any of the five samplings a

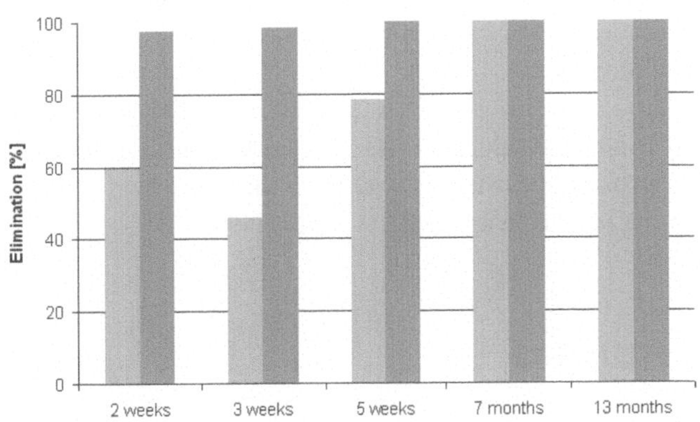

Figure 5 | Elimination of iomeprol after soil passage in the 2 °C (left) and the 20 °C (right) columns.

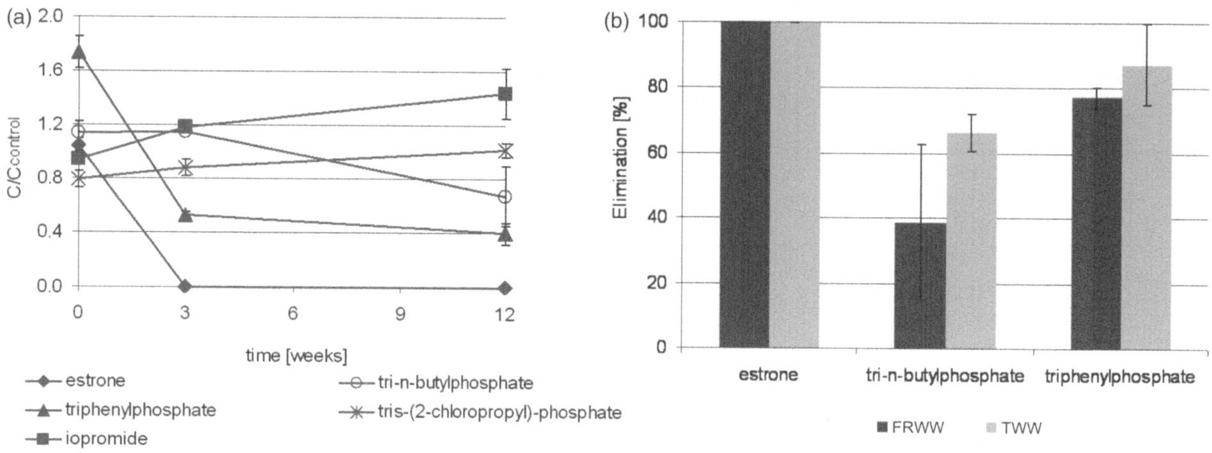

Figure 6 | Concentration of selected substances in the batch experiments with filtered raw waste water (a) and comparison of elimination rates between filtered raw (FRWW) and treated waste water (TWW) (b), error bars indicate standard deviations.

high biodegradation potential both under cold and temperate conditions is indicated for these substances. In the case of the X-ray contrast agents, elimination in the 2 °C column increased over time which indicates adaptation of the biomass to cold conditions (Figure 5). Considerable degradation rates for iopromide (Schittko *et al.* 2004; Gruenheid *et al.* 2005) and iomeprol (Schmidt *et al.* 2007) of up to more than 80% were also reported from riverbank filtration sites.

The concentrations of selected substances during the 12 weeks batch experiment are shown exemplarily for the batch assay with filtered raw waste water (Figure 6(a)). The trialkylphosphates tri-*n*-butylphosphate and triphenylphosphate showed mean elimination rates of 39 and 77%, respectively, which increased to 66 and 87% in the assays with treated raw waste water (Figure 6(b)). Estrone was eliminated to 100% regardless of the background organic load in the medium, while tri-*n*-butylphosphate and triphenylphosphate showed higher elimination rates, if treated waste water was used as a medium. Lower biodegradation in the raw waste water might be due to competition with easily degradable organic matter present in the assays. Thus the dissolved organic carbon decreased by more than 50% from 46 to 20 mg/L in the two batches set up with raw waste water, while it was stable between 7 and 8 mg/L in the batches with treated waste water.

CONCLUSIONS

Pathogenic viruses and emerging organic pollutants such as pharmaceutical residues have been assessed in the Lower Jordan Valley. Concentrations of the trace organics were in the same order of magnitude as reported for Europe and the USA. Among the studied pathogenic viruses, the rotavirus group was dominating, but the adenovirus and norovirus were also detected frequently. Intensive water reuse in arid regions as well as water recycling strategies under moderate climatic conditions can result in an accumulation of pollutants. With MBR treatment of waste water a reduction of pathogens can be achieved. The experimental results from the batch and column studies demonstrate biodegradation of most organic pollutants in waste water and during soil infiltration under aerobic conditions. However, parameters such as organic background load, temperature and adaptation of biomass have been shown to influence the degradation behaviour of the different substances tested. Future studies will continue monitoring campaigns in Palestine and Jordan, and examine pollutant removal in pilot-scale plants operated in the Jordan Valley.

ACKNOWLEDGEMENTS

The authors would like to thank all involved partners of the SMART (Sustainable Management of Available Water Resources with Innovative Technologies; www.iwrm-smart.org) project for their support. Special thanks go to: Heinz Hötzl (AGK), Ali Subah (Ministry of Water and Irrigation (MWI)) and Elias Salameh (Jordan University). Funding by the German Federal Ministry of Education and Research (BMBF, project no. 02WM1082) as well as by the International Bureau of the Ministry of Education and

Research (BMBF, project no. AUS 10/810) is gratefully acknowledged.

REFERENCES

Dillon, P., Toze, S., Page, D., Vanderzalm, J., Bekele, E., Sidhu, J. & Rinck-Pfeiffer, S. 2010 Managed aquifer recharge: rediscovering nature as a leading edge technology. *Water Science and Technology* **62** (10), 2338–2345.

Gruenheid, S., Amy, G. & Jekel, M. 2005 Removal of bulk dissolved organic carbon (DOC) and trace organic compounds by bank filtration and artificial recharge. *Water Research* **39** (14), 3219–3228.

Haramoto, E., Katayama, H., Oguma, K. & Ohgaki, S. 2005 Application of cation-coated filter method to detection of noroviruses, enteroviruses, adenoviruses, and torque teno viruses in the Tamagawa River in Japan. *Applied and Environmental Microbiology* **71** (5), 2403–2411.

Jelic, A., Gros, M., Ginebreda, A., Cespedes-Sánchez, R., Ventura, F., Petrovic, M. & Barcelo, D. 2011 Occurrence, partition and removal of pharmaceuticals in sewage water and sludge during wastewater treatment. *Water Research* **45** (3), 1165–1176.

Lipp, P., Gross, H.-J. & Tiehm, A. in press Improved elimination of organic micropollutants by a process combination of membrane bioreactor (MBR) and powdered activated carbon (PAC). *Desalination and Water Treatment – Science and Engineering.*

Lipp, P., Kreißel, K., Meuler, S., Bischof, F. & Tiehm, A. 2009 Influencing parameters for the operation of an MBR with respect to the removal of persistent organic pollutants. *Desalination and Water Treatment* **6** (1–3), 102–107.

Marklund, A., Anderson, B. & Haglund, P. 2005 Organophosphorus flame retardants and plasticizers in Swedish sewage treatment plants. *Environmental Science and Technology* **39** (19), 7423–7429.

Meyer, J. & Bester, K. 2004 Organophosphate flame retardants and plasticisers in wastewater treatment plants. *Journal of Environmental Monitoring* **6** (7), 599–605.

MWI – Ministry of Water and Irrigation 2009 Water for Life – Jordan's Water Strategy 2008–2022. Available from: http://web.idrc.ca/uploads/user-S/12431464431JO_Water-Strategy09.pdf (accessed 13 April 2012).

Page, D., Dillon, P., Toze, S. & Sidhu, J. P. S. 2010 Characterising aquifer treatment for pathogens in managed aquifer recharge. *Water Science and Technology* **62** (9), 2009–2015.

Sacher, F., Ehmann, M., Gabriel, S., Graf, C. & Brauch, H. J. 2008 Pharmaceutical residues in the river Rhine – results of a one-decade monitoring programme. *Journal of Environmental Monitoring* **10** (5), 664–670.

Sacher, F., Raue, B. & Brauch, H. J. 2005 Analysis of iodinated X-ray contrast agents in water samples by ion chromatography and inductively-coupled plasma mass spectrometry. *Journal of Chromatography A* **1085** (1), 117–123.

Schittko, S., Putschew, A. & Jekel, M. 2004 Bank filtration: a suitable process for the removal of iodinated X-ray contrast media? *Water Science and Technology* **50** (5), 261–268.

Schmidt, C. K., Lange, F. T. & Brauch, H. J. 2007 Characteristics and evaluation of natural attenuation processes for organic micropollutant removal during riverbank filtration. *Water Science and Technology: Water Supply* **7** (3), 1–7.

Tiehm, A., Schmidt, N., Stieber, M., Sacher, F., Wolf, L. & Hoetzl, H. 2011 Biodegradation of pharmaceutical compounds and their occurrence in the Jordan Valley. *Water Resources Management* **25** (4), 1195–1203.

WHO – World Health Organization 2008 Guidelines for drinking-water quality. Available from: http://www.who.int/water_sanitation_health/dwq/fulltext.pdf (accessed 13 April 2012).

Ying, G. G., Kookana, R. S. & Kolpin, D. W. 2009 Occurrence and removal of pharmaceutically active compounds in sewage treatment plants with different technologies. *Journal of Environmental Monitoring* **11** (8), 1498–1505.

Zawadsky, C. & Tiehm, A. 2009 PCR-detection of viruses in waste water and surface water of the Jordan Valley. In: *Proceedings of the 15th IWA International Symposium on Health-Related Water Microbiology*. Greece, pp. 340–341.

First received 3 January 2013; accepted in revised form 18 April 2013

Theme III

Technologies and implementation

Adapting to water scarcity: constraints and opportunities for improving irrigation management in Khorezm, Uzbekistan

B. Tischbein, A. M. Manschadi, C. Conrad, A.-K. Hornidge, A. Bhaduri, M. Ul Hassan, J. P. A. Lamers, U. K. Awan and P. L. G. Vlek

ABSTRACT

Like many irrigation schemes in Central Asia, the one in Khorezm faces a two-fold challenge: on the one side, the severe problems inherited from the past need to be remedied and on the other side, the rising supply–demand gap driven by sharpening competition for water and climate change must be dealt with. Located in the lower part of the Amu Darya basin, Khorezm irrigation and drainage scheme is particularly vulnerable to supply–demand gaps. Promising solutions towards adaptation comprise modified strategies of land and water use towards higher efficiency and flexibility in combination with measures to lessen the constraints of the system itself, which was initially designed for the management of a few, large and uniform production units and not for many diverse and small units. Solutions consist of flexible, modeling-based approaches, re-arranging institutional settings and establishing economic incentive systems. Flexible modeling allows an integrated use of surface and groundwater resources avoiding or minimizing the impact of water stress on yield. Institutional settings strengthen the position of water users via improved participation and transparency of processes in Water Consumers Associations (WCAs). Economic measures support sustainable resource use strategies and improve the functioning of WCAs. The findings could be extrapolated to other regions of Central Asia with similar conditions and challenges.

Key words | Aral sea, efficiency, irrigation water management, Khorezm, salinity management

B. Tischbein (corresponding author)
A.-K. Hornidge
A. Bhaduri
M. Ul Hassan
J. P. A. Lamers
U. K. Awan
P. L. G. Vlek
Center for Development Research,
University of Bonn, Walter Flex-Strasse 3,
D-53113 Bonn,
Germany
E-mail: *tischbein@uni-bonn.de*

A. M. Manschadi
University of Natural Resources and Life Sciences,
Vienna, Department of Crop Sciences,
Konrad Lorenz Str. 24, A-3430, Tulln,
Austria

C. Conrad
Remote Sensing Unit, Institute of Geography,
University of Wuerzburg, Am Hubland,
D-97074 Wuerzburg,
Germany

U. K. Awan
International Water Management Institute (IWMI),
Lahore-Pakistan

INTRODUCTION

The Aral Sea dilemma is a textbook example illustrating the impact of large-scale and long-term unsustainable water management. The irrigated area on the territory of the former Soviet Republics in the Aral Sea basin was expanded from 4.5 Mega hectares (Mha) in 1960 to 7.6 Mha in 1990 and 8.1 Mha in 2004 (SIC ICWC 2011a). Inefficient water management in these huge areas led to high withdrawals from its major tributaries, the Amu Darya and the Syr Darya Rivers. The inflow to the Aral Sea decreased for instance from 43 km^3 in the 1960s to 9 km^3 during the period 2001–2005 (Micklin 2007). As a consequence, the Aral Sea shrank dramatically. According to Micklin (2007), the lake lost 90% of volume and the salt concentration increased to 100 g/l. These processes resulted in enormous economic losses, disastrous impacts on the ecosystems and severe health problems for the population.

The Uzbek province Khorezm, located in the lower part of the Amu Darya basin approximately 350 km south of the current shores of the Aral Sea, exemplifies the reasons causing the Aral Sea dilemma, its consequences and future challenges to water management in Central Asia. Khorezm is equipped with a large-scale irrigation system developed during the Soviet period. Due to

doi: 10.2166/ws.2013.028

its downstream location, Khorezm is particularly vulnerable to reductions in water supply. Hydrologically caused reductions of water resources such as for instance in drought years are sharpened by water management in the upstream parts of the basin. This is illustrated by the ratio between (i) withdrawals from the Amu Darya (SIC ICWC 2011b) in the vegetation period (April–September) of a specific year in the period 1999 until 2009, and (ii) the average withdrawal for that period (Figure 1). The annual ratios estimated for irrigation schemes located upstream (Tajikistan), midstream (Karakum canal) and downstream (Khorezm) of the Amu Darya underlines that in drought years (e.g. 2000, 2001, 2008) the ratio for Khorezm had dropped sharply, whereas water supply in e.g. midstream locations was much less impacted and upstream parts of the basin were hardly influenced by these drought strikes.

It is expected that water resources available for Khorezm in the near future will become even more limited and increasingly variable due to climate change and enhancing competition for water in the upper part of the transboundary basin driven by population growth, industrialization and hydropower (Schlüter & Hirsch 2010).

The climate in Khorezm is continental and arid. The reference evapotranspiration amounts to 1,380 mm per year, whereas the annual precipitation is very low with 95 mm (Conrad *et al.* 2012). Therefore, agricultural production relies completely on irrigation and drainage systems, the latter necessary for salt management. Approximately 275,000 ha in this region are equipped with an irrigation and drainage infrastructure consisting of complex networks of 16,400 km irrigation canals and 8,000 km drainage ditches and collectors (GIS-Lab of the ZEF/UNESCO Project). Cotton, rice, winter wheat, vegetables and fodder dominate the cropping pattern in Khorezm. With the exception of 11%, irrigation canals are not lined (Ibrakhimov 2004) and water is applied to crops mainly by furrows and to a minor extent by basin irrigation (in case of leaching and rice cropping). Agriculture is regulated by the state as illustrated by the state-ordered production of cotton and wheat. Water management is dominated by hierarchically structured and centrally organized administrations (Bobojonov 2008). These administrations allocate water based on areas assigned to grow cotton and wheat and which have fixed production quota. The state-order system creates thus a strong link between the use of soil and water resources. The flexibility of water management is limited also by the lack of detailed knowledge on site-specific irrigation water demand which at present is based on static norms considering crop, soil and groundwater level in a coarse way.

Water management problems are indicated by: (i) substantial water withdrawals from the Amu Darya amounting to 4.4 km^3 per year on average (Tischbein *et al.* 2011), (ii) insufficient supply at farm level in terms of quantity and temporal appropriateness, leading to water stress shown by reduced actual evapotranspiration by 25% in tail-end regions (Conrad 2006), (iii) low depleted fraction of 0.48 (Conrad 2006), and (iv) widespread occurrence of soil salinization. Pre-seasonal leaching accounts for approximately 1 km^3 in years with average water availability (Bekchanov *et al.* 2010a).

Restructuring current water management towards efficient, appropriate and sustainable use of water resources is

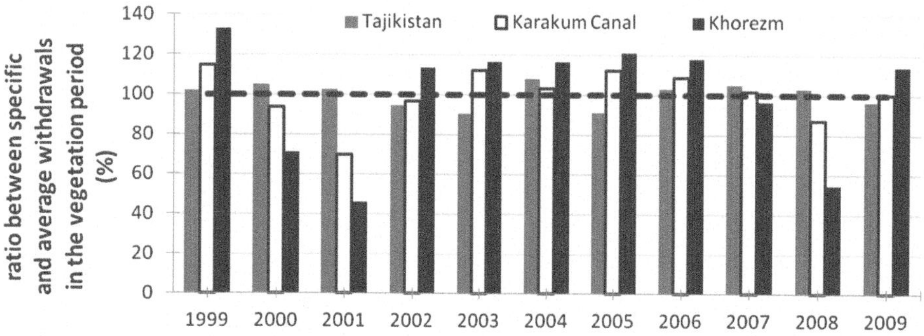

Figure 1 | Ratio between the withdrawals during the vegetation period for specific years (1999–2009) and the average of that period for upstream (Tadjikistan), midstream (Karakum canal) and downstream locations (Khorezm) (data source: SIC ICWC 2011b).

a key-intervention to improve the livelihood of approximately 1.5 million inhabitants in Khorezm especially when taking into account the unsolved problems inherited from the Soviet period: the huge expansion of irrigated areas in the Aral Sea basin; cotton monoculture; insufficient adaptation of the irrigation and drainage infrastructure and management to site-specific and dynamic conditions. This is challenged further by the present confrontations of climate change and a sharpening competition for water. Working out options to increase efficiency, appropriateness and sustainability of water management is the major objective of the research described here.

METHODS

Basic features of the approach

Interdisciplinarity

The current problems and future challenges of water management in Khorezm are caused by a mix of bio-physical, technical and socio-economical factors. Hence an adequate solution consists of an interdisciplinary approach combining technical procedures, economical methods and sociological analyses. We use therefore the bio-physical system in the case study region Khorezm as a starting-point to work out modeling approaches and to derive interventions as opportunities towards improving the current water management. These opportunities will then be embedded into the institutional and economic context by considering options to lessen constraints regarding technical, institutional and economic factors.

Focusing on field/farm level

The focus on the field level was guided by two major ideas: (i) providing the farmer with options to cope with variable water supply, as this is urgently needed and becoming increasingly relevant, and (ii) considering the field level reflects a bottom-up approach from field to system level, which has become more appropriate after the restructuring of the production units by the national administration.

Considering the links between surface and groundwater resources ('sponge phenomenon')

As a result of high recharges due to heavy leaching and huge irrigation water losses, groundwater is shallow in Khorezm. According to Ibrakhimov *et al.* (2011) groundwater level fluctuates annually between 2.2 m below ground at the end of the non-irrigation period in winter (January/February) and 1.15 m in the peak irrigation season (July). Driven by irrigation and drainage strategies, the aquifer behaves as a 'sponge', as exemplified by the analyses of a representative 850 ha sized sub-unit (Figure 2). The upper part of Figure 2 depicts the irrigation water inflow and drainage outflow. The difference between inflow and outflow (lower part) determines the groundwater dynamics (in interaction with the evapotranspiration being a further major output of the system). The groundwater level referenced to the deepest level in January (lower part), clearly follows the inflow-outflow difference with a delay in the range of a month.

While shallow groundwater tables may be beneficial for meeting crop water demand, it does negatively affect soil salinity management in two ways. Firstly, shallow groundwater accelerates the salt accumulation in the root zone. Secondly, the effectiveness of leaching is reduced due to high water levels in the drainage ditches which constrain discharging the percolated leaching water.

Modeling approaches at field level

Modeling soil water fluxes

As an alternative to the static norms presently dominating the estimation of the irrigation water distribution in Uzbekistan, a modeling approach considering explicitly the components of water and salt balances was used. Based on the HYDRUS model (Simunek *et al.* 2005) and FAO's dual crop coefficient concept (Allen *et al.* 1998), the approach allows modeling of the water fluxes and salt dynamics including the relevant capillary rise. Forkutsa (2006) applied the approach to establish water balances and to derive net irrigation amounts and irrigation timing. The dual crop coefficient enables a differentiation between evaporation and transpiration and consequently measures can be derived to reduce non-productive evaporation (by

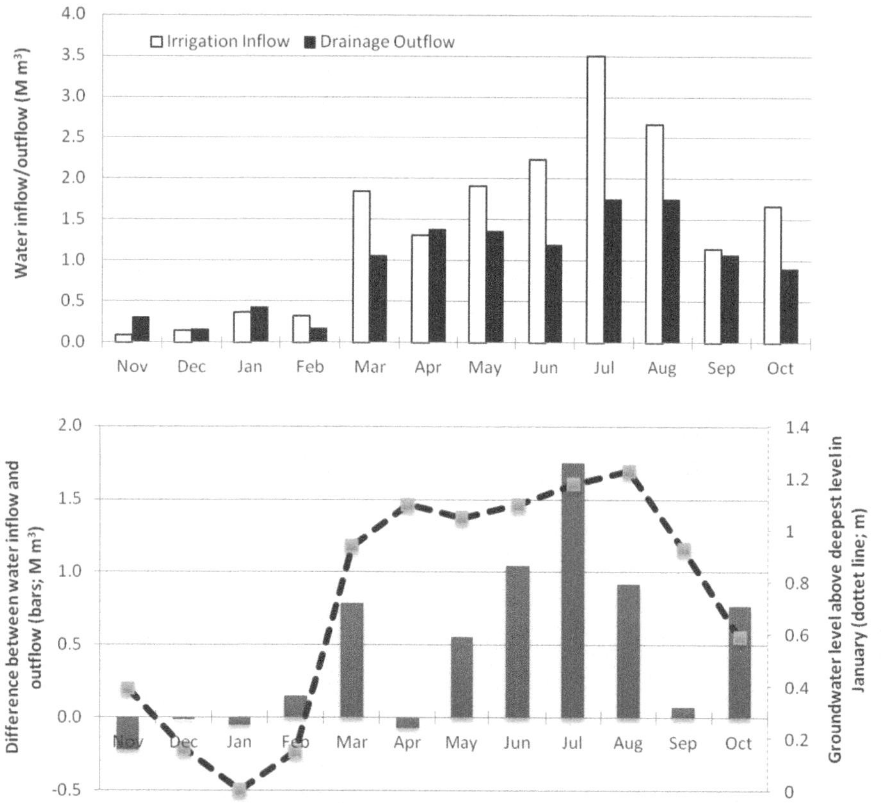

Figure 2 | Annual dynamics of groundwater level in a typical sub-unit of Khorezm (lower part) causing the 'sponge phenomenon' of the aquifer driven by the irrigation inflow–drainage outflow difference (lower part).

practices of conservation agriculture) and base irrigation scheduling explicitly on the transpiration. As the modeling describes the hydrological processes in detail, the approach has the potential to be used with a high spatial resolution considering even within-field variability. This is in particular advantageous in a situation of increased number of water users with diversified requirements (compared with the past setting in Khorezm with large and uniform production units) as emerged in the study region.

Linked irrigation scheduling–groundwater model

The implementation of flexible irrigation strategies at schemes level needs to be supported by tools taking the groundwater dynamics into account. Awan (2010) developed a tool linking the irrigation scheduling model CROPWAT (Clarke *et al.* 1998) and the groundwater model FEFLOW (Diersch 2002). To overcome the missing consideration of the capillary rise in CROPWAT, the HYDRUS model (Simunek *et al.* 2005) was used. This configuration allows establishing the field water balance, deriving optimal irrigation schedules and quantifying the spatio-temporal behavior of groundwater. In contrast to the currently, rigid irrigation scheduling based on static norms only, this tool allows us to respond adequately to temporal changes in water availability, meteorological situation and groundwater (Awan 2010). The expected increase in variability of these factors underlines the relevance of the approach. Furthermore, the tool enables us to derive site-specific solutions and to consider conjunctive use of surface and groundwater resources.

AquaCrop for deficit irrigation

To adapt to limited and non-reliable supply at farm level, a tool minimizing the impact of non-avoidable under-supply on the crop yield by controlled deficit irrigation is urgently needed to assist farmers. As the AquaCrop model (Steduto *et al.* 2009) has clear advantages in estimating the crop yield depending on water stress compared with CROPWAT, Akhtar (2011) combined the AquaCrop and HYDRUS models and developed a tool to work out deficit irrigation strategies and to compare these with full irrigation.

Salt management at field level

HYDRUS allows modeling of salt dynamics (salt accumulation and leaching). As a consequence, site-specific and flexible leaching dates and amounts can be derived aiming at highest possible leaching effectiveness and ensuring non-exceedance of crop-specific salt tolerances (Forkutsa *et al.* 2009).

Application process

According to field analyses by Forkutsa (2006) and Awan (2010), the application efficiency in Khorezm ranges between 40 and 50% only. Measures for improvement consist thus of: (i) adapting the application discharge to field size, soil conditions, slope and irrigation amount based on modeling the advance, recession and infiltration of applied irrigation water (Horst *et al.* 2005), (ii) introducing advanced handling of furrow irrigation (e.g. surge flow, alternate furrow (Pereira *et al.* 2007, Bekchanov *et al.* 2010b), and (iii) laser-guided land leveling. Due to very flat topography in Khorezm, directing the water from both ends of the furrow (double-side irrigation) provides also a feasible option (Poluasheva 2005) to increase uniformity and in turn efficiency of water application (at respective locations in Khorezm).

RESULTS AND DISCUSSION

Technical measures

Irrigation scheduling at field level

Combining HYDRUS and FAO's dual crop coefficient concept allowed modeling of water fluxes and balances with high spatio-temporal resolution. At a field typical of the study region (size of ca 4 ha; sandy loam texture), capillary rise contributes 37% to actual evapotranspiration on average over the field (Forkutsa 2006). Spatial variation of capillary rise showed a considerable variability within the field, ranging from 92 to 277 mm over the vegetation period of cotton (Forkutsa 2006). As a consequence, accumulation of soil salinity mainly driven by capillary rise is spatially variable within the field. Knowledge on the within-field variability provides the informal base to adapt irrigation and leaching amounts to site-specific conditions as a prerequisite for water saving. Furthermore, the simulations revealed a reduction of actual evapotranspiration to 65–70% of the potential level. It became obvious, that transpiration was reduced over one month before the first irrigation event was applied. As a consequence, the model allows determining appropriate irrigation timing. Optimized irrigation timing has the potential to mobilize water saving in the range of 20% without increasing water stress (Forkutsa 2006).

Akhtar (2011) simulated deficit irrigation strategies, differentiating proportional and stage-specific reduction of irrigation amounts. With respect to the impact on yield, proportionally reduced irrigation is a lower risk choice. For instance, a 20% proportional reduction proved a promising option to raise crop-water productivity of cotton. Despite a 10% loss in biomass, the yields could be kept in the same range of yields occurring at full supply of irrigation due to an improved irrigation timing, which in turn stimulated yield relevant growth components. Hence, low to moderate deficits are a first choice when conceiving effective and feasible deficit irrigation strategies. This is in-line with previous findings by Pereira *et al.* (2009) opting for mild deficits derived from field experiments in Fergana/Uzbekistan and by the application of the irrigation scheduling model ISAREG.

Rice irrigation

Rice cultivation in Khorezm is conducted in permanently flooded basins. Therefore, rice cropping demands up to 30% of the total water supply, although rice occupies only 10% of the cropped area in Khorezm (Manschadi *et al.* 2010). Improving the management of rice fields has a huge water saving potential, which in turn allows raising considerably the water productivity (Bekchanov *et al.* 2010b). With respect to fields on loamy sand and sandy loam, Devkota (2011) compared intermittent irrigation of rice fields combined with conservation agriculture practices. Intermittent irrigation consequently applied in the vegetation period and conservation agriculture practices enabled a 68–73% water saving potential. Although yield was reduced by 30–56%, water productivity was nearly doubled. These numbers indicate the maximum achievable range of the water saving potential by intermittent rice irrigation under the agro-ecological conditions predominating in Khorezm.

Favored by bio-physical (flat topography, shallow groundwater), economic (high price) and cultural (nutrition behavior) conditions Khorezm developed a rice cultivation tradition and became a major rice producing region in Central Asia. Decreasing the area cropped with rice would imply an over-proportional potential of water savings (see above mentioned numbers on the area of and the water directed to rice cropping). On the other hand, rice is the most profitable crop in the region as illustrated by its large economic benefits (Djanibekov *et al.* 2010) and restricting rice production would become a major administrative issue. More promising than completely omitting rice cultivation seems the introduction of a combination of technological, economical and institutional measures addressing for instance the selection of the most appropriate fields for rice cultivation, precise land preparation by laser-leveling, application of dry–wet strategies in periods with low sensitivity of yield and information on alternative crops (especially when anticipating seasons with low water availability). These all can lower water requirements considerably with minor impact on yield and income of farmers. Also the introduction of a rational incentive–disincentive system (see below) is an economic mechanism, of which higher impacts can be expected when combined with institutional re-arrangements mainly aiming at raising the flexibility in the decision-making of farmers.

Salt management at field level

Forkutsa *et al.* (2009) simulated the soil salinity with the HYDRUS model (Simunek *et al.* 2005), which was parameterized for representative fields in Khorezm. Figure 3 (upper part) depicts the soil salinity at characteristic times (before leaching, after leaching, end of the vegetation period) with respect to a 2 m layer differentiating the root zone (0–0.8 m) and the layer below influencing the root zone via capillary rise (0.8–2 m).

Pre-season leaching lowered the salt contents in the root zone, which accumulated during the vegetation period. The salt content in the 2 m layer reached at the end of the vegetation period the same level as before leaching. The salinity in the layer below the root zone shows ineffectiveness of leaching, because the salts were shifted from the root zone to the layer below rather than removed from the 2 m layer.

The parameterized HYDRUS model was used to assess the impact of an improved management option. The option comprises the application of crop residues (to reduce soil evaporation), lowering the groundwater table to 2 m, rescheduling leaching and irrigation (saving the non-effective part of leaching water and using it in the vegetation period).

Regarding the root zone (0–0.8 m), the improved system has the potential to increase leaching effectiveness; soil salinity decreased from 28.5 to 8 tons per layer per ha compared with 29 to 10 tons per layer per ha by the current strategy (Figure 3/lower left). The improved system is capable of nearly avoiding salt accumulation in the vegetation period. In contrast, the current strategy leads to a clear accumulation of salts in the root zone. With respect to the 0–2 m layer, the improved strategy leads to an overall decrease of salt compared with a currently more or less balanced situation (Figure 3/lower-right).

Application process

The flat topography in Khorezm allowed testing of double-side furrow irrigation as a low-cost option. Compared with 300 long conventional furrows, double-side irrigation resulted in a more uniform water distribution along the furrows, improved application efficiency and saved around 20% of the seasonal gross irrigation water input to the field (Poluasheva 2005). In addition, the sharp increase in salt accumulation, which was observed at the end of conventionally irrigated furrows, could be halved. The higher uniformity of water application and a reduced peak in salt accumulation increased the cotton yield by 30–35% (Poluasheva 2005).

Mid-term simulation ('sponge concept')

The current irrigation strategy in Khorezm includes the capillary rise from shallow groundwater, which is

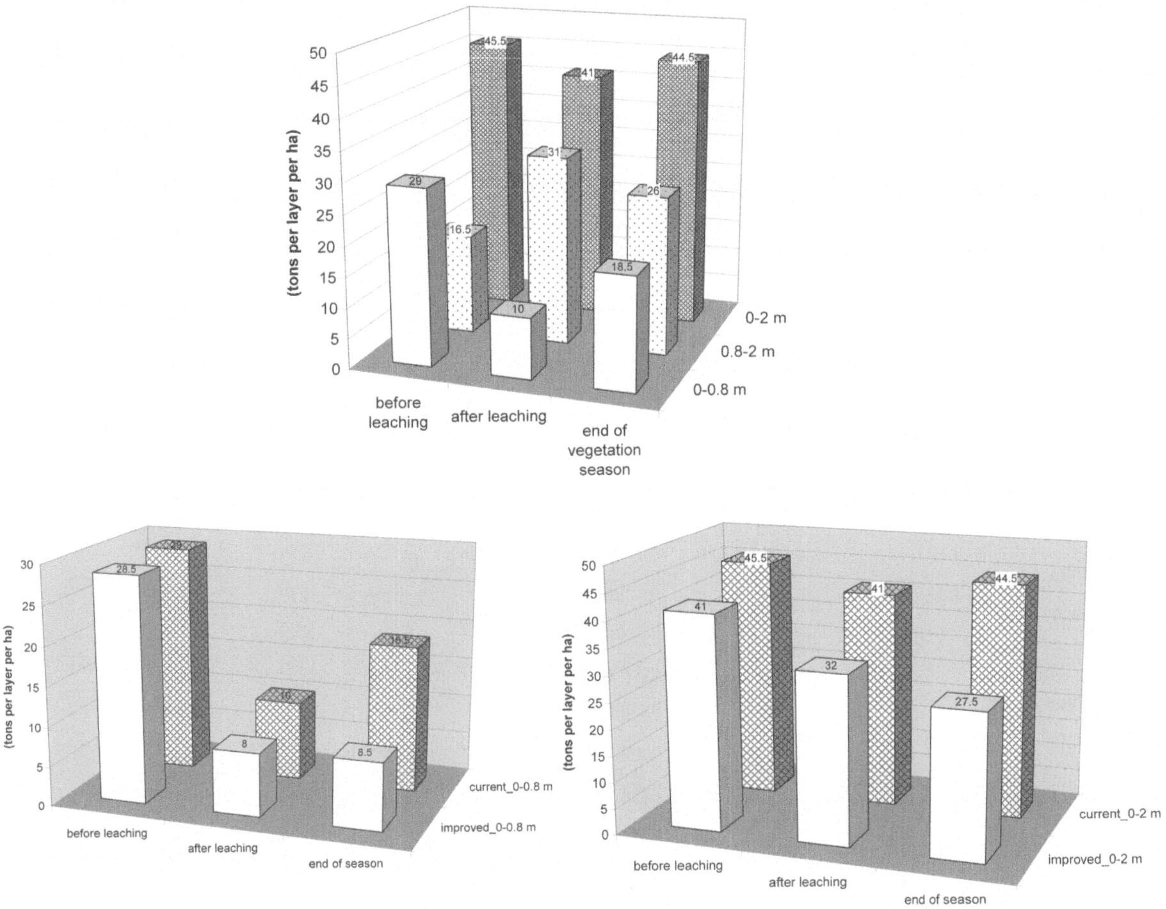

Figure 3 | Current situation of salt dynamics (upper part) and beneficial impact of an improved management on the root zone and the 2 m layer (data source: Forkutsa *et al.* (2009)).

recharged by percolation as a consequence of heavy leaching and high irrigation losses ('sponge concept'). To test the mid-term impact of the 'sponge concept' on soil moisture and salinity, a year with low water availability (dry) and one with average-high water availability (normal) were analyzed for soil moisture and salt dynamics in the root zone using the HYDRUS model. As the irrigation strategy drives the water and salt dynamics, the years were defined as 'hydrological' years: beginning with the winter period without considerable irrigation in October until the end of the irrigation and vegetation season in the following September. In the dry year (October 2000 until September 2001) the surface water supply in Khorezm was reduced to 50% of the average, whereas the hydrological year 2004/2005 was above the average (SIC ICWC 2011b). The simulations were performed for silt loam and sandy loam and referred to cotton.

The major parameters used in the simulation are shown in Table 1. Leaching is performed by three events and irrigation by four events.

For a silt loam soil texture (SL), soil moisture content does not fall below the allowable depletion (Allen *et al.* 1998) indicated by the horizontal line, ensuring that no water stress occurs (Figure 4; left part). The sufficiency of soil moisture content is not only achieved in the normal year, but also for the dry year. The findings showed that by nearly halving the surface water supplies, the soil moisture status in the root zone is still adequate to meet the crop water demand.

For sandy loam (SDL), in the normal year, farmers often are late in applying the first irrigation (Figure 4; right part). The situation becomes worse for the dry years: the first irrigation usually is too late and the second and third irrigation are insufficient. As a

Table 1 | Soil hydraulic parameters (upper part) and major input on hydrological and water management parameters (lower part) with θ_s and θ_r: saturated and residual soil water content, α and n: empirical parameters in the soil water retention function, K_S: saturated hydraulic conductivity, l: tortuosity parameter in the conductivity function

Soil type	θ_r (cm³cm⁻³)	θ_s (cm³cm⁻³)	α (cm⁻¹)	n (–)	K_s (cm/day)	L (–)		
Silt loam	0.067	0.45	0.02	1.41	10.8	0.5		
Sandy loam	0.065	0.41	0.075	1.89	106.1	0.5		
	Groundwater level (cm below ground)		Evapotranspiration of cotton (mm/day)		Irrigation (mm/month)		Leaching (mm/month)	
Month	2001	2005	2001	2005	2001	2005	2001	2005
January	229	190			0	0	150	150
February	231	195			0	0	0	0
March	157	140			0	0	250	250
April	153	121	2.75	2.72	0	0	0	0
May	182	125	2.47	2.22	0	0	0	0
June	165	117	6.61	6.12	100	225	0	0
July	159	103	8.91	8.76	200	450	0	0
August	163	102	6.65	6.99	100	225	0	0
September	173	123	4.31	4.61	100	0	0	0
October	189	154	0.31	0.41	0	0	0	0
November	197	172			0	0	0	0
December	218	191			0	0	0	0

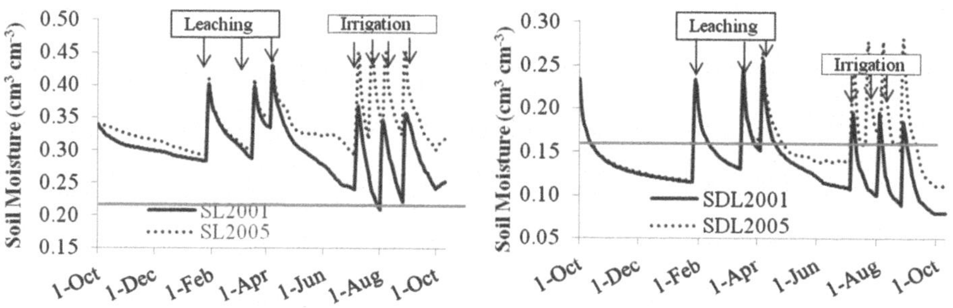

Figure 4 | Average soil moisture content for 1 m depth for silt-loam (SL; left part) and sandy loam (SDL; right part) in normal and dry years (the horizontal line indicates allowable depletion limit).

consequence, soil moisture drops considerably below the allowable depletion level, even approaching the wilting point towards the end of the season.

During normal years, an excessive amount of water is supplied to SL and SDL soils. For example, when water supply was reduced by half (dry), the 'sponge phenomenon' failed for the SDL. Soil moisture in the SL was still above the threshold value but there was significant increase in the soil salinity of the root zone.

Lessening constraints

Infrastructure and bio-physical system

The Khorezm irrigation system was designed for providing water to a small number of large and uniform production units. Because of additional insufficient maintenance and as the network is lacking appropriate hydraulic structures, water distribution to a high number of water users with diversified water requirements cannot be ensured. As a prerequisite for introducing any type of flexible irrigation schedules, the system needs thus to be equipped with structures to measure and regulate the discharge.

Implementing optimal application discharges to raise the currently low application efficiencies and realize proper dosing, adequate infrastructure for discharge control needs to become available at farm level. As irregular microtopography severely constrains uniform water distribution, laser-guided leveling has a water saving potential in the range of 20–30% (Bekchanov *et al.* 2010b). Considering the financial limitations of the farming population in Uzbekistan, Bekchanov *et al.* (2010b) concluded that less-capital demanding, but often also less water efficient measures, have the highest potential to be adopted. Although extremely effective in increasing water productivities (Bekchanov *et al.* 2012), the implementation of high efficiency irrigation technologies such as sprinkler and drip is constrained not only by the low farm capital and reserves of the farmers, but also by the overarching non-reliability of water provision which discourages long-term planning, and their attitude to investing in assets and innovations (Djanibekov *et al.* 2012). Therefore, capital intensive innovations are presently implemented at specific locations (e.g. home gardens, greenhouses, household plots) only. It is also advantageous to reflect on incentives to implement high-cost-water-wise options in the water-rich upstream regions where presently farmers are better off (Bekchanov *et al.* 2010a, b) and where the problem of temporal non-reliability of supply is comparatively much lower. Long-term strategies should aim at creating an enabling environment allowing the introduction of innovations that increase water productivity such as: (i) demonstrating modern technologies and informing local farmers on costs and benefits, (ii) ensuring a long-term planning reliability and creating thus space to unfold innovation potential of farmers (see below/subsection 'institutions'), and (iii) easing the issue of non-reliable water provision by institutional re-arrangements (e.g. strengthening WCAs) or installing decentralized reservoirs, or using the storage capacity of the aquifer.

As the current irrigation performance is characterized by temporal non-reliability of supply, indicated by a reduction of actual evapotranspiration (Conrad 2006), and by a high deviation of planned versus realized supply in the irrigation network (Awan 2010), decentral reservoirs at the level of Water Consumers Associations (WCAs) and farms are an option to assist farmers to adapt to variable supply. In particular, the implementation of deficit irrigation strategies could be supported by the consideration of such reservoirs, although the practical financial feasibility still needs to be explored.

Institutions

Implementing strategies towards the sustainable use of resources requires long-term planning reliability with respect to farmers. Assuring the currently uncertain land use rights is a prerequisite to raise farmers' willingness to invest in measures needed to introduce efficient irrigation as postulated recently (e.g. Djanibekov *et al.* 2012). Long-term planning reliability creates furthermore space for mobilizing the innovation potential of farmers (Hornidge *et al.* 2011). As the total abolition of the state-order system is not realistic in the near-future, partial transition of the state-order system is considered as a measure to enhance the innovation process: (i) developing the strict area-based production quota of cotton and wheat towards a production-based quota (keeping the production target, but leaving the decision where to produce cotton or wheat within the farm to the farmer), or (ii) defining the quota for several years (Djanibekov *et al.* 2010).

In principle, WCAs are the platform to link the water requirements by farmers with the water distribution. Practically, current performance of the processes within the WCAs needs to be improved. Transdisciplinary action research in a WCA in Khorezm proved to be an appropriate approach for raising the feeling of ownership regarding the WCA among users (Hornidge *et al.* 2011). The introduction of local water inspectors supporting the district inspectors in detecting water wastage proved to be successful as well, in particular when combining inspecting and sanctions with intensive on-site training on concrete water saving strategies (Hornidge *et al.* 2011).

Economy

Currently, many WCAs are in danger of enter a 'vicious cycle': inadequate funds limit the performance of WCAs with respect to providing farmers with appropriate water supply – delivery problems lower the willingness of water users to pay fees and in turn capacities of WCAs decrease further.

Introducing a 'Reciprocal responsibility' is a promising measure to improve the functioning of the WCA, because WCA as well as water users are addressed concurrently: in case the WCA fails to deliver water, the water user is released from water fees; water users being reluctant to pay are excluded from water delivery as recently postulated (Manschadi *et al.* 2010). Extending the service functions of WCAs towards others such as providing micro-credits or insurances in additional to conventional services is another promising approach to raise willingness to pay WCA fees (Hornidge *et al.* 2011).

The introduction of volumetric water fees has the potential to enhance water saving and improve the financial situation of WCAs. Bobojonov (2008) simulated a water saving potential of 5–6%, which can be mobilized by volumetric water fees. On the other side, a cost-covering fee, which would be in the range of 9 US $ per 1,000 m^3, would overburden many water users in Khorezm.

Instead of water pricing as an isolated measure, Rudenko & Lamers (2006) argue for an overall concept combining water fees and partial loosening of the quota system. An innovative system of incentives–disincentives by tax boni and mali is suggested to support sustainability and long-term productivity (Akramkhanov *et al.* 2010). Water users introducing strategies for water saving and improving salt management at the same time would become eligible to high boni. Interventions either for water saving or improved salt management would lead to medium boni whereas water users neglecting any interventions would be confronted with mali.

To come into effect, concepts towards water saving at the level of water users and WCAs need to be complemented by institutional and financial provisions at national and interstate/basin level (Dukhovny 2003). According to Sokolov *et al.* (2008), applying the principles on Integrated Water Resources Management facilitates reforming water management in Central Asia.

SUMMARY AND CONCLUSIONS

Intensified water use in upstream parts of the Amu Darya basin and the predicted effects of climate change for Central Asia are expected to impact water supply in Khorezm towards a lower availability and increased variability. To adapt to these changes, water management in regions located at the tail ends such as Khorezm needs in particular to become more efficient and flexible. Technical approaches (e.g. model-based irrigation scheduling), re-arranging institutional settings (e.g. strengthening the water user's participation in WCAs) and appropriate economic incentive structures (e.g. water fees) need to be conceived and implemented hand-in-hand. Partial loosening of the state-order system in combination with assuring land use rights would support more than now options for long-term planning and in turn enable the water user to unfold innovation potential and strengthen the introduction of sustainable resource management strategies.

Despite raised efficiency and flexibility, future limitations of and variability in water resources may dictate the need for controlled deficit irrigation strategies. To minimize the unavoidable impact of water stress on crop yields, the application of irrigation scheduling models could be considered which would become even more effective in combination with crop models taking the groundwater as a buffer into account. Implementing deficit irrigation strategies can be facilitated by technical approaches aiming at a higher uniformity of water application (e.g. through laser-guided land leveling) and tackling the temporal burden of the present water distribution by e.g. strengthening water users and WCAs. Integrating small-scale water storages such as lakes into irrigation concepts and considering decentral reservoirs are options, especially in periods with non-reliable water supply. Furthermore, the acceptance of deficit strategies can be improved by transparent rules governing the water allocation processes in the WCAs, enabling participation of water users.

A combination of modeling the mid-term behavior of the linked surface and groundwater system ('sponge') and mid-term forecast regarding the available water resources in the Amu Darya basin, provides an option to derive adapting strategies at an early stage.

Bundling technical interventions towards flexible irrigation/leaching scheduling and advanced handling of the water application process, institutional re-arrangements ensuring a better coordination between irrigation activities especially at the interface between farm and network level (WCA) and economic incentives are an appropriate answer to future water scarcity.

ACKNOWLEDGEMENTS

This research was carried out in the ZEF/UNESCO Project 'Economic and Ecological Restructuring of Land- and Water Use in the Region Khorezm (Uzbekistan): A Pilot Project in Development Research (www.uni-bonn.de/khorezm)'. The authors wish to thank the German Ministry for Education and Research (BMBF) for the financial support. We are grateful for the valuable comments from two anonymous reviewers on the earlier version of this manuscript.

REFERENCES

Akhtar, F. 2011 Combining AquaCrop and HYDRUS models to improve current irrigation scheduling under shallow groundwater conditions in Khorezm, Uzbekistan. MSc Thesis, University of Bonn, Germany.

Akramkhanov, A., Ibrakhimov, M. & Lamers, J. P. A. 2010 Managing soil salinity in the lower reaches of the Amudarya delta: How to break the vicious circle (Case Study 8-7). In: *Case Studies in Food Policy for Developing Countries* (P. Pinstrup-Andersen & F. Cheng, eds). Cornell University Press, Ithaca, NY, Volumes 1, 2, and 3.

Allen, R. G., Pereira, L. S., Raes, D. & Smith, M. 1998 Crop evapotranspiration – Guidelines for computing crop water requirements. FAO Irrigation and Drainage Paper, No. 56.

Awan, U. K. 2010 Coupling hydrological and irrigation scheduling models for the management of surface and groundwater resources in Khorezm, Uzbekistan. PhD Thesis, University of Bonn, Germany.

Bekchanov, M., Karimov, A. & Lamers, J. P. A. 2010a Impact of water availability on land and water productivity: a temporal and spatial analysis of the case study region Khorezm, Uzbekistan. *Water* **2** (3), 668–684.

Bekchanov, M., Lamers, J. P. A. & Martius, C. 2010b Pros and cons of adopting water-wise approaches in the lower reaches of the Amu Darya: a socio-economic view. *Water* **2** (2), 200–216.

Bekchanov, M., Lamers, J. P. A., Karimov, A. & Müller, M. 2012 Estimation of spatial and temporal variability of crop water productivity with incomplete data. In: *Cotton, Water, Salts and Soums: Economic and Ecological Restructuring in Khorezm, Uzbekistan* (C. Martius, I. Rudenko, J. P. A. Lamers & P. L. G. Vlek, eds). Springer, Dordrecht, The Netherlands, pp. 329–344.

Bobojonov, I. 2008 Modeling crop and water allocation under uncertainty in irrigated agriculture: a case study on the Khorezm Region, Uzbekistan. PhD Thesis, University of Bonn, Germany.

Clarke, D., Smith, M. & El-Askari, K. 1998 *CropWat for Windows: User Guide*. University of Southampton, Southampton.

Conrad, C. 2006 Fernerkundungsbasierte modellierung und hydrologische Messungen zur Analyse und Bewertung der landwirtschaftlichen Wassernutzung in der Region Khorezm (Usbekistan) (Remote sensing based modeling and hydrological measurements to assess the agricultural water use in the Khorezm Region (Uzbekistan)). PhD Thesis, University of Wuerzburg, Germany.

Conrad, C., Schorcht, G., Tischbein, B., Davletov, S., Sultanov, M. & Lamers, J. P. A. 2012 Agro-meteorological trends of recent climate development in Khorezm and implications for crop production. In: *Cotton, Water, Salts and Soums: Economic and Ecological Restructuring in Khorezm, Uzbekistan* (C. Martius, I. Rudenko, J. P. A. Lamers & P. L. G. Vlek, eds). Springer, Dordrecht, The Netherlands, pp. 25–36.

Djanibekov, N., Rudenko, I., Lamers, J. P. A. & Bobojonov, I. 2010 Pros and cons of cotton production in Uzbekistan. Case study #7-9. In: *Food Policy for Developing Countries: Case Studies* (P. Pinstrup-Andersen & F. Cheng, eds). Cornell University, Ithaca, New York, pp. 13.

Djanibekov, N., van Assche, K., Bobojonov, I. & Lamers, J. P. A. 2012 Farm restructuring and land consolidation in Uzbekistan: New farms with old barriers. *Europe-Asia Studies* **64** (6), 1101–1126.

Devkota, K. P. 2011 Resource utilization and sustainability of conservation-based rice–wheat cropping systems in Central Asia. PhD Thesis, University of Bonn, Germany.

Diersch, H.-J. G. 2002 *Reference Manual for FEFLOW – Finite Element Subsurface Flow and Transport Simulation System. User's Manual/Reference Manual/White Paper Release 5.0*. Wasy Ltd., Berlin.

Dukhovny, V. A. 2003 The Aral Sea basin – rumors, realities, prospects. *Irrigation Drainage* **52** (2), 109–120.

Forkutsa, I. 2006 Modeling water and salt dynamics under irrigated cotton with shallow groundwater in the Khorezm Region of Uzbekistan. PhD Thesis, University of Bonn, Germany.

Forkutsa, I., Sommer, R., Shirokova, Y., Lamers, J. P. A., Kienzler, K., Tischbein, B., Martius, C. & Vlek, P. L. G. 2009 Modeling irrigated cotton with shallow groundwater in the Aral Sea basin of Uzbekistan: II. Soil salinity dynamics. *Irrigation Science* **27** (4), 319–330.

Hornidge, A.-K., Oberkircher, L., Tischbein, B., Schorcht, G., Bhaduri, A. & Manschadi, A. M. 2011 Reconceptualizing water management in Khorezm, Uzbekistan. *Natural Resources Forum* **35** (4), 251–268.

Horst, M. G., Shamutalov, S. S., Pereira, L. S. & Gonçalves, J. M. 2005 Field assessment of water saving potential with furrow irrigation in Ferghana, Aral Sea basin. *Agricultural Water Management* **77** (1–3), 210–231.

Ibrakhimov, M. 2004 Spatial and temporal dynamics of groundwater table and salinity in Khorezm (Aral Sea Basin), Uzbekistan. PhD Thesis, University of Bonn, Germnay.

Ibrakhimov, M., Martius, C., Lamers, J. P. A. & Tischbein, B. 2011 The dynamics of groundwater table and salinity over 17 years in Khorezm. *Agricultural Water Management* **101**, 52–61.

Manschadi, A. M., Oberkircher, L., Tischbein, B., Conrad, C., Hornidge, A.-K., Bhaduri, A., Schorcht, G., Lamers, J. P. A. & Vlek, P. L. G. 2010 'White Gold' and Aral Sea disaster – towards more efficient use of water resources in the Khorezm region, Uzbekistan. *Lohmann Information* **45** (1), 34–47.

Micklin, P. 2007 The Aral Sea disaster. *Annual Review of Earth and Planetary Sciences* **35**, 47–72.

Pereira, L. S., Horst, M. G., Shamutalov, S. S. & Gonçalves, J. M. 2007 Assessing impacts of surge-flow irrigation on water saving and productivity of cotton. *Agricultural Water Management* **87**, 115–127.

Pereira, L. S., Paredes, P., Sholpankulov, E. D., Inchenkova, O. P., Teodoro, P. R. & Horst, M. G. 2009 Irrigation scheduling strategies for cotton to cope with water scarcity in the Fergana Valley, Central Asia. *Agricultural Water Management* **96**, 723–735.

Poluasheva, G. 2005 Dynamics of soil saline regime depending on irrigation technology in conditions of Khorezm oasis. In: *Proceedings of the International Scientific Conference: 'Science Support as a factor for Sustainable Development of Water Management'*, Taraz/Khazakhstan, 20–22 October 2005 (in Russian).

Rudenko, I. & Lamers, J. P. A. 2006 The comparative advantages of the present and future payment structures for agricultural producers in Uzbekistan. *Central Asian Journal of Management, Economics and Social Research* **5** (1–2), 106–125.

Sokolov, V., Dukhovny, V. & Mirzaev, N. 2008 IWRM implementation: experiences with water sector reforms. In: *Central Asian Waters* (M. M. Rahaman & O. Varis, eds). Helsinki University of Technology, Finland.

Schlüter, M. & Hirsch, D. 2010 Ten answers from the Amudarya basin. Workshop of the Global Catchment Initiative, Bonn, February 17–19, 2010.

SIC ICWC Scientific Information Center–Interstate Coordination Water Commission 2011a Portal of Knowledge for Water and Environmental Issues in Central Asia (CAWATERinfo). Available from: http://www.cawater-info.net/bk/water_land_resources_use/english/docs/table6.pdf (accessed 30 August 2011).

SIC ICWC Scientific Information Center–Interstate Coordination Water Commission 2011b Portal of Knowledge for Water and Environmental Issues in Central Asia (CAWATERinfo). Available from: http://www.cawater-info.net/amudarya/index_e.htm (accessed 30 August 2011).

Simunek, J., Van Genuchten, M. T. & Sejna, M. 2005 The HYDRUS-1D software package for simulating the one-dimensional movement of water, heat, and multiple solutes in variably-saturated media. Version 3.0, HYDRUS Software.

Steduto, P., Hsiao, T. C., Raes, D. & Fereres, E. 2009 AquaCrop-The FAO crop model to simulate yield response to water: I. Concepts and underlying principles. *Agronomy Journal* **101** (3), 426–437.

Tischbein, B., Manschadi, A. M., Hornidge, A. K., Conrad, C., Lamers, J. P. A., Oberkircher, L., Schorcht, G. & Vlek, P. L. G. 2011 Ansätze für eine effizientere Wassernutzung in der Provinz Khorezm, Usbekistan (Proposals for the more efficient utilization of water resources in the Province Khorezm, Uzbekistan). *Hydrologie und Wasserbewirtschaftung* **55** (2), 116–125.

First received 6 January 2012; accepted in revised form 24 September 2012

Sustainable water resources management in the Long Bien district of Hanoi, Vietnam

C. Stefan, T. Fröhlich, L. Fuchs, R. Junghanns, H. M. Phan, V. N. Tran and P. Werner

ABSTRACT

The accelerated industrialization of Hanoi, Vietnam, coupled with a high population growth rate and changing climatic conditions create increasing pressure on the local water balance. Despite abundant precipitations, overexploitation endangers the groundwater resources, which are not able to sustain an adequate water supply. The present paper presents a sustainable approach for balancing the lowering of groundwater levels by increasing the rainwater percolation rates through enhanced infiltration. The efficiency of the method was assessed by a scenario analysis based on hydrological and hydrogeological models. Multi-criteria simulations revealed the optimum infiltration sites by considering technical and site-specific aspects and the positive impact of artificial recharge on seasonal groundwater budget.

Key words | drainage, groundwater recharge, Hanoi, stormwater, sustainable, Vietnam

C. Stefan (corresponding author)
R. Junghanns
H. M. Phan
P. Werner
Institute of Waste Management and Contaminated Site Treatment, Technische Universität Dresden, Pratzschwitzer Str. 15,
01796 Pirna,
Germany
E-mail: *catalin.stefan@tu-dresden.de*

T. Fröhlich
L. Fuchs
Institute for Technical and Scientific Hydrology,
Engelbosteler Damm 22,
30167 Hanover,
Germany

V. N. Tran
Institute for Environmental Science and Engineering,
Hanoi University of Civil Engineering,
55 Giai Phong Road,
Hanoi,
Vietnam

INTRODUCTION

Motivation and challenges

The management of water resources in Hanoi, Vietnam, is endangered by two main factors. On one side, the population growth rates reached recently record levels of 3.5% (VIETNAMNET 2008) with densities of about 3,490 inhabitants per km^2 (VIETNAMNET 2007). At the same time, the country's industrialization accelerated dramatically with insufficient protective measures for the environment. The downside of this rash development is the difficulty to provide a safe water supply and sustainable urban sanitation. In addition, the expansion of residential areas by merging adjacent provinces and districts into the metropolitan zone increases the challenges posed by the infrastructure-related works. On the other side, the extreme regional weather conditions, with high precipitations in a rather short time, lead to high loads of the existing urban drainage system which is often unable to work properly and flooding events inevitably occur. And since the system is mainly designed for a quick disposal of rainwater and domestic wastewater, the natural water resources located in the city's subsurface are not sufficiently replenished naturally and tend to decrease in volume at worrying rates (Nguyen & Helm 1995; Nguyen & Nguyen 2002).

Description of the study area

Long Bien district is situated on the western side of Hanoi, Vietnam, between Red River and Duong River (20°57′36″ N21°05′00″ N and 105°49′05″ E–105°57′45″ E). With a total surface of about 60 km (Figure 1), Long Bien is a new model district which is expected to become one of the city's centres for trade, services and transportation that will connect Hanoi with other provinces in the north of Vietnam.

doi: 10.2166/ws.2013.049

Figure 1 | Location of Long Bien district, Hanoi (modified after Jusseret *et al.* 2008).

In 2008, the district's population was about 200,000 inhabitations with predicted 350,000 inhabitants by 2020. Over the years, both rivers brought fine alluvial sediments such as clay, silt, sandy clay and fine sand into the study area, making the subsurface of Long Bien poorly permeable. As a flood prevention measure, the district was almost completely surrounded by a dike system along both riverbanks that also divides the area into three distinct zones: flat area inside the dike system, area outside the dikes and the alluvial terrain along the riverbanks.

The study area is also characterized by numerous irrigation and drainage ditches that are diverting the waters in the Cau Bay River which then flows into the Red River in the south of the district. During heavy rain events, the district's drainage is impaired by Red River overflow and channelling the water out of the district is almost impossible. This leads to severe flooding events not only in the areas near the riverbanks but also in areas inside the dike system, as the district does not possess sufficient surface storage facilities to attenuate the high flow peaks.

Despite large volumes of water resulting from precipitations, the groundwater levels in the district decrease considerably (Fischer *et al.* 2011). The lowering rate of piezometric heads in Long Bien is yet not as alarming as in other districts of Hanoi (e.g. Thanh Cong) but this is considered to be the major cause for an average land subsidence of about 15 mm per year (HTA 2003).

Objectives

To overcome these challenges, a sustainable concept oriented mainly at naturally occurring processes was developed for the optimization of the existing master plan for the urban drainage system (Stefan & Werner 2009; Werner *et al.* 2011). The concept was tailored to match the hydrological, hydrogeological and socio-economic conditions of Long Bien district. The specific objective of this concept is the establishment of adequate water retention spaces (above surface and subsurface) in the Long Bien urban landscape. The approach includes measures for the delay of surface drainage in order to allow the decrease of flow peaks and to improve the subsurface reservoirs' replenishment rate.

METHODS

A conceptual hydrologic model was combined with a groundwater model to assess the natural and enhanced infiltration potential of rainwater in Long Bien. The hydrological rainfall-runoff model KOSIM 7.3 (Itwh 2006) was used for the simulation of the system load in the catchment area and its general drainage behaviour.

For the runoff calculations, Hanoi's yearly precipitation (1,682 mm) and evaporation (1,262 mm) rates were extended by their monthly distribution and supplemented by hourly precipitation values for 2009 collected from Lang Weather Station (Long Bien, Hanoi). Additionally, daily precipitation data from 1961–2004 were obtained from Aphrodite's Water Resources Project (2010). The degree of soil surface sealing was calculated for existing constructed areas as well as for residential and industrial zones included in the Master Plan of the district by land use analysis of available city plans. For the infiltration calculations, the average water percolation rate through the upper soil layer was estimated from the sedimentological data at 1×10^{-6} m/s.

The hydrological model was coupled with a transient groundwater hydraulic model run with Visual MODFLOW 4.2 software (Waterloo Hydrogeologic Inc. 2006). The model assessed the seasonal variations of groundwater piezometric levels of the water-bearing layer of Pleistocene origin considering both natural and anthropogenic factors (lateral recharge and withdrawal for above-surface use). The following data were used for the setup of the model: 30 drillings up to 150 m deep, daily water level stages measured from 1996 to 2003 by three gauging stations located on Red River (two stations) and Duong River (one station), monthly piezometric heads collected from 13 observation wells (1996–2003) and 23 municipal and private pumping wells (Figure 2). Three more boreholes were drilled in the study area for verification of the data set and for the calculation of hydrogeological coefficients. The Red River was chosen as model hydraulic boundary and the area on the eastern side (city centre) was considered as 'no flow' region.

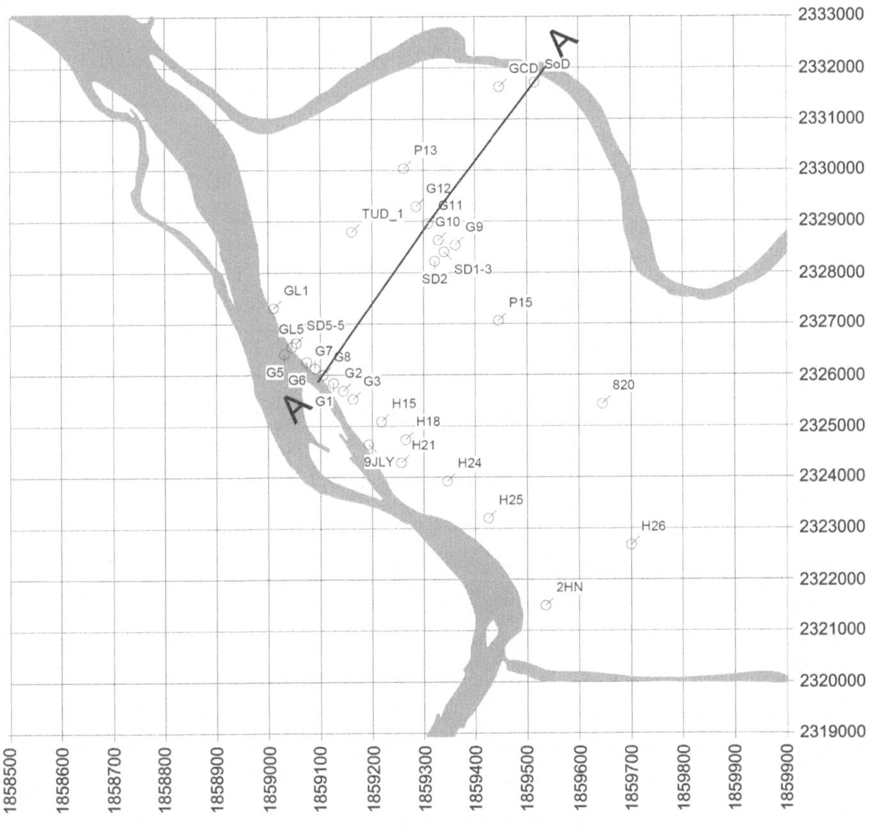

Figure 2 | Location of observation wells, pumping wells and geological cross-section.

A scenario analysis was done for the assessment of the impact of rainwater infiltration on the local groundwater budget, the potential infiltration sites being selected using a multi-criteria analysis. Two infiltration approaches were considered for the simulation of deep-wells injection of rainwater into the deep aquifer as compensation of the groundwater withdrawal.

RESULTS AND DISCUSSION

Under natural conditions, the water balance is rather stable and sustainable. Man-made structures influence this equilibrium by changing the hydraulic conditions, i.e. reduce the natural percolation rates and increase the surface drainage by sealing more and more pervious surfaces. In addition, the increase in water demand worsens the fragile balance even more by increasing almost simultaneously the groundwater abstraction rates and the introduction of pollutants into the subsurface. However, the situation can be restored. For this it is only necessary to learn from and reproduce naturally occurring processes such as the creation of natural retention facilities, green spaces, avoid sealed surfaces, etc. The concept presented in this paper reflects the impact of a nature-oriented approach of the drainage system on the local water cycle. The approach is based on the increase and sustainable exploitation of infiltration potential of rainwater and it is shown from two complementary perspectives: (a) the impact of decentralized rainwater management systems on the surface rainwater drainage and (b) the subsurface rainwater infiltration and storage.

Surface rainwater drainage

Precipitation and evaporation are the guiding principles for the design and determination of the basic conditions for the planning of rainwater management systems. The study started with the classification and characterization of precipitation events (intensity-duration-frequency analysis) and preparation of data as input for the simulation of the rainfall-runoff process. Recurrence intervals for the daily precipitation rates were calculated for 2009 data using Cunnane plotting formula (Cunnane 1978), several 1-hour values enabling a detailed analysis in terms of total amount, length, change in intensity, flow peaks, etc. Additionally, hourly precipitation rates for 2009 were used for measuring rainwater management units (Figure 3).

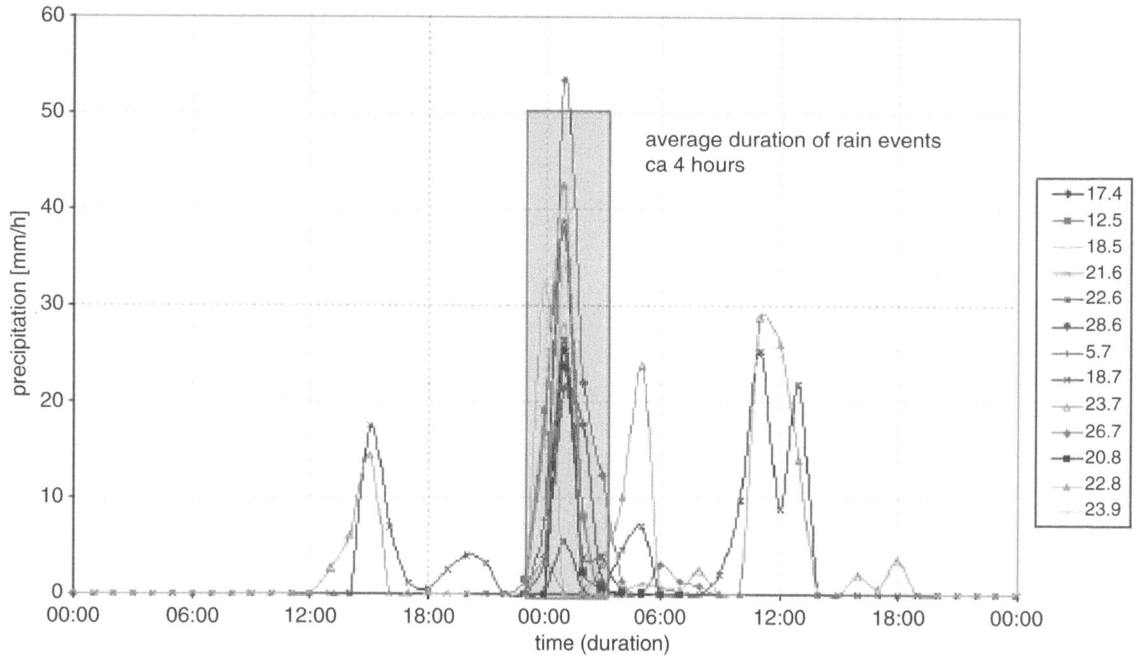

Figure 3 | Precipitation hydrograph for Hanoi, Vietnam (2009).

The next step in data preparation was the soil and land use characterization. The degree of soil surface sealing was necessary to be estimated for the calculation of the runoff resulting from precipitations. Representative land use types were selected among the existing areas together with new areas from the master plan. For each type of construction, building and streets' surface data were collected and contrasted with the total land surface. The water system was divided into various catchment areas whose shape was determined by the structure of the drainage system. The simulated hydrograph was then analysed using simple statistical methods. Beside the various total sum and average calculations, the overflow events at the discharge facilities in particular were examined in terms of amounts and frequency and their monthly distribution was represented. As expected, the percolation percentage increases with the decrease of impervious areas (an inflow of $5\,\mathrm{l/(s \times ha)}$ would be the safeguarding limit for rainwater treatment when about 10% from the total land is occupied by decentralized rainwater management facilities while still providing a high percolation percentage) and with the increase in the degree of soil sealing.

Based on the previous estimations, it was assumed that, given the existing soil characteristics in Long Bien, 20% of the total pervious area can be made available for decentralized rainwater management measures. For this, the area flow restriction level was set at $10\,\mathrm{l/(s \times ha)}$. Under these conditions, approximately 11% of rainwater can percolate via decentralized facilities (e.g. rainwater retention ponds). Up to 50% of impervious surfaces can be connected to decentralized facilities without overloading them (estimation using precipitation data from 2009). The other 50% shall, however, be channelled unchecked into the municipal central collection system. Overall, it is estimated that the implementation of decentralized rainwater management units would reduce the daily flow peak in case of a system load of 200 mm/day by almost 50%.

Subsurface rainwater infiltration and storage

If the runoff is higher than the area restriction level, the above-ground decentralized systems for rainwater management (retention ponds) may get overloaded and fail. Moreover, in case of heavy rain during the monsoon season, the units must be emptied within a very short time (maximum 12 hours). Under the given soil characteristics, about 400 $\mathrm{m^3}$ of retained precipitation requires approximately 1 ha and 12 hours for percolation into soil. Additional measures were sought for increasing the percolation rates and the one selected was the use of subsurface storage by means of managed groundwater recharge. This will not only reduce the load on the rainwater management system but also replenish the groundwater reservoir and compensate the subsidence of groundwater levels (Fischer *et al.* 2011).

Characterization of local stratigraphy

Stratigraphic and sedimentological data indicate a heterogeneous sedimentary architecture with two main types of deposits: the medium – coarse sands with rather high specific yields form aquifer layers and the alluvial silts and clays form low-permeable units between aquifers. The spatial distribution shows that upper permeable layers of Holocene origins are scattered only along the two rivers and are partially overlapping with the area of the district that is exposed to frequent flooding events. The layers are locally about 15–20 m deep, they are unconfined and consist of sandy sediments. Because of their local distribution, the Holocene layers were neglected in the groundwater model and their water permeability value was estimated as fitting parameter. The next water-bearing layers underlying the Holocene aquifer have Pleistocene origin and could be differentiated in: qp2 – upper Pleistocene (semi-confined layer, 20–40 m deep, medium sands) and qp1 – lower Pleistocene (confined, 30–40 m deep, coarse sand mixed with clay and gravel). The layers are separated from the Holocene aquifer by low-permeable clay and silt lenses (aquitard-type formations) that also inherit the direct recharge from the Red River during the dry season. Figure 4 shows the simplified sedimentological cross-section used in the groundwater model.

After model calibration, the water permeability values for the lower Pleistocene layer were set to values from 6×10^{-5} to 5.5×10^{-3} m/s and for the upper Pleistocene layer from 2.5×10^{-4} to 5.5×10^{-3} m/s. The specific storage and specific yield of both layers were estimated at values from 1×10^{-2} to 1×10^{-4} m/s and the soil's effective porosity has values from 0.1 to 0.2.

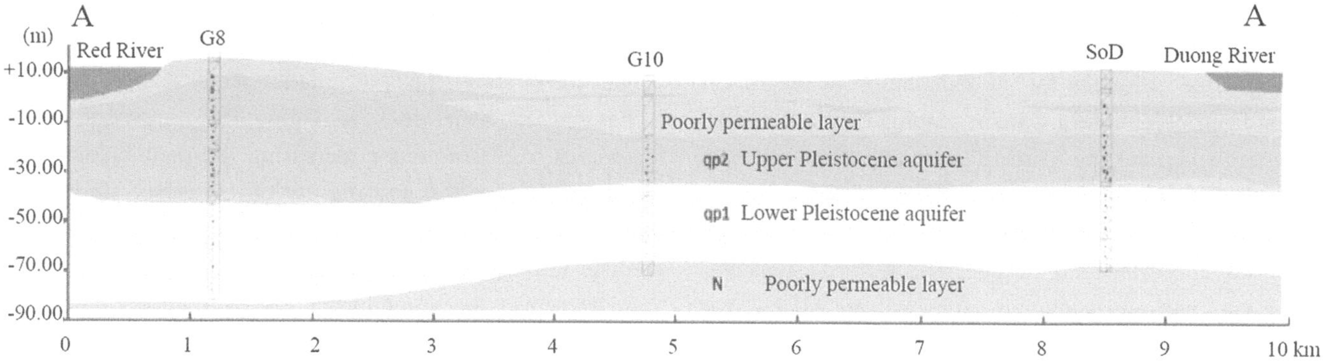

Figure 4 | Simplified sedimentological cross-section.

Seasonal variations of piezometric heads and groundwater infiltration characteristics

The next step is the simulation of natural variations of piezometric heads in order to set the optimum infiltration schedule (Figure 5).

The depression cones in the centre of the district correspond to the well fields of Hanoi Water Company and had the lowest variations in depth during the season change. This can be explained by the little lateral recharge due to the long distance to both adjacent rivers. The difference between the piezometric heads in dry and rainy seasons can reach up to 7 m. However, many authors reported a severe decline in piezometric heads of Hanoi's groundwater over the past years (Nguyen & Helm 1995; Thu & Fredlund 2000; Nguyen & Nguyen 2002; HTA 2003). This drawdown of piezometric heads produced an increasing stress in the underlying aquifers caused by the decrease of pore water pressure from an unbalanced recharge–discharge mechanism. This stress has been coupled with the weak geological structure of the upper layers (mud, clay, peat, silt) and, as a direct consequence, land subsidence cases occurred and were reported in the literature (Nguyen & Helm 1995; Thu & Fredlund 2000; Nguyen & Nguyen 2002; Luu & Pham 2008; Go *et al.* 2010).

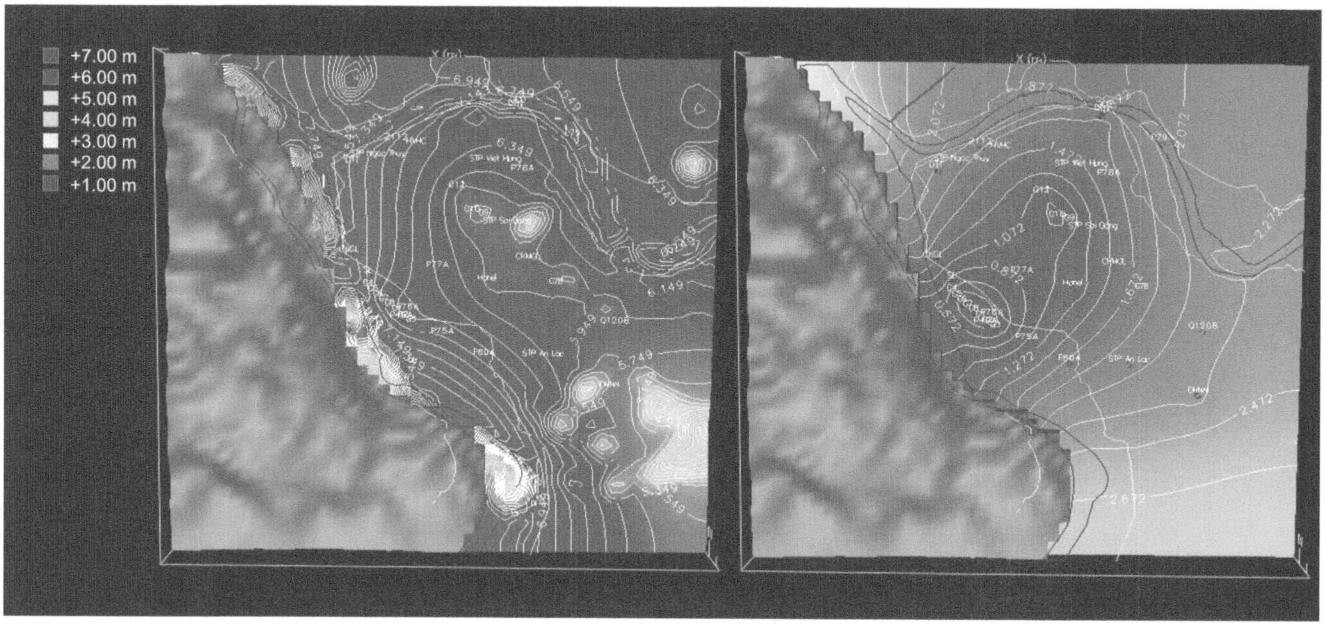

Figure 5 | Seasonal variations of groundwater levels expressed in metres above mean sea level (left: highest level in the rainy season; right: lowest level in the dry season) (after Werner *et al.* 2011).

Selection of infiltration site

GIS-based multi-criteria decision analysis (MCDA) is frequently used for the selection of potential sites for managed groundwater recharge (Masciopinto *et al.* 1991; Kallali *et al.* 2007). MCDA was used in the present study for the identification of the most suitable locations of groundwater infiltration wells in Long Bien district. The analysis included environmental, geological, hydrogeological, and urban criteria, which are assumed to be constraints of the same weight. The acceptability ranges were determined by a critical threshold value (Table 1).

Based on these acceptability ranges, the results were graphically expressed for each criterion individually using a simple Boolean mapping system, where the excluded areas are represented in white and the areas considered for infiltration in black (Figure 6).

Table 1 | Criteria for the selection of groundwater infiltration wells

Selection criteria	Acceptability range	Comments
A. Environmental	500 m around main groundwater extraction wells	Sanitary protective area
B. Geological	Thickness of poorly permeable soil layer below ground surface less than 10 m	Land subsidence can be influenced by the thickness of the alluvial material right below surface
C. Hydrogeological	Groundwater levels higher than 2 m below surface	Infiltration potential increases with groundwater drawdown
D. Urban/land use	Outside residential and densely constructed areas	Available open spaces (no constructions)

The MCDA was done by abstracting the acceptable ranges from the overlaid Boolean maps (Figure 6(a)–(d)). After overlaying, it was concluded that about 30% of the district's surface (about 18 km^2) is suitable for the installation of managed groundwater recharge units (Figure 6(e)). The potential infiltration sites are mostly located in the centre-east of the district, with smaller areas scattered in the southwest.

Infiltration scenarios

The lower Pleistocene layer (up to 80 m deep) consists of coarse sand and gravel and offers good conditions for use as a primary source of drinking water. With a good specific yield, this was also the layer selected for the managed infiltration with filters installed at about 50 m below mean sea level. In this case, it was estimated that the direct input by precipitation is very low and therefore the natural groundwater recharge rates from precipitation were neglected.

Four scenarios were simulated taking into consideration the best-case locations and the objectives above (Figure 7). The base-line scenario ('S-0') simulates the natural fluctuations of piezometric levels in the complete absence of any anthropogenic influence (neither extraction, nor managed infiltration). In this case, the piezometric heads reach 7–8 m above mean sea level during the rainy season and 3–4 m during the dry season. In case of 'Sn' scenario, the lateral recharge has been neglected and only the effect of groundwater extraction is represented. In this case, the piezometric heads were following the same time profile but they reach only 4–6 m above mean sea level during the rainy season and 1–2 m during the dry season. The aim of the infiltration is therefore 'to lift' the piezometric heads as close as possible to the 3–4 m lower boundary during the dry season without passing the upper limit of 7–8 m above mean sea level.

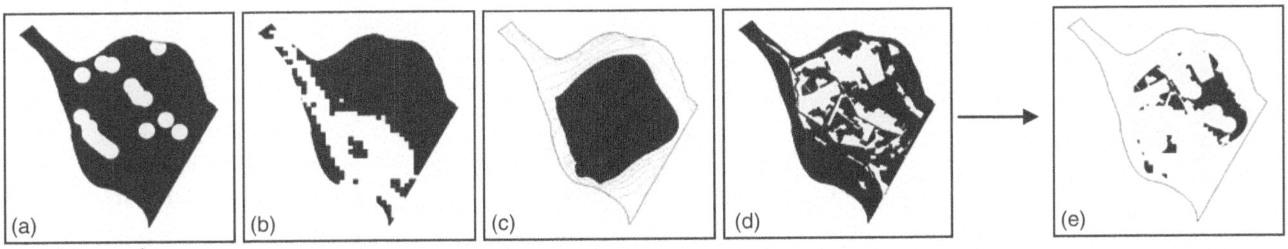

Figure 6 | The Boolean mapping system used for assessment of optimum infiltration sites using selection criteria from Table 1.

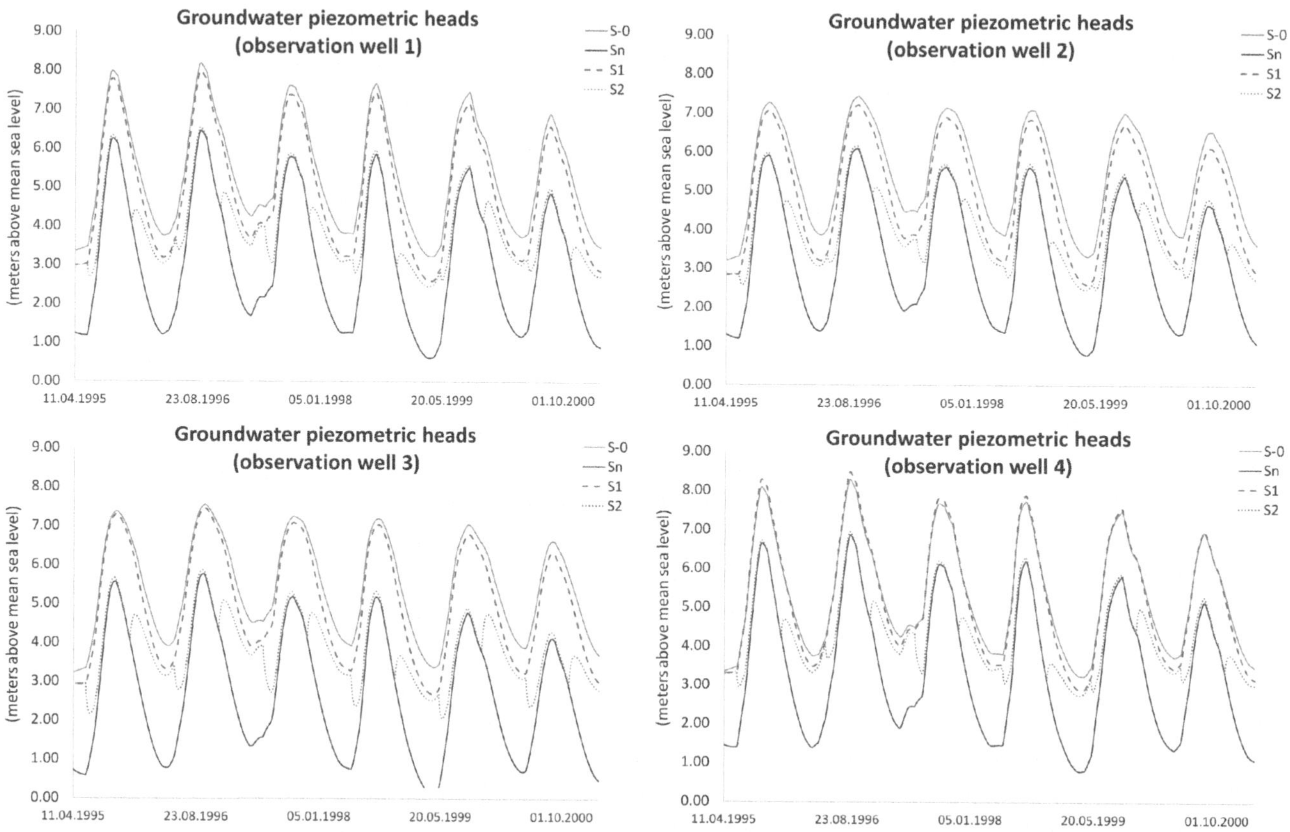

Figure 7 | Infiltration scenarios and their effect on the groundwater piezometric heads measured at four arbitrary observation wells.

Two approaches were used for the infiltration: (1) Scenario 'S1' implies the infiltration of rainwater through five infiltration fields, each well field located in the vicinity of the main groundwater extraction sites (but not closer than 1,000 m due to sanitary reasons). Choosing this location can provide a fast compensation of local groundwater depletion and helps avoid major land subsidence issues but with strong qualitative considerations for the infiltration water; (2) Scenario 'S2' implies the infiltration of rainwater using only three well fields that are located in the vicinity of the future wastewater treatment plants. This scenario takes into account the planned drainage infrastructure, e.g. the construction of wastewater treatment units that can be partially adapted to handle only rainwater. The optimum number of wells for each scenario was estimated iteratively by running the simulation of groundwater recharge for arbitrary values and comparing the resulting piezometric heads with the values from scenarios S-0 and Sn.

The best results were obtained for the scenario S1, where the calculated piezometric heads reproduce the conditions of the base-line scenario S-0. In this case, the enhanced rainwater infiltration may compensate almost 100% of the whole groundwater extraction in Long Bien and therefore provides sustainable groundwater replenishment in Long Bien district. The scenario S2 also promotes a significant increase of piezometric heads but mostly during the dry season, a fact which makes this approach suitable as a flood mitigation measure by providing subsurface storage space for peak flow reduction.

CONCLUSIONS

The objective of the present paper is to provide a nature-oriented concept for the sustainable urban drainage in Long Bien district, Hanoi, Vietnam. The example presented aims at strengthening the city's identity by urban development on and around the water and the integration of water bodies into the city's social life. The decentralized rainwater management measures will be integrated into the urban landscape tackling the problems at origins while it will relive the drainage system and improve the general aspect of the area. In the meantime, the measures for managed groundwater recharge will reap two rewards: taking over partial load from the drainage system and by counterbalancing the lowering of piezometric heads caused by groundwater overexploitation. The two scenarios were simulated using a groundwater model and showed that infiltration of rainwater is also possible during the rainy season but without significant increase of groundwater piezometric heads. The first method allows additional infiltration of rainwater during the whole year and represents a sustainable solution for the management of rainwater in fast developing areas such as Hanoi, Vietnam.

For the sake of simplification, it was assumed that all the water required for direct infiltration into the subsurface is collected and provided to the infiltration sites either by decentralized rainwater management units or by other storage and delivery means. The quantitative approach presented in this paper assumes therefore that rainwater meets the quality criteria at the time and site of infiltration requested by national regulations. Further work is necessary to elucidate the exact qualitative criteria imposed by the implementation of enhanced rainwater infiltration with consideration to groundwater protection and sustainable management.

ACKNOWLEDGEMENTS

This work was funded by the German Federal Ministry of Science and Education as part of the initiative International Water Research Alliance Saxony – IWAS (project no. 02WM1028).

REFERENCES

Aphrodite's Water Resources Project 2010 Asian Precipitation – Highly Resolved Observational Data Integration Towards Evaluation of the Water Resources. Available from: http://www.chikyu.ac.jp/precip/ (accessed 31 August 2011).

Cunnane, C. 1978 Unbiased plotting positions – a review. *Journal of Hydrology* **37** (3), 205–222.

Fischer, A., Stefan, C., Silabetzschly, K., Werner, P. & Hoc, B. 2011 Soil aquifer treatment as a tool for sustainable groundwater use in Hanoi/Vietnam. *Journal of Environmental Protection* **2**, 882–887.

Go, Y., Ayako, F., Susumu, N., Raghavan, V., Shoichi, O., Masumoto, S., Mamoru, S. & Yumio, S. 2010 3D Urban Model of Hanoi, Vietnam. In *Proceedings of GIS-IDEAS*. Hanoi, Vietnam, December 9–11, 2010.

HTA 2003 *Investigation Results of Assessing Underground Water Resources. Project for Increasing the Capacity of Gia Lam water works to 60,000 m³/d*. Hanoi Transportation and Infrastructure Agency, Water Supply Company No. 2. Main report and Annexes, Hanoi, Vietnam (in Vietnamese).

Itwh 2006 *Kontinuierliche Simulation zur Bemessung von Speicherbauwerken in urbanen Entwässerungssystemen – Modelbeschreibung – KOSIM 7.1*. Institut für technisch – wissenschaftliche Hydrologie GmbH, Germany.

Jusseret, S., Vu, T. T. & Dassargues, A. 2008 Groundwater flow modelling in the central zone of Hanoi, Vietnam. *Hydrogeology Journal* **17**, 915–934.

Kallali, H., Anane, M., Jellali, S. & Tarhouni, J. 2007 GIS-based multi-criteria analysis for potential wastewater aquifer recharge sites. *Desalination* **215**, 111–119.

Luu, X. H. & Pham, Q. T. 2008 Some problems on underground works for Hanoi metro projects. Report Hanoi People's Committee, Hanoi Metropolitan Rain Transport Project Board. Hanoi, Vietnam.

Masciopinto, C., Palmisano, V., Tangorra, F. & Vurro, M. 1991 A decision support system for artificial recharge plant. *Water Science and Technology* **24** (9), 331–342.

Nguyen, T. Q. & Helm, D. C. 1995 Land subsidence due to groundwater withdrawal in Hanoi, Vietnam. In *Proceedings of the Fifth International Symposium on Land Subsidence*. The Hague, The Netherlands.

Nguyen, V. D. & Nguyen, T. D. 2002 Groundwater Pollution in the Hanoi Area, Vietnam. ESCAP – IWMI Seminar on Environmental and Public Health Risks Due to Contamination of Soils, Crops, Surface and Groundwater from Urban, Industrial and Natural Sources in South East Asia. Hanoi, Vietnam, 10–12 December 2002.

Stefan, C. & Werner, P. 2009 How to protect water resources in Hanoi. International water alliance saxony in Vietnam. *Technikfolgenabschätzung – Theorie und Praxis* **2** (18), 77–81.

Thu, T. M. & Fredlund, D. G. 2000 Modeling subsidence in the Hanoi City area, Vietnam. *Canadian Geotechnical Journal* **37**, 621–637.

VIETNAMNET 2007 VietNamNet – Vietnam's population density nearly doubles that of China. Available from: http://english.vietnamnet.vn/social/2007/07/716861/ (accessed 29 August 2011).

VIETNAMNET 2008 VietNamNet – Census reports Hanoi population at 3.4 million. Available from: http://english.vietnamnet.vn/social/2008/01/765257/ (accessed 29 August 2011).

Waterloo Hydrogeologic Inc. 2006 *Visual MODFLOW 4.2 User's Manual for Professional Applications in Three-Dimensional Groundwater Flow and Contaminant Transport Modeling*. Waterloo Hydrogeologic Inc., Canada.

Werner, P., Röstel, G., Fuchs, L. & Stefan, C. (ed.) 2011 *Integrated Water Resources Management in Vietnam – Handbook for a Sustainable Approach*. Forum für Abfallwirtschaft und Altlasten, Dresden, Germany.

First received 1 January 2013; accepted in revised form 30 April 2013

Stakeholder participation and capacity development during the implementation of rainwater harvesting pilot plants in central northern Namibia

M. Zimmermann, A. Jokisch, J. Deffner, M. Brenda and W. Urban

ABSTRACT

This paper summarises the lessons learned during the planning and construction of four rainwater harvesting (RWH) pilot plants in the village of Epyeshona (400 inhabitants) in central northern Namibia. The main problem of the region is that the water demand of its population exceeds the local natural resources. The rainfall in the project region is extremely variable (50–990 mm per year), evaporation rates are high, perennial rivers do not exist, and groundwater aquifers are saline due to low soil permeability and high evaporation. The project's activities were prepared and accompanied by stakeholder participation and capacity development. The village community chose the techniques and pilot sites in several participatory workshops. Three roof catchment systems with differing tank designs (polyethylene, ferrocement, and concrete bricks) for individual households and a concrete-lined ground catchment facility for six households were built. The collected water is supposed to be used for horticulture. Hence, gardening plots and drip irrigation systems were created. Selected trainees were trained to build, operate and maintain the RWH systems and gardens. Finally, the pilot plants were monitored and evaluated to examine their social and technological feasibility.

Key words | capacity development, demand-responsive approach, horticulture, northern Namibia, rainwater harvesting, stakeholder participation

M. Zimmermann (corresponding author)
A. Jokisch
J. Deffner
M. Brenda
W. Urban
Technische Universität Darmstadt,
Department of Civil Engineering and Geodesy,
Institute IWAR,
Chair of Water Supply and Groundwater Protection,
Petersenstr. 13,
64287 Darmstadt,
Germany
E-mail: *m.zimmermann@iwar.tu-darmstadt.de*

INTRODUCTION

Description of the study area

Namibia is located in the south-west of Africa and is the driest country south of the Sahara. The study deals with the four northern regions of the country, namely Oshana, Oshikoto, Ohangwena and Omusati (Figure 1). An estimated 40–50% of the Namibian population lives here (Araki 2005; Kluge *et al.* 2008). The region is almost exclusively inhabited by the largest ethnic group of Namibia, the Owambo (Niemann 2000). Oshakati is the major city in the region.

In the semi-arid region, up to 96% of the rainfall occurs in the summer months from October to April with a peak in January and February. The winter months are almost completely dry. The rainfall is unpredictable and varies from approximately 50–990 mm per year (Sturm *et al.* 2009). The potential evaporation rate is as high as 2,500 mm per year (Kluge *et al.* 2008). Besides limited rainfall, the area faces another serious water problem due to its shallow topography. The Cuvelai basin stretches from southern Angola to the central north of Namibia and is named after a system of intermittent rivers with a surface area of approximately 100,000 km². Its headwaters are located in southern Angola. Depending on the amount of rainfall in that area, central northern Namibia can experience heavy floodings towards the end of the rainy season. The floodings were exceptionally high between 2009 and 2011.

The water supply system in the region is fed by the Namibian-Angolan border river Kunene and consists of a 150 km long open canal and a pipeline grid with an overall length of about 2,000 km (Zimmermann & Urban 2009). This is due to the fact that perennial rivers are missing and

doi: 10.2166/ws.2013.024

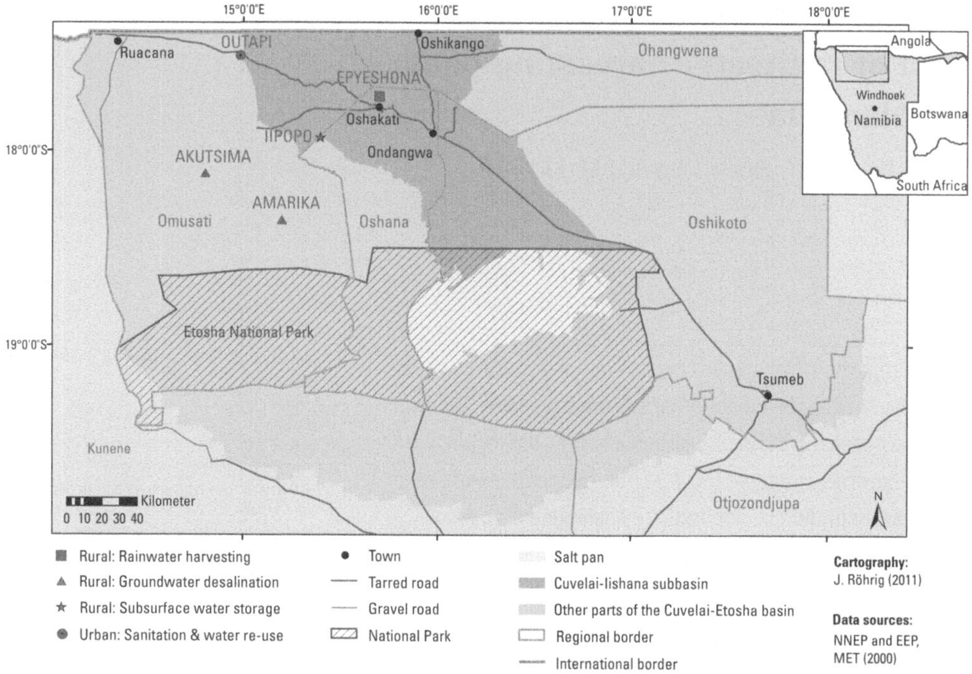

Figure 1 | The Cuvelai-Iishana sub-basin in central northern Namibia and the pilot plant locations of the CuveWaters project (Eisold & Benzing 2011).

the groundwater is predominantly saline. The salinity mainly stems from the low water permeability of the soil and high evaporation rates. Further problems can be seen in the population growth of up to 2.8% and migration into the urban centres of the region (Kluge *et al.* 2008). These factors will probably enhance the demand for water. Since the regional water demand exceeds the local natural resources, new sources for the water supply have to be developed.

Rainwater harvesting pilot plants

One of the water supply techniques which is examined in the CuveWaters research project is rainwater harvesting (RWH). Rainwater is hardly collected in the region by technical means (Zimmermann 2010). It can be harvested from several kinds of surfaces with high runoff coefficients. Gutters and downpipes lead the water into a reservoir (Sturm *et al.* 2009). When using roof catchments, corrugated iron roofs with high runoff coefficients of 0.8–0.85 are preferred over thatched roofs (Gould & Nissen-Petersen 2003). For the construction of ground catchments, concrete lined surfaces have comparably high runoff coefficients of 0.73–0.76 (Gould & Nissen-Petersen 2003). Regarding the reservoirs, attention has to be paid to the local availability of construction materials and workforce. Selected designs should be cost-effective and robust. The rainwater harvested is expected to be of medium quality and thus not potable, particularly with regard to the ground catchment. This is why it is mainly intended to be used for micro-scale gardening in order to increase self-sufficiency or to generate a small income.

To determine the potential locations of RWH pilot plants in the project region, essential selection criteria were defined by the project team. Potential sites had to fulfil the following prerequisites:

- Peri urban area to have options for the marketing of horticultural products
- Availability of rainwater
- Availability of tap water
- Availability of suitable roof catchments and potential ground catchment sites
- Local availability of construction materials
- Demand and interest of users for RWH technology
- Willingness of potential users to participate in the construction, operation, maintenance and monitoring of the plants.

The village of Epyeshona (Okatana constituency, Oshana region) fulfilled the selection criteria and was jointly selected as the pilot site by the community, the Department of Rural Water Supply (DRWS) of the Oshana region and the project team. The village is situated approximately 10 km north of Oshakati (Figure 1) and has roughly 400 inhabitants in 80 households (Deffner *et al.* 2008). The people mainly live from subsistence agriculture and are thus often not employed. Traditionally, pearl millet (Mahangu) is cultivated. The average annual rainfall is around 472 mm (Sturm *et al.* 2009). The drinking water supply is provided by the regional pipeline system. In the following sections, the lessons learned during the planning and construction of the RWH pilot plants are described. Special attention will be paid to the topics of stakeholder participation and capacity development.

METHODOLOGY

The methodology of this study consists of a demand-responsive approach (Deffner *et al.* 2010), combining qualitative social research methods on the one hand and participatory planning methods on the other. This allows for the collection of empirical data as well as for the integration of stakeholders. Methods of qualitative social research comprise a large number of approaches and are often applied in development and health studies (Patton 1990; Thomas *et al.* 1999). Participatory assessment and planning methods (PLA) are widely used (Chambers 1994), e.g. in urban and environmental planning, transport and zone planning, tenant participation or vision development (Davidoff 1965; Arnstein 1969; Bischoff *et al.* 1995). PLA are being applied in development projects and water resources management as well (Lammerink *et al.* 1999; Poolman & van de Giesen 2006).

With this approach, the social realities of local community members can be grasped and considered. This is important in terms of the development and implementation of the proposed RWH options. In our case, workshops were conducted with the inhabitants of Epyeshona, including the future users of the pilot plants. Interactions and discussions among the participants as well as with the facilitators were fostered. In doing so, the stakeholders were informed and, thus, enabled to take knowledgeable decisions. This was already one part of capacity development. Additionally, these decisions were taken jointly by the project team as well as the stakeholders in a participatory manner (stakeholder participation).

The demand-responsive approach consists of several stages which can be roughly subdivided on one hand into a planning stage and on the other into an implementation stage (Figure 2). In the planning stage, the local living conditions are examined and assessed to identify and understand the community's (water supply) problems. Furthermore, the proposed RWH options are introduced and discussed to get an impression of their acceptance as well as to receive feedback in terms of proposals for (technical) modifications. After the incorporation of change requests, the proposed options are once again presented and discussed with the community. This step links the technical decision-making process with the following participatory development of the pilot plant's institutional setups, as well as the capacity development, monitoring and evaluation processes. These following steps were carried out in a similar manner. In the implementation stage of the pilot facilities, several threads of the approach were applied, such as monitoring generally the use and acceptance of the facilities, monitoring the technical status but also making interventions for technical capacity development and their monitoring. Several training courses were conducted and can be seen as such an intervention which was then monitored for its effects on the maintenance and management of the pilot plants. Quantitative and qualitative data were collected by researchers as

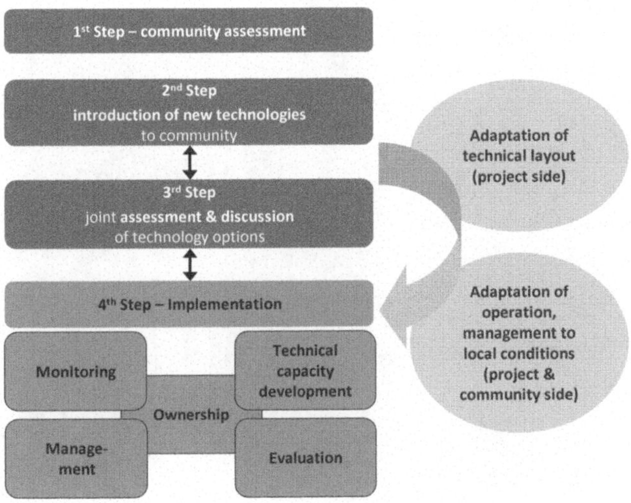

Figure 2 | Steps of the demand-responsive approach (community level).

well as the users themselves in the implementation stage (e.g. water quantities, routines, incidents).

The methods applied in the steps presented had to be modified to the conditions of the rural area in the developing country. This refers to the language, the participants' educational level and other socio-cultural issues such as traditional hierarchies. The handling of these aspects was considerably improved by the mixed research team which comprised German and Namibian members of different scientific disciplines, i.e. engineers and social scientists. The workshops took place in the village of Epyeshona and were carried out in the local language, Oshiwambo. Furthermore, all activities were coordinated with formal political and administrational institutions, such as the Ministry of Agriculture, Water and Forestry (MAWF), the Directorate of Rural Water Supply (DRWS), the Regional Council, the Community Development Committee (CDC), and the Local Water Committee (LPC). Care was taken to ensure that the workshop participants were representative in gender and age. Around 20–40 community members took part at each of the five workshops. One half to two thirds of the attendants were women. Both young and elderly people participated at the workshops.

RESULTS AND DISCUSSION

The results of the study are presented and discussed in three sections, namely according to the planning, the construction, and the operation, monitoring and evaluation phases.

Planning phase

During several community workshops and social-empirical inquiries in 2007 and 2009, the local water supply situation and water usage were recorded and assessed in Epyeshona (Figure. 3). The participants consisted of males and females who actively took part in the workshop processes and gave valuable inputs (CuveWaters 2009). RWH techniques were introduced to the participants using drawings and models of the plants. Advantages and disadvantages were extensively discussed by and with the community. Generally, the people gave a positive response on the concept of RWH. On the one hand, the participants valued advantages such as saving money for the public water points or private taps, using less water, and having water available in the proximity of their own houses as well as for different

Figure 3 | Social-empirical community workshop in the village of Epyeshona in 2007.

purposes (CuveWaters 2007). On the other hand, they worried that mosquitos could breed in the tanks or children might fall into them, both of which are unfounded fears. During the discussions, some participants mentioned that they wanted to store the total amount of the rainwater harvested in the rainy season in order to use it only in the dry season. This kind of usage pattern, however, would require a reservoir which would be three times as large as that implemented later on. This was why the suggestion was rejected and the reservoirs were sized according to a usage pattern with withdrawals during the dry as well as the rainy season. This was one of the few cases when recommendations by the participants were finally overruled for technical or economic reasons.

After the workshop attendees agreed to build four RWH pilot plants (three roof catchment facilities and one ground catchment system), the community selected three households for the installation of private roof catchment facilities. The criteria which served as a guideline for choosing the beneficiaries were jointly selected and comprised, among others, the availability of corrugated iron roofs of sufficient size and the willingness of the future users to contribute to the construction of the facilities. Three different tank types which had been preselected by the project team were installed at the three pilot plants (Table 1). The criteria for the preselection were that the designs of the facilities should be reproducible and materials or prefabricated parts should be locally available. Among the beneficiaries were the headman's household as well as the vice headman's household. Both families are highly respected in the local community. The whole participatory process would not have been possible at all without the contributions of the traditional authorities and their family members. Apart from that, the community also selected the site for the ground catchment system from a number of identified potential sites. The six households which would jointly use the ground catchment facility were chosen by the participants as well. In this regard, the main criterion for the selection was the immediate vicinity to the pilot plant. The institutional setup of the ground catchment system was solely developed by the community members, which is why it can be regarded as one of the main outcomes of the participatory workshops.

Most participants expressed their willingness to help with the construction of infrastructure by volunteering. The community discussed potential members who would have to be trained in the skills required for the construction (Deffner *et al.* 2008). The main criteria for the selection were: potential trainees should be willing to stay in the village for a longer period of time so that they might apply their skills even when the project was over. They also had to be able to speak and write English (CuveWaters 2009). The language barrier might be seen as a constraint but the trainees are supposed to train others in their local language Oshiwambo in the future. Eventually, six trainees were identified by the community to be trained in the construction phase. The trainees were local young men and women who were unemployed at that time and who were not members of the traditional authority's family. During the construction phase, further labour was needed and additional trainees were chosen by the community.

Construction phase

The RWH plants were constructed right after the participatory planning process during two phases. The first one started in October 2009 and ended in February 2010, whereby all four mentioned RWH facilities were

Table 1 | Main features of the four RWH pilot plants

Catchment type	Catchment size (m²)	Reservoir material	Reservoir size (m³)	Expected yield (m³ per year)	No. user households	Construction days
Roof catchment	100	Polyethylene	30 (6*5)	28–40	1	9
Roof catchment	100	Ferrocement	30	28–40	1	11
Roof catchment	87	Concrete bricks	30	24–35	1	13
Ground catchment	480	Concrete bricks and ferrocement	120	118–170	6	20

Figure 4 | RWH system with ferrocement reservoir (30 m³) and irrigated garden in the village of Epyeshona.

constructed, including accompanying infrastructures (Figure 4). The main features of the four RWH systems are shown in Table 1. The building sites were supervised and managed by project team members, but the plants were constructed by the selected trainees under the guidance of Namibian and Kenyan consultants. The second construction phase lasted from February to March 2011 and mainly comprised irrigation facilities.

As part of the technical capacity development, appropriate training was intended to enable the trainees to construct, operate and maintain the RWH facilities (Figure 5). The technical instructors taught the trainees in all areas of guttering and piping, including soldering, ferrocement and brick tank construction, concrete brick making, and also in the organisation of construction sites, e.g. building material supply, calculations, etc. At the beginning, only the six selected trainees were trained at the first construction site (which was the roof catchment system with polyethylene tanks at the headman's homestead). But the team of trainees increased to 25 people during the simultaneous works at the underground reservoir for the ground catchment and the ferrocement cistern. Mainly volunteers, such as local inhabitants from the surrounding households or students who were visiting their families, joined the team. This not only shows the interest that arose from the activities but also the motivation of the community. The trainees' skills and educational backgrounds were very diverse, ranging

Figure 5 | Technical capacity development during the construction phase.

from people without any handicraft experience to good skilled masons. Nevertheless, the trainees' work assignment, commitment and attitude to work were very good to excellent.

During the first weeks of construction, the trainees were not paid but received food and drink as well as workwear. However, for the construction of the last RWH system, which was the concrete brick tank, and during the second construction phase, they received a payment according to local salaries. This is due to the fact that the trainees had already acquired some skills at that point. These trainees were primarily those who were originally selected by the community. After the completion of the pilot plants, certificates of attendance were handed over to all trainees.

Other community members were also involved in the construction process, particularly members of the headman's family. They gave organisational support, for instance by providing food and drink for the whole team. In doing so, it could be promoted that as many inhabitants of the village as possible felt part of the project. This is expected to lead in general to a better acceptance of the RWH systems but particularly also to better operation and maintenance by the users. Furthermore, a sense of ownership for the facilities can be created and fostered.

During the second construction phase in February and March 2011, small-scale irrigation systems and gardens were installed at the RWH pilot sites, since the harvested water is mainly supposed to be used for horticulture. Interestingly, the rural population did not apply any kind of artificial irrigation at all until then. All four sites were provided with lo-tech drip irrigation systems. Additionally, a greenhouse was built at the ground catchment. Therefore, the trainees received training in the installation of drip irrigation systems and in the construction of greenhouses as well as ponds for additional water storage. Furthermore, the user households and some of the trainees were trained in the preparation of gardens and received intensive gardening training.

Operation, maintenance and monitoring phase

Right after the end of the construction works, the responsibility for the operation of the RWH pilot plants including the gardens was handed over to the users. The roof catchment systems are operated by the respective household members. In most cases, there is one responsible person within each user household who was elected by the family or by the community. In terms of the ground catchment, the six selected households operate the plant jointly. Each one of them has the same right to use the harvested rainwater as well as one of six gardening plots of the same size.

The user households had already received training in the basic operation and maintenance of the pilot plants during the construction phase. Generally, RWH systems are comparably easy to operate and require relatively low maintenance (One World Consultants 2011). Operational activities consist of, for instance, diverting the first flush and checking the catchment surface for dirt. Maintenance comprises the annual cleaning of the catchment surface, the guttering and the reservoir. Apart from that, the trainees were instructed to carry out minor repairs such as fixing gutters and pipes or sealing tank leakages with cement if necessary. All described capacity development measures are supposed to ensure the reliable operation of the RWH systems as well as the fertility of the gardens in the long term. This is of particular importance when the project has ended and the responsibilities for the facilities will be completely handed over to the user households.

Monitoring measures have been divided into the monitoring carried out by the users, further technical monitoring and socio-cultural research. Many technical monitoring activities are being undertaken by the users of the RWH systems themselves. Parameters such as water quantity and quality as well as technical functionality of the tanks and adjoining infrastructures are monitored. Additionally, socio-cultural inquiries are being made every year in which the users of the RWH pilot plants are interviewed to determine the role of this additional source of water in their everyday lives. The same applies to the monitoring of the gardens watered with harvested rainwater. Here, the users are responsible for keeping records of ploughing, seeding, harvesting and manure application. The users are also asked to document the garden products they consume, share or sell at local markets. In doing so, the RWH systems' contribution to income generation and livelihood improvement can be measured and assessed.

CONCLUSIONS

The project CuveWaters aims to develop endogenous water resources in central northern Namibia. With groundwater desalination, subsurface water storage, and sanitation and water re-use, RWH is one of four technological options which is implemented and tested in the project region as part of CuveWaters' multi-resource mix. The regional hydrological situation is a major constraint for irrigated agriculture and gardening. Hence, the additional water resource provided by RWH is a considerable improvement. With it, local subsistence farmers are enabled to cope with the dry season and to achieve higher yields. The horticultural harvest can either be used for self-consumption or to generate income by selling it at local markets. This can be seen as a crucial contribution to improving the livelihoods of rural communities. Apart from that, the harvested rainwater has already proven to be an essential water source. During the floods in 2011, several households in Epyeshona were cut off from the public water supply. However, they could be supplied with harvested rainwater by one of the pilot plant users during these times of need.

All activities in the research project include stakeholder participation and capacity development measures. The project's understanding of participation not only relates to the decision-making process, but also to the processes of planning, construction, operation, maintenance and monitoring. This high degree of participation is supposed to guarantee the acceptance of the proposed techniques. The experiences of many development projects have shown that users often are not satisfied with the measures provided if, for instance, the technique does not meet their expectations. The consequences are that the pilot plants might be operated improperly and maintained poorly. Without the users' acceptance and sense of responsibility, even well designed and constructed systems are neither useful for the community nor the researchers. Another important aspect of participation is the adaptation of the proposed technique and its embedding into local conditions. To achieve this, an understanding of not only the physical but also the social living conditions is necessary. This comprises the water usage patterns of the users as well as their attitudes, needs and preferences regarding the proposed technique. Hence, stakeholder involvement and participation was encouraged from the beginning of the project.

Stakeholder participation and capacity development are interdependent. The one cannot be carried out without the other. Without explaining the techniques to the community and discussing advantages and disadvantages with them, the users would not be able to participate in the activities and to take informed decisions. Apart from that, the selected trainees received technical training. In doing so, not only could job opportunities be created by capacity development but access to the regional labour market could also be improved for those who participated in the training. This is of particular importance in a region where unemployment and emigration of young and educated people is a serious problem. Even though only a limited number of participants can be trained by the project, it is expected that those who received training are able to spread the knowledge they gained.

The whole community of Epyeshona participated in the planning process and more than 30 inhabitants were directly involved in the construction process. Another 10 people are now responsible for the operation of the facilities, mainly as farmers of the gardens irrigated with harvested rainwater. The work commitment of all participants was outstanding and the trainees have already successfully proved their skills in reservoir construction by building another ferrocement tank for an external client. During a 2 year monitoring process until 2012, the RWH pilot plants are currently being evaluated in terms of their general operability, cost-benefit ratio and viability. In the third phase of CuveWaters, it is planned to spread those techniques in the region which have proven to be sustainable in the pilot phase.

ACKNOWLEDGEMENTS

CuveWaters is a joint research project by the Institute for Social Ecological Research (ISOE, Frankfurt) and the Darmstadt University of Technology (TUD). The project partners in Namibia are the Ministry of Agriculture, Water and Forestry (MAWF), the Desert Research Foundation of Namibia (DRFN), the Gesellschaft für Internationale Zusammenarbeit (GIZ) and the Federal Institute for Geosciences and Natural Resources (BGR). The project is funded by the German Ministry for Education and Research (BMBF). Thanks are due to all community members of Epyeshona, trainees, volunteers, as well as to the consultants Walter Holch, Isaac Kariuki, John Ndongo, Beatrice Wambui and Joseph Machura.

REFERENCES

Araki, S. 2005 Change in population and land-use intensities in several villages of the four northern regions of Namibia. *African Study Monographs* **Suppl. 30**, 77–88, March 2005.

Arnstein, S. 1969 A ladder of citizen participation. *Journal of the American Institute of Planners* **35** (4), 216–224.

Bischoff, A., Selle, K. & Sinning, H. 1995 *Informieren, Beteiligen, Kooperieren. Kommunikation in Planungsprozessen. Eine Übersicht zu Formen, Verfahren, Methoden und Techniken.* Dortmunder Vertrieb für Bau- und Planungsliteratur, Dortmund.

Chambers, R. 1994 The origins and practice of participatory rural appraisal. *World Development* **22** (7), 953–969.

CuveWaters 2007 CuveWaters Project Situation Assessment Summary Report, Rural and Urban Assessment I & II, Final Summary Report, Internal project report. ISOE/DRFN/TUD, Frankfurt.

CuveWaters 2009 Rural Assessment IV: Rainwater harvesting technologies. Epyeshona, Okatana Constituency. Internal project report. ISOE/DRFN/TUD, Frankfurt.

Davidoff, P. 1965 Advocacy and pluralism in planning. *Journal of the American Institute of Planners* **31** (4), 331–338.

Deffner, J. & Mazambani, C. 2010 Participatory empirical research on water and sanitation demand in central northern Namibia: A method for technology development with a user perspective. CuveWaters Papers, No. 7, Institute for Social-Ecological Research (ISOE), Frankfurt/Main, Germany. Available from: http://www.cuvewaters.net/ftp/cuve_7_deffner.pdf (accessed 17 August 2011).

Deffner, J., Mazambani, C., Klintenberg, P., Nantanga, K., Moser-Norgaard, P. & Seely, M. 2008 Selected results from socio–ecological participatory situation assessments in two sites in Central Northern Namibia. CuveWaters Papers, No. 3, Institute for Social–Ecological Research (ISOE), Frankfurt/Main, Germany. Available from: http://www.cuvewaters.net/ftp/cuve_3_deffner.pdf (accessed 17 August 2011).

Eisold, J. & Benzing, C. 2011 *From Concept to Tap – Integrated Water Resources Management in Northern Namibia.* edts. Institute for Social-Ecological Research (ISOE) GmbH, Frankfurt/Main, Germany.

Gould, J. & Nissen-Petersen, E., 2003 *Rainwater Catchment Systems for Domestic Supply – Design, Construction and Implementation.* ITDG Publishing, London.

Kluge, T., Liehr, S., Lux, A., Moser, P., Niemann, S., Umlauf, N. & Urban, W. 2008 IWRM Concept for the Cuvelai Basin in Northern Namibia. *Physics and Chemistry of the Earth* **33**, 48–55.

Lammerink, M. P., Bolt, E., de Jong, D. & Schouten, T. 1999 Strengthening community water management. *PLA Notes (IIED)* **35**, 21–28.

Niemann, S. 2000 Wasserversorgung und Wasserverwendung in Namibia: Nutzungstraditionen als Grundlage eines nachhaltigen Ressourcenverbrauches im ehemaligen Ovamboland (Water supply and usage in Namibia: Traditions of usage as a basis of a sustainable resource use in the former Ovamboland). *Hamburger Beiträge zur Afrika-Kunde, Vol. 61.* Institut für Afrika-Kunde im Verbund der Stiftung Deutsches Übersee-Institut, Hamburg, Germany.

One World Consultants Ltd 2011 *Operation and Maintenance Training Manual on Rainwater Harvesting Tanks.* Internal Project paper, CuveWaters.

Patton, M. Q. 1990 *Qualitative Evaluation and Research Methods.* Sage, Newbury Park, CA.

Poolman, M. I. & van de Giesen, N.C. 2006 Participation: rhetoric and reality. The importance of understanding stakeholders based on a case study in Upper East Ghana. *International Journal of Water Resources Development* **4**, 561–573.

Sturm, M., Zimmermann, M., Schütz, K., Urban, W. & Hartung, H. 2009 Rainwater harvesting as an alternative water resource in rural sites in Central Northern Namibia. *Physics and Chemistry of the Earth Parts A/B/C* **34** (13–16), 776–785.

Thomas, A., Chataway, J. & Wuyts, M. 1999 *Finding out fast: Investigative skills for policy and development.* Sage, Thousand Oaks, London & New Delhi.

Zimmermann, M. 2010 The coexistence of traditional and large-scale water supply systems in central northern Namibia. *Journal of Namibian Studies, History – Politics – Culture* **7** (2010), Otjivanda Presse, Essen, Germany, 55–84.

Zimmermann, M. & Urban, W. 2009 Wasserversorgungstechnische Lösungsansätze innerhalb des IWRM im Norden Namibias (Water supply solutions within the IWRM in northern Namibia). In: *1. Darmstädter Ingenieurkongress – Bau und Umwelt*, (H. J. Linke, ed.). 14–15 September 2009, Conference proceedings, Technische Universität Darmstadt, Darmstadt, Germany.

First received 25 December 2011; accepted in revised form 19 March 2013

The situation of sanitary systems in rural areas in the Miyun catchment, China

C. Kröger, A. Xu, S. Duan, B. Zhang, H. Eckstädt and R. Meissner

ABSTRACT

The Miyun Reservoir provides most of Beijing's drinking water. Despite its importance, the Miyun reservoir suffers from decreasing water quality caused by uncontrolled wastewater discharges, inadequate land use and over fertilization, which increase the pressure on soil and water resources. The major pollutants are nitrogen and phosphorus which emanate to some extent from untreated sewage. So far there is little data about the existing wastewater quantity and quality in rural settlements in northern China. This study was conducted in typical villages situated along upstream rivers in the catchment of the Miyun Reservoir. The main objective was to determine the current situation and efficiency of the wastewater treatment system in rural settlements.

Key words | China, Miyun, rural areas, sanitary systems, wastewater treatment

C. Kröger (corresponding author)
A. Xu
H. Eckstädt
Prof. Hydromechanics and Sanitary Engineering, University of Rostock,
Satower Str. 48, 18059 Rostock,
Germany
E-mail: *christina.kroeger@uni-rostock.de*

S. Duan
B. Zhang
Beijing Soil and Water Conservation Center,
No.5 Yuyuantan South Rd.,
Haidian District, 100038 Beijing,
China

R. Meissner
Helmholtz Centre for Environmental Research – UFZ, Department Soil Physics, Lysimeter station,
Dorfstrasse 55, 39615 Falkenberg,
Germany

AIM OF THE REGIONAL SURVEY

Background

Beijing is situated in the dry north-east of the North China Plain bordering the Mongolian Plateau in central Hebei province. In the 1950s drinking water was mainly supplied by surface water, whereas the Guanting Reservoir, 80 km northwest of Beijing, was designed in 1954 to provide water for industrial use and urban waterways. Severe pollution by untreated sewage from Zhangjiakou and Datong and effluents from rural counties ended up in the 1970s in the Guanting Reservoir. In the 1990s water quality deteriorated sharply to class V standard and a part of the drinking water supply was prohibited for Beijing city in 1997. The most critical resource in China is not land or food, but water (Brown 2001).

Since the disaster with the Guanting Reservoir the Miyun Reservoir has become the main source of drinking water for Beijing. The Miyun Reservoir was finished in 1960 during China's Great Leap Forward and its purpose was mainly flood control and irrigation of rural areas. The water supply capacity of Miyun is 600 million m^3 of urban use per year and it provides 75% of Beijing's drinking water (Peisert & Sternfeld 2004). Beijing has experienced a drought in recent years whereby the water level in the Miyun Reservoir dropped nearly 8 m and the ground water level dropped about 20 m. Whereas Guanting Reservoir has experienced frequent eutrophication, Miyun Reservoir has been mesotrophic with only a tendency for eutrophication (Wang *et al.* 2001; Zhengjun *et al.* 2008). However, the water quality category of the Miyun Reservoir has degraded from class II to III and still conforms to the Chinese drinking water standard, but treatment is required (Enders 2005; Zhengjun *et al.* 2008). In China's system of surface water quality classification, class I is the highest quality, suitable for headwaters and national protected areas. Class III is the lowest quality still considered suitable for drinking water. Classes IV and V are considered to be suitable for agricultural use or for normal landscape needs. Lower than class V is considered to be highly polluted, in which water system functionality has been severely degraded (MEP 1997).

Situation in the Miyun catchment

The major pollutants are nitrogen (TN) and phosphorus (TP) which emanate to some extent from untreated sewage and TN has a larger pollution risk compared with TP in most years (Zhang *et al.* 2010). In the Miyun catchment,

doi: 10.2166/wst.2013.296

there are mainly small traditional agricultural villages characterized by little infrastructure. Seventy per cent of the Miyun Reservoir catchment belongs to the province of Hebei where two-thirds of the 860,000 inhabitants (2003) live while the Miyun Reservoir is situated in the Beijing district; most of the residents in the catchment live in rural villages (UNESCAP 2002; Peisert & Sternfeld 2004; Hongliang 2008). In 1995, the Beijing government issued a detailed administrative regulation that divided the Miyun catchment into three watershed protection zones and limited activities within them. The water protection zones enclose 4,500 km² and are home to 140,000 residents (Enders 2005). The innermost zone, covering the land located within approximately 300 m of the reservoir restricts all agriculture and residential activities (Regele 2008). The inflows to the reservoir are problematic as buildings are often built right up to the river bed and also adjacent farmland.

The role of villages in water pollution becomes important in the context of the very large percentage, about 47% (Berlin-Institut für Bevölkerung und Entwicklung 2007) of the Chinese population that lives in small villages across the country (Ongley *et al.* 2010). Wastewater in rural areas was mostly discharged without any treatment until 2006 (Zhang 2009) or disposed into soakaways and partly to sewer systems, which lead to simple multiple-compartment settling tanks where solids are removed by sedimentation. Most settling tanks are made of bricks or stone with concrete and are gradually permeable for nutrients. The supernatant percolates into soil or is discharged directly into receiving waters. Further biological treatment stages are rarely conducted. Zhang (2005) proposes that because of water shortage and low water quality in water bodies the treatment capacity must be strongly increased, especially in northern China. Furthermore water pollution from small rural industries and large scale intensive animal raising (feedlots) is a serious problem throughout China (Ongley *et al.* 2010). Rural industries mean businesses located in rural areas which are involved in non-agricultural activities, classified into TVE (Town and Village Enterprises) and private ownerships (Wang 2008). There is a lack of well trained personnel as well as a shortness of monetary incentives for wastewater treatment (Abele 2010). According to the State Environmental Protection Agency (Zhenhua 2003), only 24% of sewage was treated to the national standard. However, Zhao & Wong (2002) report that 76% of wastewater caused serious pollution and only 5% was treated to reach the required standard. So far, there is little data about the wastewater quantity and quality in rural settlements in China (Ongley *et al.* 2010).

Aim

Our objective was to collect data about wastewater from households, public institutions, rural industries and other typical water pollution sources in rural villages to develop a sanitation concept for the catchment area of the Miyun reservoir.

MATERIALS AND METHODS

Spatial survey

The data acquisition to determine the prevailing sanitation systems in selected village types resulted from questionnaire interviews with residents and local authorities. Furthermore, samples of the various wastewater sources and sewage treatment plants are investigated to determine the composition of the effluent according to the wastewater origin and to assess the treatment capacity of the prevalent wastewater treatment systems. For the experiments, three typical but different types of villages were used to detect differences in living standards (Table 1). The village Shicheng is situated

Table 1 | Overview of the research villages in the catchment of the Miyun Reservoir to determine the prevalent sanitary systems and wastewater pathways in rural areas

	No. of inhabitants	Village type/infrastructure	Main income sources	Main research objects
Shicheng	700	High developed infrastructure with primary school, police station, hospital, hotels	Tourism, retail industry	Wastewater from hotel (50 beds), restaurant, capacity of membrane-bioreactor plant, three-chamber settling tank
Huangyukou	400	Less developed	Forestry, horticulture, animal husbandry	Poultry farm, three-chamber settling tank
Xiwanzi	100	Less developed	Forestry, horticulture, animal husbandry	Public bath house

close to the Miyun reservoir and has a relatively highly developed rural infrastructure with a primary school, hospital, police station and hotels. The main source of income is tourism. The villages Huangyukou and Xiwanzi are located in valleys of the highlands in the north of the Miyun reservoir along the river Sheyuchuan, annually supplying water to a volume of 517 m^3 from the Miyun reservoir. The main income sources are forestry, horticulture and animal husbandry.

In the villages, wastewater samples were analyzed for pH, temperature, chemical oxygen demand (COD) N_{total}, P_{total}, NH_4-N, NO_3 using photometric cuvette tests (Hach-Lange Company). The analyses were carried out in April. The wastewater samples were stored in a cool place and analyzed at least 6 h later. The time delay between sampling and analysis may lead to slight changes in the composition of nitrogen parameters, but does not effect the overall result significantly. The sample containers were filled completely to prevent oxidation. Before analysis, samples were homogenized, partly according to the ranges of the cuvette diluted with tap water. A blank sample of the tap water showed the correction value. For the investigation of NO_3 and NO_4-N, the sample was centrifuged at 2,000 rpm for 10 min and then filtered through a 0.45 µm membrane filter. The pH meter was calibrated daily.

One region of Shicheng was examined by testing the water supply of a typical family hotel with 50 beds. The grey, black and kitchen wastewater was collected in a three-chamber tank and then was discharged into an artificial lake or infiltrated into soil. Furthermore, a membrane bioreactor (MBR) was investigated in Shicheng.

As in many villages in the highlands, there is no biological treatment plant in Huangyukou. Any wastewater was collected in multi-chamber septic tanks and then infiltrated into soil directly. Here, the runoff from a family of 4 persons and a livestock farm (poultry farm) was examined.

In Xiwanzi, a village in the highlands with similar conditions as Huangyukou, the wastewater of a public bathhouse was analyzed. Residents without private bath, use this facility.

Economic status in investigation villages

The main source of income with 38% of the total income of 700 residents in Shicheng comes from tourism. The second largest income sector is the construction and retail trade. The average wastewater flow is about 118 l/(p*d) with the largest volume from fishery (39%) and bathing waters (34%) (Jie 2009).

Huangyukou and Xiwanzi are mainly rurally characterized. The cultivation of chestnuts and other nut trees, horticulture and animal breeding are prevalent sources of income. The average per capita net income in 2005 was about 7.200 RMB/yr (for comparison: the average net income of urban population is 15.600 RMB/yr) (CIIC 2005). Our survey showed that the average net income of a family in Shicheng is 4.700 RMB/yr. The average wastewater flow is about 26 l/(p*d), approximately 60% originating from bathing waters (Jie 2009).

LEGISLATION FRAMEWORK IN CHINA

The values of Chinese water guidelines for the discharge of treated municipal wastewater (GB 18918–2002 (MEP 2003)) are more sophisticated than the European guidelines (91/271/EWG) but are quite similar to German national standards (AbwV (Bundesministerium 2004)). However, the Chinese standards are categorized depending on the classification of the receiving water bodies, whereas in European guidelines the limit values are also classified to the population equivalents.

Chinese control parameters contain standard chemical parameters, as well as physical and microbiological parameters. The subdivisions of class I into IA and IB are assigned to drinking water protection zones to avoid further deterioration of water quality.

RESULTS AND DISCUSSION

Wastewater pathways in rural areas

The wastewater treatment in rural areas is limited almost exclusively to mechanical processes. Only in larger settlements are biological processes applied. As Regele (2008) points out, the rural upstream areas remain poor, preliminarily relying on traditional agriculture practices and manufacturing jobs to sustain themselves and cannot be compared with larger downstream communities which have benefitted from the past half century of economic growth. The most common wastewater pathways in rural settlements are shown below.

Rainwater

In our study area, four different wastewater paths are prevalent. Rainwater is usually percolated locally or fed separately through gutters and sewer systems with wastewater to the next receiving water and is subjected to no further purification.

Wastewater

Three-chamber settling tank – biological wastewater treatment plant This type of sanitation is often well developed in terms of infrastructure in places with a high population density. In general, there is hardly any differentiation of wastewater in rural settlements, making no distinction between industrial, commercial, municipal and wastewater from households, but therefore all kinds of wastewater is collected and treated together. The above-named system is the most common in the Miyun reservoir catchment. In Shicheng, the pre-cleaned wastewater is fed to a large treatment plant, which operates on the principle of a MBR after collection and treatment (sedimentation of solids) in the sedimentation tank.

Three-chamber settling tank – infiltration into soil In places with little infrastructure, such as Huangyukou and Xiwanzi with poorly developed sewage systems, three-chamber pits for mechanical and anaerobic wastewater treatment are used and the treated wastewater infiltrates directly into the soil. These places often lack funds for the expansion and maintenance of existing drains, which is also costly because of the low population density. Statements about the condition of the sewers are simply very daring. It can be assumed that the structural condition is poor and determines a high proportion of exchange between wastewater, soil and groundwater.

Soakaway pit – infiltration into soil In areas where there is no sewage system at all, domestic wastewater is collected locally in soakaway pits without outlet and seeps directly into the ground. Wastewater treatment takes place only minimally. The population density is the lowest and the settlements often stretch along a small road.

Irrigation and evaporation Partially grey water is applied directly onto the soil surface where it evaporates and infiltrates. In Huangyukou, the excess rinse water from cleaning at a chicken farm is disposed in this way because the quantity cannot be discharged through the sewage system.

The wastewater systems are generally in poor condition. The share of infiltration is difficult to determine. The infiltrating wastewater is mostly untreated and contributes to nutrient enrichment in soil and groundwater.

Composition of wastewater and efficiency of treatment systems

In Shicheng the wastewater is treated by the MBR plant with a capacity of 60 t/d. The capacity of sewage treatment plants in China is usually given in tons per day (Duan 2011). The indication of mass is only of limited significance, as the freight or the actual loading of the treatment plant is not recorded. Specification in volume would be more appropriate. Furthermore there is little distinction between the different types of wastewater. Industrial wastewater from restaurants or livestock is not separately treated in advance but runs directly into municipal sewage treatment plants or is discharged into the receiving waters or percolates into the ground. A division into direct and indirect discharge is not common which results in varying COD values of the inflow, as was also reported by Gao *et al.* (2011).

The plant in Shicheng treats the wastewater first mechanically by grit removal, followed by an aerated MBR basin and finally disinfectant and freshwater for dilution is added. The cleaning performance of the system was very limited, only the COD concentration was reduced on an average by 87%. For nitrogen and phosphorus there could not be any measured decreases. In two out of eight discharge value measurements the COD exceeded the admissible standards as shown in Table 2; the discharge values for N_{tot}, P_{tot}, and NH_4-N exceeded the standards in any case many times over (Figure 1). The effluent is discharged into the dry river bed of the Jiudaowan river, which should contribute to further purification. The river has been dry for several years, probably even before treatment construction, so in case of flooding due to heavy rain or snow melt the expected cleaning performance will be worse. In Shicheng there is a second MBR water treatment plant, which was at the time of investigation, however, damaged by overloading and had to be decommissioned.

A general problem is that there are hardly any experts who can explain the function of these systems. Employees of the water authorities could often give only insufficient information. The discharge values are measured by private companies that are also responsible for planning and constructing the plant. A further control of effluent values by government is not scheduled.

In Huangyukou the grey water from households has been investigated. The laundry and kitchen wastewater was collected in a container daily at a 1:1 ratio. The concentration of each parameter in the effluent was very different, so it is difficult to make a statement about the

Table 2 | Chinese guidelines (GB 18918–2002) for the discharge of treated municipal wastewater (Chang 2010)

Parameters	Unit	Class I A	Class I B	Class II	Class III
COD_{Cr}	mg/l	50	60	100	120
Biochemical oxygen demand (BOD_5)	mg/l	10	20	30	60
Suspended solids	mg/l	10	20	30	50
Anionic surfactants	mg/l	0.5	1	2	5
N_{total}	mg/l	15	20	–	–
NH_4-N[a]	mg/l	5 (8)	8 (15)	25 (30)	–
P_{total} (after 1st Jan 2006)	mg/l	0.5	1	3	5
pH	–	6.....9			
Fecal coliforms	mg/l	10^3	10^4	10^4	–

[a]If water temperature <12 °C the value numbers in brackets are allowed.

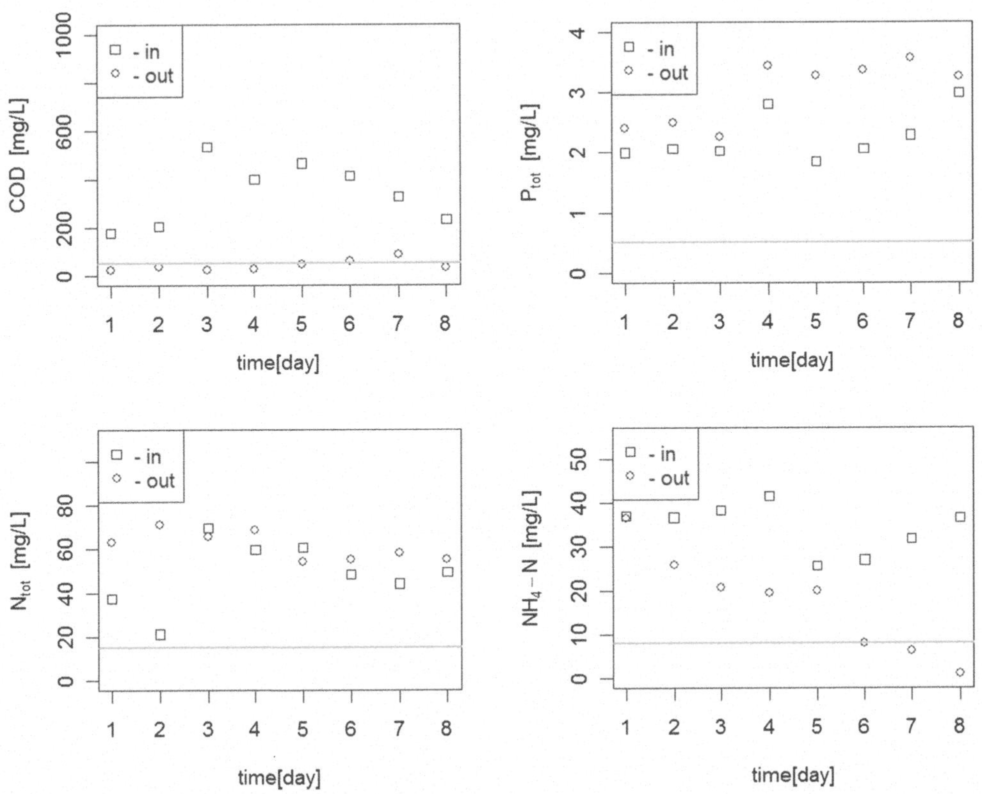

Figure 1 | Effluent quality of the MBR treatment plant in Shicheng. The permissible limits (grey line) (GB 18918–2002) are exceeded in most cases.

concentrations. The ratio of COD:N_{tot}:P_{tot} was, however, similar in each case and averaged at 438:23:1.

A widely consumed food in northern China is poultry bred on farms and sold after a period of time. After selling the livestock, the barn is cleaned. According to the DLG (2009), breeding 3,000 chicken in three cycles of 50 days would produce approximately 360 kg nitrogen and 74 kg phosphorus per year. The poultry farm has no special wastewater treatment. The rinsing water is discharged into a three-chamber settling tank together with wastewater from households. Presumably a lack of understanding of wastewater impacts on the surrounding and water bodies as well as economic interests of operators and lack of control and implementation of government regulations is essential as in many rural industries (Wang 2008).

The effluent of the family hotels and restaurants in Shicheng is collected in a three-chamber settling tank and then seeps into the soil. In Xiwanzi the effluent of the public bath house, consisting of treated grey water and toilet wastewater is also treated in a three-chamber settling tank. Up to 60 people can simultaneously use the bathroom. In Shicheng only the incoming wastewater could be analyzed, as the position of the last chamber was unknown. However, in Xiwanzi, samples from several chambers could be inspected. The degrees of degradation between the chambers were very different. On some days, no degradation could be detected at all but rather an accumulation. On other days the degree of degradation of COD was at 96%, 88% for N_{tot}, 59% for P_{tot} and for NH_4-N at 23%. A constant cleaning is not performed in the multi-chamber pits (Figure 2). A comparison of daily discharge values with the legally permissible limits (Table 2) revealed that only one of the 36 values fell below the standard. The effluent values for COD were on average 406 mg/l, for N_{tot} at 32 mg/l, for P_{tot} at 0.77 mg/l NH_4-N at 17 mg/l.

In conclusion, the statutory critical values (Table 2) were exceeded by all plants. It can be assumed that a large amount of nutrients passes through the inflow into the reservoir. Therefore point sources contribute significantly to nutrient enrichment.

Partly, there exist planning errors or inappropriate detailed solutions, e.g. the effluent of the MBR plant in the dry part of the river bed for further purification, without considering the consequences in case of sudden flooding. Furthermore, a sewer with a manhole was placed transversely through the river bed in Shicheng, which has been dry in recent years. Probably it was not considered, that floods can occur.

Especially in the highland villages, wastewater is discharged into soakaways without drains, so the nutrients can pass directly to a large extent into the groundwater. Many nutrients are also dumped by rural industries (fishery, poultry farm), because they have no adequate sanitation facilities.

Condition of plants

Even noticeable is the poor quality of building materials. A sewage treatment plant built in 2005 in Shicheng shows,

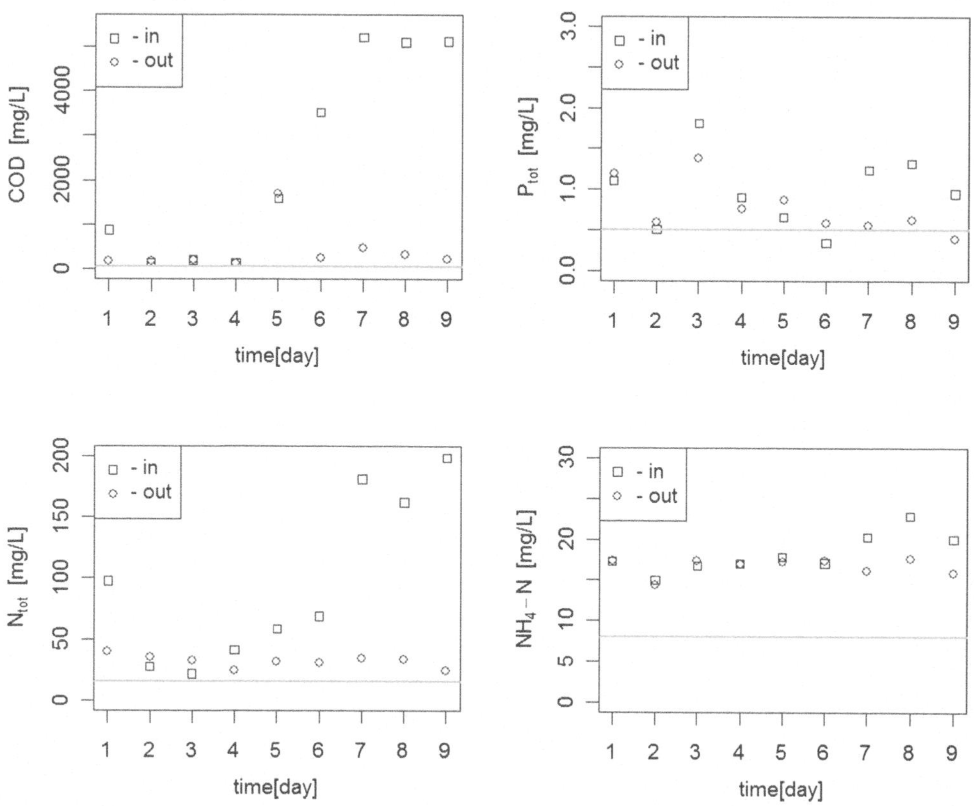

Figure 2 | Effluent quality of the three-chamber settling tank in Xiwanzi, the prevailing wastewater treatment system in rural areas. The permissible limits (grey line) (GB 18918–2002) were exceeded in most cases.

as early as 2010, major damage to concrete and metal parts (broken and rusted manhole covers, cracks in concrete). A few septic tanks are built of bricks, which are presumably permeable to some degree. Furthermore, it can be assumed that the underground pipes are in a similar condition and there is a sufficient interaction with groundwater.

Enforcement of legislation framework and guidelines in rural areas

The on-going deterioration of water quality could not be stopped, as the penalties regularized by laws could only be enforced by intensive controls (Chang 2010). China's environmental laws are comparatively strong but not effective in rural areas due to legislative shortcomings, poorly designed policy instruments and unsupportive work for environmental regulators (Wang 2008). Furthermore, planners and developers of wastewater treatment plants are also often responsible for the control of the effluent values of their own constructed plants and forward the values to the superior government agencies. We assume that it lacks top-to-bottom-mechanisms of regularly state control.

In general, technical solutions are often preferred, while operation and maintenance of such facilities is not considered. This is particularly problematic because the staff often have little knowledge about the treatment processes. Cornel & Wagner (2005) reported that there is no special training in China for wastewater professionals. The information about the municipal sewage systems of authorities and employees is often contradictory or incomplete. A general problem seems to be the lack of state control over the valid critical discharge values. Many of the big flagship facilities are poorly maintained and overloaded, so that the cleaning efficiency is extremely low. The disposal of surplus sludge from wastewater treatment plants and multi-chamber tanks is not regulated at all.

CONCLUSIONS

Situation of sanitary systems in the rural areas of the Miyun catchment

Most of the wastewater is treated only mechanically with settlement tanks, but the sludge treatment or disposal is not regulated. Furthermore the maintenance of most plants is worse, and larger plants with biological treatment stages are not running in an optimal operation mode, which could be due to the fact that the personnel has little knowledge. In the majority of our measurements the critical values of the Chinese guidelines were exceeded. Beside this, a lot of plants are made of poor quality building material, so the amount of untreated wastewater leaking into waterbodies is even larger. All in all, point sources seriously contribute to nutrient enrichment in the groundwater, the upstream rivers and finally the Miyun reservoir.

ACKNOWLEDGEMENTS

The study is funded by the German Federal Ministry of Education and Research (BMBF; FK 02WM1048). We would like to thank all colleagues at the Chinese and German partner institutions for their cooperation and support during the investigations.

REFERENCES

Abele, C. 2010 IFAT China bedient wachsenden Umweltmarkt (China serves growing environmental market). Germany Trade and Invest. Available from: http://www.gtai.de/GTAI/Navigation/DE/Trade/maerkte,did=67558.html (accessed 04 July 2011).

Berlin-Institut für Bevölkerung und Entwicklung 2007 Ländliche Räume Chinas. Available from: http://www.berlin-institut.org/online-handbuchdemografie/bevoelkerungsdynamik/regionale-dynamik/laendliche-raeume-chinas.html (accessed 13 July 2011).

Brown, L. 2001 How water scarcity will shape the new century. *Water Science and Technology* **43** (4), 17–22.

Bundesministerium 2004 Verordnung über Anforderungen an das Einleiten von Abwasser in Gewässer - Abwasserverordnung – AbwV. (German wastewater ordinance).

Chang, Y. 2010 *Greywater Treatment within Semi-Centralised Supply and Treatment Systems by the example of the People's Republic of China.* PhD Thesis. TU Darmstadt, IWAR, Germany.

CIIC 2005 China Internet Information Center: Chinas Provinzen, regierungsunmittelbare Städte und autonome Gebiete (China's provinces, municipalities directly under the government and autonomous regions). Available from: http://german.china.org.cn/german/204507.htm (accessed 27 July 2011).

Cornel, P. & Wagner, M. 2005 *Anforderungen an die Abwassertechnik in anderen Ländern – China (Requirements for Waste Water Technology in other Countries – China).* Abschlussbericht zum BMBF-Vorhaben 02WA0452, TU Darmstadt, Germany.

DLG 2009 Deutsche Landwirtschafts-Gesellschaft e.V. Bilanzierung der Nährstoffausscheidun-gen landwirtschaftlicher Nutztiere – Hähnchenmast (Assessing nutrient excretions of farm animals – broiler). *Arbeitskreis Futter und Fütterung – Band 199*. DLG, Frankfurt am Main.

Duan 2011 Personnel communication with Prof. Shuhuai Duan, Beijing Soil and Water Conservation Center, 2011.

Enders, S. 2005 *Miyun – Integrated Watershed Management for the Protection of Beijing's Drinking Water*. Research Experience for Undergraduates, National Science Foundation and Michigan State University. Available from: http://forestry.msu.edu/China/New%20Folder/S.Enders-Miyun.pdf (accessed 14 March 2011).

Gao, D., An, R., Tao, Y., Li, J., Li, X. & Ren, N. 2011 Simultaneous methane production and wastewater reuse by a membrane-based process: evaluation with raw domestic wastewater. *Journal of Hazardous Materials* **186** (1), 383–389.

Hongliang, L. 2008 Miyun Reservoir. International Lake Environment Committee. Available from: http://www.ilec.or.jp/database/asi/asi-08.html (accessed 15 April 2011).

Jie, Y. 2009 Discharge features of rural domestic wastewater from different types of villages in water source protection area in Miyun reservoir of Beijing. *RDA Journal of Agro-Environment Science* **28** (6), 1200–1207.

MEP 1997 Ministry of Environmental Protection of the PRC: The National Standards of the People's Republic of China - Environmental Quality Standards for Surface Water. Available from: http://english.mep.gov.cn/SOE/soechina1997/water/standard.htm (accessed 20 May 2011).

MEP 2003 Ministry of Environmental Protection of the PRC: Discharge standard of pollutants for municipal wastewater treatment plant – GB 18918–2002. Available from: http://english.mep.gov.cn/standards_reports/standards/water_environment/Discharge_standard/200710/t20071024_111808.htm (accessed 20 May 2011).

Ongley, E., Xiaolan, Z. & Tao, Y. 2010 Current status of agricultural and rural non-point source pollution assessment in China. *Environmental Pollution* **158** (5), 1159–1168.

Peisert, C. & Sternfeld, E. 2004 Quenching Beijing's thirst: the need for integrated management for the endangered Miyun reservoir. *China Environmental Series* **7**, 33–46.

Regele, A. 2008 *Managing Beijing's Water Resources with Payment for Environmental Services*. College of Natural Resources, University of California, Berkeley.

UNESCAP 2002 United Nations Economic and Social Commission for Asia and the Pacific – Hebei. Available from: http://www.unescap.org/esid/psis/population/database/chinadata/hebei.htm#basic (accessed 15 May 2011).

Wang, M. 2008 Rural industries and water pollution in China. *Journal of Environmental Management* **86** (4), 648–659.

Wang, X., Li, T., Xu, Q. & He, W. 2001 Study of the distribution of non-point source pollution in the watershed of the Miyun reservoir. *Water Science and Technology* **44** (7), 35–40.

Zhang, C. 2009 Decentralized sewage remediation in drinking water source protection zones of rural areas in Beijing. China University of Mining and Technology. Available from: http://www.ecosan.org.cn/wp-content/uploads/Zhang_descentralised_sewage_Beijing_October_20091.pdf (accessed 14 August 2011).

Zhang, W. 2005 *Ökologische Siedlungswasserwirtschaftliche Konzepte für urbane Räume Chinas unter Berücksichtigung deutscher Techniken und Erfahrungen (Ecological Urban Water Management Concepts for Urban Areas in China, taking into Account German Experiences and Techniques)*. PhD Thesis. TU Darmstadt. IWAR, Germany.

Zhang, W., Sun, D., Li, H. & Zhou, L. 2010 Surface water quality assessment in Miyun reservoir watershed, Beijing in the period 1980–2003. *Environmental Science* **31** (7), 1483–1491.

Zhao, S. & Wong, K. 2002 The sustainability dilemma of China's township and village enterprises: an analysis from spatial and functional perspectives. *Journal of Rural Studies* **18** (3), 257–273.

Zhengjun, W., Jianming, H. & Guisen, D. 2008 Use of satellite imagery to assess the trophic state of Miyun Reservoir, Beijing, China. *Environmental Pollution* **155** (1), 13–19.

Zhenhua, X. 2003 *Report on the State of the Environment China 2002*. Available from: http://english.mep.gov.cn/SOE/soechina2002/water.htm (accessed 15 August 2011).

First received 12 January 2013; accepted in revised form 20 April 2013

A mathematical approach to find long-term strategies for the implementation of resource-orientated sanitation

I. Kaufmann Alves

ABSTRACT

In the present discussion of sustainability centralised water infrastructures are exposed to new challenges, which may cause a conceptual alteration in urban water management. If technologies for closing urban water and nutrient cycles are to at least partially replace existing systems, then intensive reconstruction work becomes essential. The paper presents the development and implementation of a mathematical approach to minimise environmental impact and economic costs on the way to more source-controlled future states in urban water management. To find an optimal transformation strategy, a simultaneous project scheduling and network flow problem was defined as a bi-criteria mixed-integer program. An optimal solution is found by minimising two objective functions concurrently – the economic costs and 'ecologic costs' for the period of consideration. This paper discusses the influence of the weighting of these two costs on optimal transformation strategies for a real catchment in Germany. The results show that the approach can very well support decision makers when showing all impacts of transformation processes in detail. All in all, the developed model can be seen as a first step in strategy-finding for transformations in existing urban water systems.

Key words | decision support, mathematical optimisation, mixed-integer linear program, resource-orientated sanitation, transformation strategies

I. Kaufmann Alves
Institute of Urban Water Management,
University of Kaiserslautern,
Paul-Ehrlich-Str. 14,
67663 Kaiserslautern,
Germany
E-mail: *inka.kaufmann_alves@bauing.uni-kl.de*

NOMENCLATURE

$\mathcal{A}$	Set of arcs
A	Flow type 'flow contributing area'
bf_j	Factor for dimensioning of arc j
b_i^K	Demand of a flow K in a node i (e.g. wastewater components, rainwater runoff)
$B_{jpt} \in \{0,1\}$	Variable: begin of construction of arc j in mode p at time t
bk_{jp}	Investment costs of arc j in mode p
bud_t	Budget at time t
COD	Chemical oxygen demand
d_{jp}	Construction time of arc j in mode p
$\mathcal{F}$	Set of flow types A, R, T, FR, FT
fk_j^K	(Fictious) flow dependent costs for flow K on arc j
FR	Flow type 'pollution load in stormwater runoff', here for parameter COD
FT	Flow type 'pollution load in dry weather flow', here for parameter COD
g^A	Annually mean rainfall
g^T	Annually mean domestic wastewater flow in present state
H_i	Predecessor arcs
i	Number of node
j	Number of arc
K	Flow types in mathematical model
$K_i(1)$	Parts of economic costs, $i \in \{1,\ldots,4\}$
$K_i(2)$	Parts of ecologic costs, $i \in \{1,\ldots,5\}$
l_{jp}^K	Lower bound of flow K for arc j in mode p
$LB_{jpt} \in \{0,1\}$	Variable: compliance with minimum flow velocity
M_j	Set of modes for an arc j
mk_j^K	Flow dependent costs for flow K on arc j

doi: 10.2166/wst.2013.691

MV	Set of arcs for which flow velocity is tested
nd_{IST}	Remaining useful life span for elements at time $t = 0$
nd_{jp}	Useful lifespan of arc j in mode p
nds_{jp}	Useful lifespan of arc j in mode p after rehabilitation
NI	Set of arcs not existing in present state
p	Mode
r_{jpi}	Necessary quantity of resources i for arc j in mode p
R	Flow type 'design rainfall runoff'
ra_{it}	Available resources of resource type i at time t
res	Resources
$\mathcal{RK}$	Set of resource types
RT	Sum of flow types 'design rainfall runoff' and 'dry weather flow'
$S_{jpt} \in \{0,1\}$	Variable: rehabilitation of arc j in mode p at time t
sk_{jp}	Costs for rehabilitation and reinvestment respectively of arc j in mode p
spk_j	Higher sewer flushing costs for arc j
T	Flow type 'dry weather flow'
T_i	Successor arcs
$\mathcal{T}$	Time of consideration
$\mathcal{T}_{BY}^{(jp)}$	Timespan, in which an arc j in mode p exists
$\mathcal{T}_{B}^{(jp)}$	Timespan, in which an arcs j in mode p can be built
$\mathcal{T}_{\mathrm{B}}$	Time for transformation
u_{jp}^{K}	Upper bound of flow K for arc j in mode p
u_{jp}^{RT}	Maximum flow RT in arcs j in mode p representing transport elements
uk_{jp}	Operational costs of arc j in mode p
$\mathcal{V}$	Set of nodes
$X_{jt}^{K} \geq 0$	Variable: flow K on arc j at time t
$X_{jt}^{A} \geq 0$	Variable: flow A on arc j at time t
$X_{jt}^{R} \geq 0$	Variable: flow R on arc j at time t
$X_{jt}^{T} \geq 0$	Variable: flow T on arc j at time t
$X_{jt}^{FR} \geq 0$	Variable: flow FR on arc j at time t
$X_{jt}^{FT} \geq 0$	Variable: flow FT on arc j at time t
X_{gew}^{R}	Flow R discharged to receiving waters
X_{gew}^{T}	Flow T discharged to receiving waters
X_{gew}^{FR}	Flow FR discharged to receiving waters
X_{gew}^{FT}	Flow FT discharged to receiving waters
X_{rs}^{T}	Actual values of resources protection components
X_{wh}^{A}	Actual values of water balance components
$Y_{jpt} \in \{0,1\}$	Activation variable: activation of arc j in mode p at time t
Z	Interest rate
ZF	Objective function
zf_j	Factor for calculation time-dependent operational costs on arc j
Z_{rs}	Goal values for resources protection components
Z_{wh}	Goal values for natural water balance components
$\mathcal{Z}_{gew}$	Set of arcs containing receiving water values
$\mathcal{Z}_{r}$	Set of arcs containing resources protection values
$\mathcal{Z}_{w}$	Set of arcs containing natural water balance values
ΔT	Time step
$\gamma_i(1)$	Weights for portions of economic costs, $i \in \{1,\ldots,4\}$
$\gamma_i(2)$	Weights for portions of ecologic costs, $i \in \{1,\ldots,5\}$
λ_j^K	Flow diversion factor of flow K for arc j

INTRODUCTION

Background

In industrialised countries complex water infrastructure systems have developed over decades and centuries. In the present discussion of sustainability they are exposed to new challenges. Especially, problems resulting from demographic and climate change can cause a conceptual alteration in urban water management. If resource-orientated sanitation techniques are to replace or complement existing centralised systems, then extensive reconstruction work becomes necessary. This system change is superposed by a high demand for rehabilitation in existing water supply and sewer networks. Currently, changes or rehabilitation measures are carried out rather as a result of locally limited and 'spontaneous' decisions.

The methodical and optimal procedure of an extensive or complete conversion is a challenging question. Such a retrofitting can only be realised successively over a long period and requires new strategies for transition. At all stages of a transition reliable water supply and disposal of wastewater have to be guaranteed. To ensure that every step of reconstruction ecologically and economically benefits the future, an optimised strategy should be investigated. As these two aspects – financial efforts and impact on the environment – have to be minimised, the development of a mathematical optimisation model seems obvious.

State-of-the-art and aim of the study

The use of linear optimisation models (linear programs – LPs) in urban water management history goes back almost 50 years. Since the 1960s operations research methods have been used for planning alternatives in regional wastewater systems (listed for example in de Melo & Câmara (1994)). In recent applications in wastewater disposal, LPs are used to find cost-efficient combinations of wastewater systems under pollution load constraints (Refsgaard 2001), to evaluate the sustainability of decentralised sanitation and reuse (DESAR) concepts (van der Vleuten-Balkema 2003) or to analyse the potential of wastewater reuse under physical and economic constraints (Chu *et al.* 2004). Furthermore, decision support approaches for the selection of sustainable urban drainage systems (SUDS) or DESAR concepts have been investigated in a number of studies including assessment methods for comparing different options of sanitation (e.g. Icke *et al.* 1999; Niederer *et al.* 2007; Thévenot 2008; Pearson *et al.* 2010). Only a few decision support approaches exist, which comprise technical option selection for integrated urban water management (Mitchel & Diaper 2006; Makropoulos *et al.* 2008). Most published decision support approaches are used to choose an appropriate future system solution. Research on a complete transformation of water infrastructure systems seems to be rather rare. Most studies (e.g. Kluge & Libbe 2006; Duffy & Jeffries 2011; Schiller 2010) focus on organisational and financial aspects, whereas the technical part is not investigated profoundly.

Therefore this paper will present the implementation of a mathematical tool to find an optimised strategy for the realisation of resource-orientated drainage and sanitation concepts in existing urban areas. Two components of mathematical optimisation are of interest for the mentioned question – project scheduling problems (PSP) and network-flow problems (NFP). The aim of project planning is to schedule time- and resource-restricted activities in such a way that an objective function dependent on starting times of the activities is optimised. In NFP a flow function is implemented on a network of nodes and arcs. A flow-dependent objective function has to be optimised regarding flow conservation and capacity constraints. A simultaneous-constrained consideration of both problems has not been worked on so far. However, for a realistic modelling of long-ranging retrofitting and construction works in urban water management systems, this would be necessary. The presented approach will enable the evaluation and minimisation of time-dependent impacts not only on financial costs but also on the operability of systems and on the environment.

METHODOLOGY

Development and formulation of the model

The mathematical model is formulated as a bi-criteria mixed-integer program with two objective functions (financial effort and environmental impact). For the first time a simultaneous project scheduling and network flow problem, *ProNet*, has been defined. Details about the model structure, a complex arc-node-network, are given in Kaufmann *et al.* (2007). Within this network, nodes and arcs have to be valid for PSP and NFP simultaneously. Nodes in the activity network of the PSP represent construction activities (implementation of new devices, rehabilitations and adaptations). In the flow network of the NFP nodes represent at the one hand elements of urban settlements (surface and wastewater types) and on the other hand elements of water infrastructure (e.g. network nodes, storage or treatment facilities or sinks and sources such as ground, receiving waters or air). The arcs connect the nodes in direction of 'flows'. Arcs are weighted, i.e. they are associated with some values: costs (bk_{jp}, uk_{jp}) and lifespans or installation periods (d_{jp}, nd_{IST}, nd_{jp}, nds_{jp}) for the activity network and capacities and diversion factors for the different flows in the flow network (l^K_{jp}, u^K_{jp}, λ^K_j). The defined flow variables X^K_{jt} of the model are: maximum design rainfall (X^R_{jt}), mean dry weather flow (X^T_{jt}) with its components, runoff contribution area (X^A_{jt}) and pollution loads in dry weather flow (X^{FT}_{jt}) and stormwater runoff (X^{FR}_{jt}). These flows are 'led' through the network, which is expanded as new elements are built and reduced if elements are shut down respectively.

In the PSP, specifications for starting times of activities, construction periods, useful life spans as well as budget and resource restrictions are formulated in linear (un)equations.

The 'building variable' B_{jpt} equals 1 when a device is implemented; variable S_{jpt} (rehabilitation or reinvestment of devices) has to become 1 when the useful lifespan is over. The indices mark the number of the respective arc (j), the mode in which an arc is activated (p) and the time (t). (The mode represents different outcomes of characteristics of an arc, e.g. different plant sizes or different diameters of sewers.) With these variables, main constraints for the PSP are:

- Compliance with resource restrictions

$$\sum_{j\in\mathcal{A}}\sum_{p\in M_j}\sum_{k=\max\{0,t-d_{jp}+1\}}^{t} r_{jpi}B_{jpk} \leq ra_{it} \quad \forall i\in\mathcal{RK},\ t\in\mathcal{T} \tag{1}$$

- Compliance with budget restrictions

$$\sum_{j\in\mathcal{A}}\sum_{p\in M_j} bk_{jp}B_{jpt} \leq bud_t \quad \forall t\in\mathcal{T} \tag{2}$$

- Construction of new devices, in a way they are finished in time for transformation T_B

$$B_{jpt}=0 \quad \forall j\in\mathcal{A},\ p\in M_j{:}d_{jp}>0,\quad t\in\mathcal{T}_B^{(j,p)}=\{T-d_{jp}+1,\dots,T\} \tag{3}$$

The combination to *ProNet* is accomplished by the implementation of a variable Y_{jpt}, which marks the date, onward from which a new element can be used after completion. The main constraints for this variable are:

- Activation of arcs after installation period until time T (end of period under consideration)

$$Y_{jpt}=\sum_{r=0}^{t-d_{jp}} B_{jpr} \quad \forall(j,p)\in NI,\quad t\in\mathcal{T}_{BY}^{(j,p)}=\{d_{jp},\dots T\} \tag{4}$$

- Reinvestment/rehabilitation at the end of useful lifespan

$$\sum_{r=0}^{t} Y_{jpr} \geq nd_{jp}\sum_{r=0}^{t} S_{jpr} \quad \forall j\in\mathcal{A},\ p\in M_j,\ t\in\mathcal{T} \tag{5}$$

$$\sum_{r=0}^{t} Y_{jpr}-nd_{jp} \leq \sum_{r=0}^{t} nds_{jp}S_{jpr} \quad \forall j\in\mathcal{A},\ p\in M_j,\ t\in\mathcal{T} \tag{6}$$

$$\sum_{t\in\mathcal{T}} Y_{jpt} \leq nd_{jp}\sum_{t\in\mathcal{T}} B_{jpt}+\sum_{t\in\mathcal{T}} nds_{jp}S_{jpt} \quad \forall j\in\mathcal{A},\ p\in M_j \tag{7}$$

For the NFP, flow conservation and flow diversion constraints were formulated. These constraints guarantee the feasibility of the systems and allow the evaluation of the environmental impact. By this, also the suitable size and associated installation costs for measures are ensured (mode p of an arc). The main sets of constraints are:

- Flow conservation constraints for every node

$$\sum_{j\in T_i} X_{jt}^K-\sum_{j\in H_i} X_{jt}^K=b_i^k \ \forall K\in\mathcal{F},\ i\in\mathcal{V},\ t\in\mathcal{T} \tag{8}$$

- Flow diversion constraints (e.g. infiltration rates, runoff coefficients, removal efficiencies)

$$X_{jt}^k \leq \lambda_j^k\left(b_i^K+\sum_{e\in H_i} X_{et}^k\right) \quad \forall K\in\mathcal{F},\ i\in\mathcal{V},\ j\in T_i,\ t\in\mathcal{T} \tag{9}$$

- Activation of arcs in adequate mode p (variable capacity boundaries)

$$bf_j\left(X_{jt}^R+X_{jt}^T\right) \leq \sum_{p\in M_j} u_{jp}^{RT}Y_{jpt} \quad \forall_j\in\mathcal{A},\ t\in\mathcal{T} \tag{10}$$

$$X_{jt}^K \leq \sum_{p\in M_j} u_{jp}^K Y_{jpt} \ \forall K\in\{T,A,RT\},\ j\in\mathcal{A},\ t\in\mathcal{T} \tag{11}$$

$$X_{jt}^K \geq \sum_{p\in M_j} l_{jp}^K Y_{jpt} \quad \forall K\in\mathcal{F}=\{R,T,A,FR,FT\},\ j\in\mathcal{A},\ t\in\mathcal{T} \tag{12}$$

An optimal solution represents a set of variables in every time step, which fulfils constraints and minimises objective functions. The objective functions economic costs $K(1)$ and ecologic costs $K(2)$ have to be minimised concurrently for the period of consideration. Ecologic costs as present values with an interest rate Z consist of:

- Investment and reinvestment costs

$$K_1(1) := \sum_{t\in\mathcal{T}} \frac{1}{(1+Z)^t} \left(\sum_{j\in\mathcal{A}} \sum_{p\in M_j} bk_{jp} B_{jpt} + sk_{jp} S_{jpt} \right) \quad (13)$$

- (Time-dependent) operational costs

$$K_2(1) := \sum_{t\in\mathcal{T}} \frac{1}{(1+Z)^t} \left(\sum_{j\in\mathcal{A}} \sum_{p\in M_j} uk_{jp} Y_{jpt} + zf_j\, uk_{jp} \left(\sum_{r=0}^{t} Y_{jpr} - nd_{jp} \sum_{r=0}^{t} S_{jpr} \right) \right) \quad (14)$$

- Additional operational costs by reduction of minimal flow velocity

$$K_3(1) := \sum_{t\in\mathcal{T}} \frac{1}{(1+Z)^t} \left(\sum_{j\in MV} \sum_{p\in M_j} spk_{jp} LB_{jpt} \right) \quad (15)$$

- Flow-dependent monetary costs (e.g. sewage tax)

$$K_4(1) := \sum_{K\in\mathcal{F}} \sum_{t\in\mathcal{T}} \frac{1}{(1+Z)^t} \sum_{j\in\mathcal{A}} mk_j^k X_{jt}^k \quad (16)$$

and result in:

$$K(1) := \sum_{i=1}^{4} \gamma_i(1) K_i(1) \quad (17)$$

Ecologic costs $K(2)$ are calculated as sum of:

- Costs for distance from goal values
  for adaption of a natural water balance (infiltration, evaporation and runoff rate, rate of rainwater use)

$$K_1(2) := \sum_{w\in Zw} \sum_{t\in\mathcal{T}} \left| Z_{wh} - \frac{100}{g^A} X_{wh}^A \right| \quad (18)$$

for adaption of favoured resources protection values (rates of direct reduction of potable water demand, nutrient and energy recovery and greywater recycling)

$$K_2(2) := \sum_{h\in Zr} \sum_{t\in\mathcal{T}} \left| Z_{rs} - \frac{100}{g^T} X_{rs}^T \right| \quad (19)$$

- Costs for emissions in receiving waters

$$K_3(2) := \sum_{g\in Zgew} \sum_{t\in\mathcal{T}} \left(X_{gew}^{FR} + X_{gew}^{FT} \right) \quad (20)$$

- Costs for immissions

$$K_4(2) := \sum_{g\in Zgew} \sum_{t\in\mathcal{T}} \left(X_{gew}^{R} + X_{gew}^{T} \right) \quad (21)$$

- (Fictitious) flow-dependent costs

$$K_5(2) := \sum_{K\in\mathcal{F}} \sum_{t\in\mathcal{T}} \sum_{j\in\mathcal{A}} fk_j^K X_{jt}^K \quad (22)$$

and result in:

$$K(2) := \sum_{i=1}^{5} \gamma_i(2) K_i(2) \quad (23)$$

The different ecologic cost portions can directly be accounted as differences or sums as formulated in Equations (18)–(21) or can be scaled to an interval [0,1], as shown for two components in Figure 1, wherein 1 represents the highest detriment (can be chosen to characterise present state) and 0 being the matching of a target value (no detriment).

The solution of the LP represents an optimal strategy towards the future state, which could not be enhanced in both criteria (economic and ecologic costs). Generally, not only one solution of the optimisation problem exists but numerous reasonable Pareto-optimal solutions (see for example Ehrgott (2005)). Pareto solutions are those for which improvement in one objective can only occur with the worsening of the other objective. Therefore, a 'compromise' has to be found.

Procedure of model application

The implementation of the developed program follows the flowchart presented in Figure 2.

Data analysis and preparation

For the implementation of the mathematical model for real catchment areas the main work is data analysis and preparation. An important step is to subdivide the entire

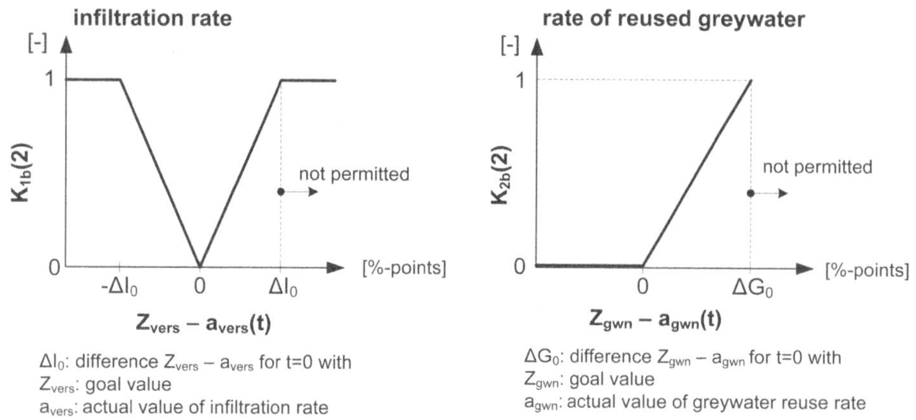

Figure 1 | Examples of transformation functions for criteria of ecologic costs (distance from natural infiltration rate (left) and rate of reused greywater (right)).

catchment in subcatchments since the scale of single houses or lots would be too detailed. This step is dependent on numerous factors in the fields of subcatchment data, pollution data, topography parameters as well as sewer network and receiving water data. For these subcatchments then, appropriate activities for SUDS, DESAR and drainage of remaining flows can be prepared. These feasible, necessary or infeasible activities are arranged in a specific catalogue of activities. Here, for each catchment's characteristics the expedient installation and operational costs for the measures – prepared for different possible sizes of the device depending on connected surfaces or dry weather flow components (cf. mode p) – are stored.

Beside the generation of this catalogue a simplification of existing sewer networks is advisable. For the resulting 'computational network', capacity and flow velocity constraints are tested. At the same time possible new transport elements such as foul sewers, vacuum pipes, stormwater sewers or surface drains have to be prepared with possible flow paths and discharge points. Within, all possible dimensions are provided with allocated capacities and costs.

In a detailed pre-processing routine the combination of parameters, activities, costs, connections and interrelations to the input formats for the mathematical model takes place.

Mathematical modelling

Subsequent to data preparation the model *ProNet* can be applied. Indeed, many of the implicated constraints or evaluations are dependent on the specific catchment. Goal values for a natural evaporation rate or infiltration rate and initial-state parameters are different for each application. Mean values for runoff coefficients or infiltration rates for surface types are dependent on precipitation characteristics and should be evaluated for the specific case if possible. Optimisation parameters have to be chosen and weighted.

To solve the multi-criteria problem two methods for constructing a single aggregate objective function are implemented in the optimisation tool: the weighted sum method and the ε-constraint method. At the first method, for the two objective functions, economic cost $K(1)$ and ecologic cost $K(2)$, scalar weights are specified and the sum of both has to be minimised. At the second method one objective function is limited to a specific value and the second objective function has to be minimised without exceeding the limit of the other objective. The problem itself is solved with the software ILOG CPLEX® 9.0 and 11.0 (ILOG 2005). The mathematical programming engine CPLEX® uses Branch-and-Cut-methods for solving mixed-integer linear programs.

Interpretation

In the solution of the mathematical optimisation the results for all elements of the mathematical network are documented for every calculated time step. For larger catchment areas with numerous subcatchments, this number of representations becomes too complex to review. Moreover, in different applications, different interpretations of the result are of interest: in some cases only objective functions' values are important, in other cases specific points of the drainage system are of interest. Therefore in a substantial post-processing the desired representations of results are automatically generated.

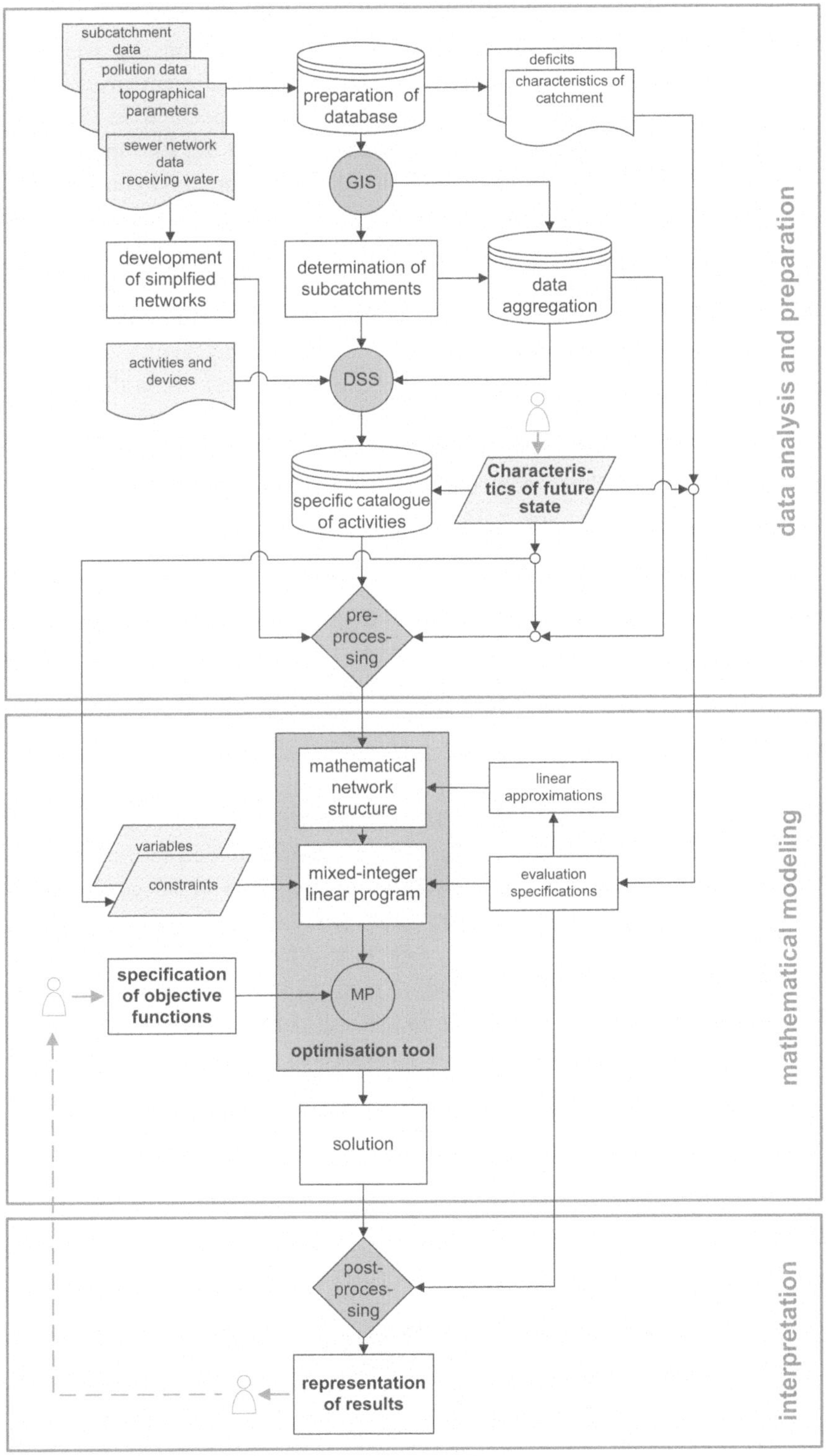

Decision maker's involvement
GIS Geographic Information System
DSS Decision Support System
MP Mathematical Programming Engine, here ILOG CPLEX®

Figure 2 | Flowchart for model application.

APPLICATION IN A CASE STUDY

Investigated catchment

The model has been applied to a suburb of Kaiserslautern in Germany, a rural catchment of about 3,000 inhabitants. Conditions of this town are representative for numerous rural structured communities in Germany. The entire catchment has a drainage area of about 90 ha and implies 35 ha of paved area. About 70% is drained by combined sewer systems whereas newer developed areas are drained by separate sewer systems. Two combined sewer overflow devices and one final sewer overflow tank are installed in the sewer system. Twenty hectares of the catchment is a business park, the effluent of which shows the characteristics of domestic wastewater. Dry weather flow amounts to 11.4 L/s and consists of 8.2 L/s foul sewage and 3.2 L/s infiltration water. The pollution of chemical oxygen demand (COD) in dry weather flow is 560 mg O_2/L. Wastewater is led to the central wastewater treatment plant (WWTP) of Kaiserslautern. For this study the catchment area was subdivided into six subcatchments according to numerous parameters such as land use, topography, subsurface conditions or population density.

Future state and boundary conditions

The approach presented here allows finding an optimised *strategy* to reach a favoured future state. The future state itself with the general decision for a sanitation concept is known. Moreover, goal values for a more natural water balance or a resource protection have to be defined. For this paper the chosen specifications for a possible sustainable future state are summarised in Table 1.

In this example different weights of $K(1)$ and $K(2)$ in objective function are examined by using the weighted sum method. Table 2 shows the six different investigated scenarios to reach the future state specified in Table 1. Among the two extreme cases S0 and S5, four meaningful combinations are calculated. For example, S0 characterises an optimal strategy for reaching the future state with minimal economic costs, regardless of environmental impact. In S5 reverse preferences in reaching the same future state are investigated: the strategy should have minimal ecologic impact while economic costs are non-relevant.

Another important definition is the criteria that should contribute to the objective function of ecologic costs $K(2)$. So far, 11 different criteria in four fields are implemented in the model (cf. $K_1(2)$ – $K_4(2)$ in Equations (18)–(21)). Depending on specific conditions they have to be chosen in a balanced combination appropriate for the application case. For this example, the three fields of adapting a more natural water balance $K_1(2)$, resources protection $K_2(2)$ and emission $K_3(2)$ were chosen and weighted equally with 1/3. Every criteria $K_i(2)$ of $K(2)$ is scaled to an interval from 0 to 1 (see Figure 1). The ecologic costs per time-step will then result as the weighted sum of the chosen criteria, which will sum up to a maximum of 1.

In this example a total period of consideration of 80 years is chosen. Within, the transformation to the favoured future state is calculated for a period of 50 years, and a period of 30 years afterwards is investigated to evaluate the 'maintenance costs' of the future state. The economic costs are calculated as total project costs with an interest rate of 3% relating to starting point of consideration $t = 0$. This interest rate is recommended for cost calculations in water management by LAWA (2005), where also neglecting of inflation is proposed. In addition, a budget of €2.5 million € per time step ($\Delta T = 5$ years) is given.

Table 1 | Specifications for favoured future state

Overall specifications	Stormwater	Stormwater runoff and wastewater should not be mixed any more; near-nature and source-controlled management	
	Greywater	Should completely be treated centrally at WWTP	
	Blackwater	Completely decentralised treatment	
Specific goal values	Evaporation rate	As a percentage of mean precipitation height	55%
	Infiltration rate		30%
	Rate of rainwater utilisation		10%
	Rate of greywater recycling[a]	As a percentage of mean domestic wastewater flow	0%
	Rate of fertiliser and energy recovery		20%
	Rate of direct reduction of use of potable water		40%

[a]Not chosen for objective function as greywater should be treated centrally at WWTP

Table 2 | Specifications for objective functions in the different scenarios

	Weights in weighted sum method		
	Economic costs $K(1)$	:	Ecologic costs $K(2)$
Scenario S0	1	:	0
Scenario S1	1	:	0.2
Scenario S2	1	:	0.4
Scenario S3	1	:	1
Scenario S4	1	:	2
Scenario S5	0	:	1

RESULTS AND DISCUSSION

Objective functions' values

The economic and ecologic costs for the different transformation scenarios S0 to S5 are shown in Figure 3, where every scenario is characterised by a point in the $K(1)$–$K(2)$ graph. With increasing weight on ecologic costs $K(2)$ for the optimal strategy of transition, the economic costs will rise. The conditions of this specific case result in €32 million as minimum economic cost $K(1)$ to achieve and maintain the future state (scenario S0). This high value result is because numerous measures are required anyway to reach the future state. If only the ecologic costs $K(2)$ contribute to the objective function (scenario S5), a minimum of 7.2 results for these costs in the calculated 80 years. As sum of $K(2)$, a maximum of 16 (for 16 time steps in 80 years) could result, if no new elements were implemented. The environmental impact thus can be reduced to a great extent in optimal strategies.

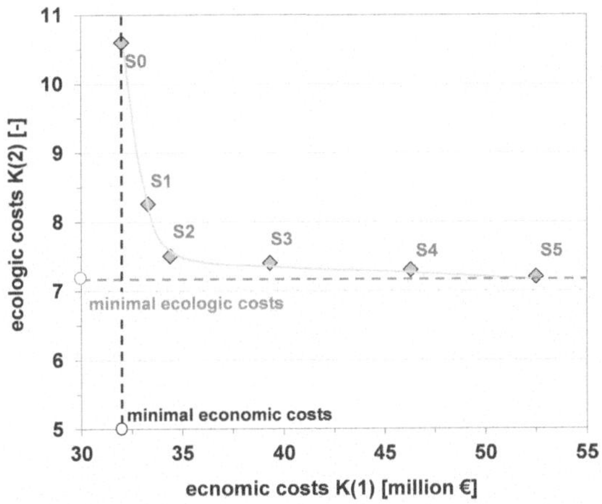

Figure 3 | Results for economic costs $K(1)$ and ecologic costs $K(2)$ for the six scenarios.

A comparison of the strategies resulting in S0 and S2 shows, obviously, that with few additional financial efforts environmental impacts can be reduced efficiently. On the other hand it can be seen that higher weights than 1 for $K(2)$ (S3, S4 and S5) allow only a marginal reduction of ecologic costs with high financial effort. A more detailed comparison of scenario S2 and S5, differing by only 0.3 in ecologic costs but by 50% in economic costs, is done in the following.

Figure 4 allows a closer look at those aspects of the ecologic costs $K_i(2)$ that are chosen to contribute to the objective function. For the total period of consideration, in every axis of the radar chart the sum of costs for each criterion is shown for the both scenarios S2 and S5 in comparison to S0. That means for example that the 'costs' for adaptation to a near-nature infiltration rate in scenario S2 sum up to 0.6. Although the criteria $K_i(2)$ are weighted equally they are not minimised in the same way in the different scenarios. The difference in the sum of $K(2)$ in scenario S2 and S5 results in a lower impact on the water balance criteria in S5, lower emissions in S2 and nearly the same values for resources protection for both scenarios. So in sum, S5 produces lower ecologic costs when the impacts on criterion 'rainwater utilisation' are lower. The influence of this criterion is very high, as a high rate of rainwater should be used for substitution of potable water and these installations are more expensive than other natural rainwater management devices. In fact, the reduction of costs for this criterion can only be achieved by worsening other criteria, e.g. costs for emissions.

In Figure 5 it is shown that the results for $K_i(2)$ are dependent on the definition of ecologic costs, where criteria result from the evaluation of the distance from favoured goal values. That means the actual value of, for example, infiltration rate is compared to the more natural goal value of infiltrated percentage of rainwater. Figure 5 illustrates the minimisation of the three criteria for water balance. The values in the temporal sequence are shown for the discussed scenarios S2 and S5.

The corresponding costs for the implementation of devices in the field of SUDS, which will cause changes in the water balance, are shown as an area diagram in Figure 5. The costs are calculated at the beginning of implementation of devices. Due to installation periods and the used time step of 5 years, the effects can appear later. In S5 the three distances of water balance components are reduced more efficiently and right from the beginning. The early implementation of a bigger part of SUDS devices and the higher number of devices cause economic costs to be €7.5 million

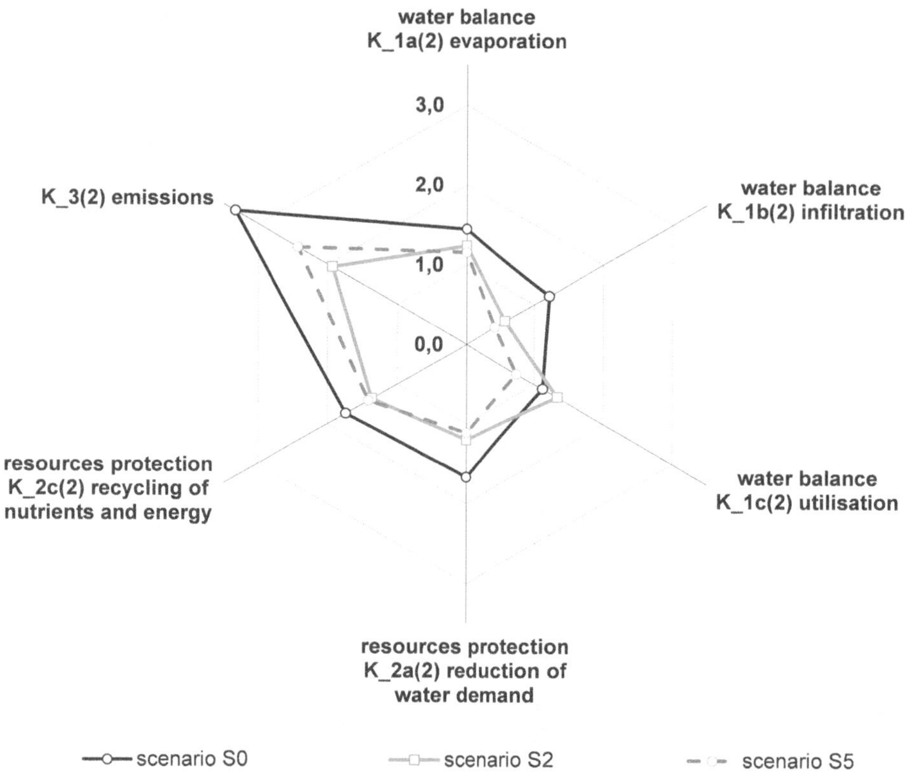

Figure 4 | Ecologic costs $K_i(2)$ [–] for the scenarios S0, S2 and S5 in a radar chart.

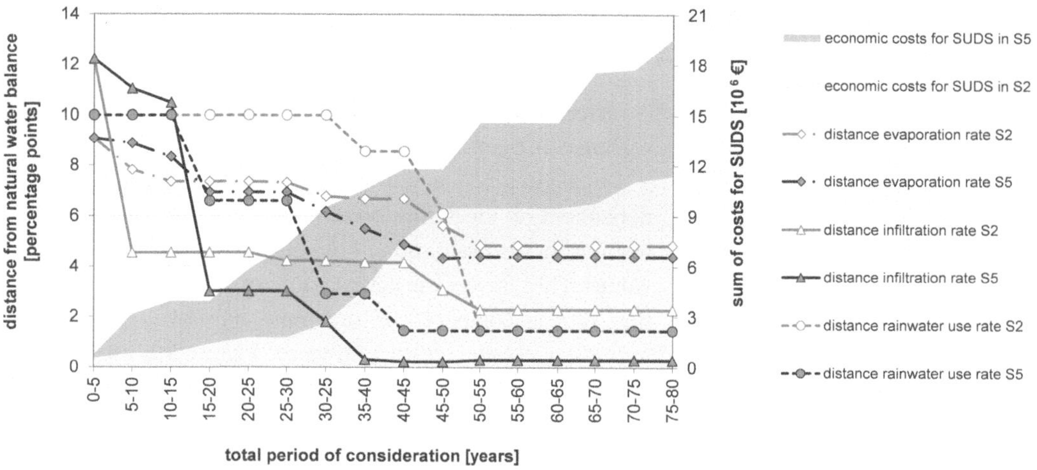

Figure 5 | Criteria for adaption of a natural water balance for total period of consideration.

higher. If devices would have been implemented later, as done in S2, they would have caused lower costs $K(1)$ due to the interest rate. However, the distance from goal values then stays high for a longer period producing higher total costs $K(2)$. Even in S5, with highest weight on ecologic costs, the goal values are not reached exactly. On the one hand, the regeneration of a natural water balance in existing developments is difficult and on the other hand in the shown scenarios economic costs are restricted by a budget.

Decision on a solution

The choice of an appropriate solution out of the Pareto-optimal solutions is difficult. Recommendations for the definition of optimisation criteria result from the respective objective of a system transformation (Kaufmann Alves 2012). So if new devices are to be implemented in a rather short time, then ecologic costs should have a high weight. This can be the case if, for example, receiving waters are very weak, the existing sewer system is surcharged or other regulations are not fulfilled. The discussion with local deciders will lead to the definite choice of an optimal strategy for system transformation. In this application, S2 could be a good preference as, with rather moderate costs, a considerable reduction of environmental impact during the transformation process is possible. It has to be kept in mind that other criteria and weights $\lambda_i(2)$ for components of the ecologic costs $K_i(2)$ will have caused other strategies.

For decision-making processes many more result representations can be considered. At every point of interest, e.g. resulting pollution loads, changes in dry weather flow components connected to centralised networks or quantity and quality of WWTP inflow can be illustrated.

The most important result, if having decided for one strategy, will be a construction schedule, where construction and rehabilitation periods for the implemented devices are demonstrated in temporal and spatial sequence for every subcatchment. Figure 6 shows a small section for one subcatchment for scenario S2. In this solution, many devices of SUDS are implemented. The former combined sewer system is used to drain remaining stormwater runoff. Long installation periods for implementation of decentralised blackwater treatment are due to the fact that here all components, house installations as well as biogas plant and connections are included.

Discussion

To sum up these results, it can be seen that different optimal solutions are possible to reach one future state. The introduced approach can support local decision makers by making it possible to show all impacts in detail when calculating different scenarios.

The data analysis and preparation to formulate the mathematical network is very expensive. But for complex systems an optimal solution for transformation processes cannot be found manually if, besides the financial costs, aspects of the urban water and nutrient cycle have to be taken into account. Even the investigation of small settlements results in a difficult optimisation problem, as many interactions in water systems are modelled. Thus, simplifications are necessary. However, the developed optimisation tool produces feasible solutions for a succession of retrofitting measures, which meet functional and flow constraints. The objective functions represent the financial frame and quality of adaptation to target values in an adequate way. A problem is to specify the weights of the two introduced costs. Different dimensions and scales make it difficult to choose a predefined value. Also, diverse characteristics of ecological criteria complicate choice and weighting of cost portions $K_i(2)$. In the result, a lower value of $K_i(2)$ can imply on the one hand a better adaptation to goal values and a lower emission rate respectively. On the other hand, it can stand for a 'faster' adaptation with the same final values in a future state, as the sum of $K(2)$ has to be minimised. The definition of ecologic costs for transformation phases thus remains difficult. Nevertheless, new knowledge about time-dependent demonstration of environmental impact and effects on infrastructures is of great value.

Varied strategies for transformation (different weights, different optimisation criteria) do not differ much in choice of implemented devices, as with a given future state (e.g. complete decentralised treatment of blackwater) many retrofitting steps are necessary anyway. In fact, the dates of implementation during the transformation period and the plant sizes, especially for sustainable drainage elements, are the significant differences when choosing different optimisation criteria. With high weight on ecologic costs, effective measures are implemented at the beginning

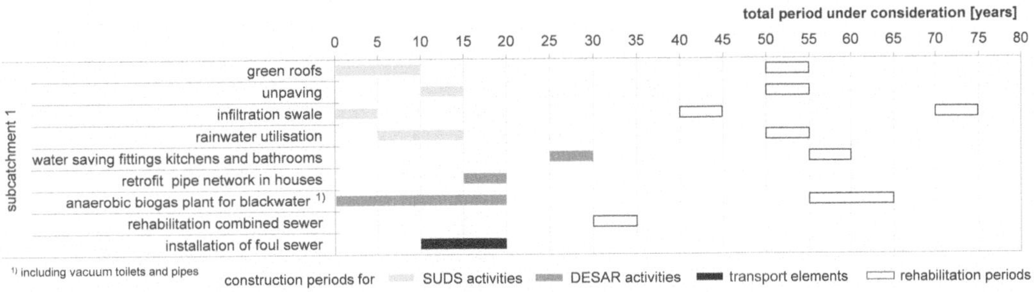

Figure 6 | Construction schedule for a subcatchment (27 ha, 1,200 inhabitants) for scenario S2.

of the transformation period even though investment costs are high. Strategies that should result in low financial efforts (high weight on economic costs) initially show an implementation of measures with low expense.

The calculated results cannot be verified in reality at the moment, as so far an extensive transformation of larger settlements has not been undertaken. The trend of conclusions about transformations is in agreement with the literature: the further use of infrastructures – here sewers and WWTP – is considered as important in system transformations (e.g. Hiessl *et al.* 2003; Kluge & Libbe 2006; Hillenbrand 2009; Meinzinger 2010). Starting points of transformations are seen in short-term realisable measures with low technical and financial effort. Later areas with higher transformation effort could follow. Such strategies result in the model when a high weight is put on economic costs $K(1)$.

At a transformation of urban water systems, a bigger part of retrofitting has to be done in private households. This could make it difficult to determine spatial and temporal distribution of measures in terms of optimal strategies. Communities and public policy-makers must ensure an aim-orientated way towards sustainable future systems (e.g. promotion and support programmes).

CONCLUSIONS

An extensive integration of new systems for drainage and sanitation in existing infrastructures requires, besides a suitable choice of techniques, a strategy for implementation. For development of such strategies mathematical optimisation methods can be an adequate instrument. The approach presented in this paper showed that:

- the overall aim of the developed model, to find and optimise a long-range transformation strategy, could be reached in different implementations,
- solutions resulting from calculations with the model are Pareto-optimal transformation strategies, at which weighting of introduced costs has a high influence,
- changes in optimisation criteria generally lead to an earlier or later implementation of respective measures and to different plant sizes,
- recommendations for the definition of optimisation criteria result from the respective objective of a system transformation,
- the interaction with local deciders is important and will lead to the definite choice of an optimal strategy for application.

All in all, the developed model can be seen as a first step in a strategy for finding transformations in existing urban water systems. It cannot replace engineering design studies but gives an important indication for strategies for extensive transformations. Until now, several important data have not been included in the model, e.g. variable boundary conditions. So far the work has focused on technical aspects (functioning, dimensioning and fulfilment of regulations); in future work different planning and policy levels should be included. Nevertheless, it could be shown that an extension of existing assessment methods in terms of a detailed temporal and spatial evaluation of the economic and ecologic situation is essential. The exclusive planning and consideration of a favoured sustainable future state of urban water systems is inadequate for assessment of effects of extensive implementation of resource-orientated wastewater management.

REFERENCES

Chu, J., Chen, J., Wang, C. & Fu, P. 2004 Wastewater reuse potential analysis: implications for China's water resources management. *Water Res.* **38** (11), 2746–2756.

de Melo, J. J. & Câmara, A. S. 1994 Models for the optimization of regional wastewater treatment systems. *Eur. J. Oper. Res.* **73**, 1–16.

Duffy, A. & Jeffries, C. 2011 A new framework to guide urban water transitioning. *Proceedings 12th International Conference on Urban Drainage*, Porto Alegre, Brazil, 10–16 September 2011.

Ehrgott, M. 2005 *Multicriteria Optimization*. 2nd. ed. Springer, Berlin.

Hiessl, H., Toussaint, D., Becker, M., Dyrbusch, A., Geisler, S., Herbst, H. & Prager, J. 2003 *Alternativen der kommunalen Wasserver- und Abwasserentsorgung - AKWA 2100 (Alternatives of municipal water supply and wastewater disposal – AKWA 2100)*. Pysika-Verlag, Heidelberg.

Hillenbrand, T. 2009 *Analyse und Bewertung neuer urbaner Wasserinfrastruktursysteme (Analysis and Assessment of New Urban Water Infrastructures)*. Dissertation. Schriftenreihe SWW der Universität Karlsruhe, Vol. 134, Verlag Siedlungswasserwirtschaft, Karlsruhe.

Icke, J., von der Boomen, R. M. & Aalderink, R. H. 1999 A cost-sustainability analysis of urban water management. *Wat. Sci. Technol.* **39** (5), 211–218.

ILOG 2005 ILOG CPLEX 9.0 User's Manual, ILOG, S.A., San Francisco, USA.

Kaufmann, I., Meyer, T., Kalsch, M., Schmitt, T.G. & Hamacher, H.W. 2007 Implementation of sustainable sanitation in existing urban areas – long-term strategies for an optimised solution. *Wat. Sci. Technol.* **56** (5), 115–124.

Kaufmann Alves, I. 2012 *Strategieentwicklung zur Strategieentwicklung zur Integration ressourcenorientierter Abwasserbewirtschaftung durch mathematische Optimierung (Development of Strategies for the Realisation of Resource Orientated Drainage and Sanitation Concepts in Existing Systems by Mathematical Optimisation)*. Dissertation at the faculty of Civil Engineering, University of Kaiserslautern.

Kluge, T. & Libbe, J. (eds) 2006 *Transformation netzgebundener Infrastruktur – Strategien für Kommunen am Beispiel Wasser (Transformation of Network Infrastructures – Strategies for Municipalities Using the Example of Water)*. netWorks: Difu-Beiträge zur Stadtforschung, Vol. 45, Berlin.

LAWA (Länderarbeitsgemeinschaft Wasser) 2005 *Leitlinien zur Durchführung dynamischer Kostenvergleichsrechnungen (Guidelines for Dynamic Cost Comparison Methods)*. Kulturbuchverlag, Berlin.

Makropoulos, C.K., Natsis, K., Liu, S., Mittas, K. & Butler, D. 2008 Decision support for sustainable option selection in integrated urban water management. *Environ. Modelling Software* **23** (2008), 1448–1460.

Meinzinger, F. 2010 *Resource Efficiency of Urban Sanitation Systems: A Comparative Assessment Using Material and Energy Flow Analysis*. Dissertation, Hamburger Berichte zur Siedlungswasserwirtschaft, Vol. 75, Gesellschaft zur Förderung und Entwicklung der Umwelttechnologien an der Technischen Universität Hamburg-Harburg e.V, Hamburg, Germany.

Mitchel, V. G. & Diaper, C. 2006 Simulating the urban water and contaminant cycle. *Environ. Modelling Software* **21** (2006), 129–134.

Niederer, S., Kriz, K. & Pucher, A. 2007 Development of a spatial decision support system for sustainable water management. *Proceedings MEDA WATER International Conference*, Sustainable Water Management, Tunis, 21–24 March 2007.

Pearson, L. J., Coggan, A., Proctor, W. & Smith, T. F. 2010 A sustainable decision support framwork for urban water management. *Water Resources Manage.* **24** (2010), 363–376.

Refsgaard, K. 2001 Cost efficient strategies for handling of wastewater from households to recipient – a modelling approach. In: *Evaluation of Sustainability in Farm and Wastewater Systems – Decision Theoretical Applications*. Doctor Scientiarum thesis 2001, No. 5, Agricultural University of Norway, Ås, Norway. pp. 73–98.

Schiller, G. 2010 *Kostenbewertung der Anpassung zentraler Abwasserentsorgungssysteme bei Bevölkerungsrückgang (Cost Evaluation of Adaptations of Central Waste Water Disposal Systems due to a Decline of Population)*. IÖR Schriften, Vol. 51, Rhombus-Verlag, Berlin.

Thévenot, D. (ed.) 2008 *DayWater: an Adaptive Decision Support System for Urban Stormwater Management*. IWA Publishing, London.

van der Vleuten-Balkema, A. 2003 *Sustainable Wastewater Treatment – Developing a Methodology and Selecting Promising Systems*. Dissertation, TU Eindhoven, Eindhoven University Press, Eindhoven, The Netherlands.

First received 2 January 2012; accepted in revised form 30 October 2012

Theme IV

Water management indicators at different scales

Risk and monitoring based indicators of receiving water status: alternative or complementary elements in IWRM?

J. Völker, S. Richter, D. Borchardt and V. Mohaupt

ABSTRACT

The European Water Framework Directive (WFD) was enacted in the year 2000 with a stepwise approach. After legal implementation in the various member states large efforts were undertaken for the initial characterization of water bodies, risk assessment, to implement extensive monitoring schemes and to develop management plans at different aggregation levels by the year 2010. The initial characterization process and risk assessment had to be finalized by 2004 and delineated water bodies including a typological classification and identified the significant pressures and impacts in a screening procedure. In parallel, monitoring programmes and new biological indicator systems were developed in order to proof and refine the results of the risk assessment with an ecological indicator based assessment in a subsequent step which was finalized in 2009. Although the risk assessment for Germany was based on existing data that were originally collected for other purposes and came from a large variety of environmental or economical sectors, the results differ only slightly from the monitoring and indicator based information with respect to classifications of the 'ecological status' and 'chemical status'. From this result we conclude that a risk assessment based on a careful application and intelligent combination of existing data sources with proven quality allows the recognition of trends and the identification of priorities for action of measures already at an early stage of a management process. However, monitoring schemes and advanced sets of ecological indicators are essential in later management steps both for narrowing uncertainties remaining from the risk assessment and to allow for effect controls of implemented measures. Moreover, these monitoring indicators should differentiate the effects of multiple stressors more factor specific and with respect to ecosystem states and functions. In conclusion, we see risk and indicator based assessments as complementary elements in Integrated Water Resources Management (IWRM), which have to be linked in systematic and phased procedures.

Key words | ecological status of water bodies in Germany, indicators, river basin management strategies, Water Framework Directive

J. Völker (corresponding author)
S. Richter
D. Borchardt
Helmholtz Centre for Environmental Research – UFZ Department,
Department Aquatic Ecosystems Analysis and Management,
Brückstraße 3a,
39114 Magdeburg,
Germany
E-mail: *jeanette.voelker@ufz.de*

V. Mohaupt
Umweltbundesamt,
Wörlitzerplatz 1,
06844 Dessau-Rosslau,
Germany

THE WATER FRAMEWORK DIRECTIVE

The purpose of the European Water Framework Directive (WFD) (European Parliament and Council 2000) is to attain and protect an almost natural status for water bodies both for human usage and ecological functions. The main objective of the WFD is to achieve a 'good status' for all surface water bodies by 2015. This encompasses a 'good ecological status' and a 'good chemical status'. The achievement of these objectives is dependent on a shift in strategy from traditional water quality focused approaches towards ecologically orientated management. In many cases the existing ecological deficits in running waters are determined by point sources, but in addition to this a range of complex anthropogenic pressures exist, including diffuse pollution and morphological alterations.

The WFD lays down a clearly defined timeline (Figure 1). The first part of this included the characterization process, which was completed in 2004 and consisted of the first assessment of water bodies based on WFD requirements that had ever been carried out. Monitoring programmes were then established and implemented on the basis of the results obtained. Monitoring results and programmes of measures were documented in the first river basin

doi: 10.2166/wst.2013.526

Figure 1 | Implementation timeline for the Water Framework Directive (Richter & Völker 2010, modified).

management plans presented in late 2009. The first management period will be completed by 2015. Two more 6-year periods will follow. A status check is to be carried out at least once within these 6 years, so that comprehensive information about the status of water bodies will be available Europe-wide every 6 years.

Whereas existing data and methods were used to assess water body status during the risk assessment phase, several new assessment methods for biological quality elements and supporting elements have been developed to assess the status of surface water bodies for the monitoring programmes and the first river basin management plans. These methods have been designed to permit conclusions to be drawn regarding the importance of a multitude of different stressors, mainly loss of hydromorphology, the absence of river continuity and the issue of nutrient input from diffuse and point sources. These indicators will also be used to evaluate improvements by comparing the status of waters now and following the implementation of measures in the first, second and third management plan periods.

CHARACTERIZATION PROCESS AND RISK ASSESSMENT

Procedure and methods

The characterization of water bodies and the anthropogenic pressures to which water bodies are subjected represented one of the most important steps toward implementation of the WFD in the Member States. The Directive required that the effects of anthropogenic activity on the statuses of surface and ground water bodies have to be assessed. Once all testing has been completed and the relevant data have been processed, a decision is then made as to whether the environmental objectives of the WFD can be met without further measures.

The risk assessment revolves around the following three questions:

- Which water bodies have failed to satisfy the environmental WFD objectives?
- Which chemical and non-chemical pressures are responsible for the failure to meet these objectives?
- Which mechanisms and effects should be placed at the focus of operative monitoring?

The initial characterization process for surface waters consisted of three phases. In the characterization and typology definition phase, each water body was identified as belonging to a given water body category, and a decision was then made as to which organisms and substances would occur in each of these habitats in a state largely uninfluenced by anthropogenic activity. In the second phase, an assessment was made of the anthropogenic pressures to which the water body in question is currently being subjected and the impact of these pressures on the status of the water body (risk assessment). To this end, Annex II(1)(4) of the WFD requires that pressures from substances, water abstraction, water flow regulation and morphological alterations are to be identified and that their impact on water bodies is to be assessed. This process has been described in various Guidance Documents and the LAWA guidelines (CIS 2002a, b: LAWA 2003). The assessment of pressures and impacts was based on existing data as well as assessment methods and indicators that have been used in Germany for decades. Threshold criteria for several indicators were recommended by LAWA (2003) (Table 1). The results were used as a basis for determining whether a water body meets its objectives as defined by the WFD or whether additional measures are required. For the assessment, the worst-case scenario was applied, i.e. a water body was considered to be at risk or possibly at risk of failing the WFD objectives if it was not compliant with even one assessment criterion.

In the third step of the characterization process, economic issues were addressed such as cost and the efficiency of measures.

Table 1 | Assessment indicators and criteria recommended by LAWA (2003) to assess the impact of water body status within the characterization process of the WFD (Source: Borchardt *et al.* 2005).

Indicator	Criteria resulting in failure to meet the WFD objectives
Pressures on oxygen balance (saprobia)	Ecomorphological status lower than class II (LAWA 2000) for more than 70% of watercourse length. If more than 30% of the water body is affected, it is identified as being possibly at risk of failing to meet the objectives.
Nutrient pressures (trophic status)	Eutrophication worse than class II (LAWA 2000) or annual mean concentration of orthophosphate-P >= 0.2 mg/l and nitrate-N > =6.0 mg/l.
Specific pollutant levels	Exceed environmental quality standards.
Warming	Exceeds upper limit for temperatures or temperature increases as required under the Freshwater Fish Directive.
Salinization	Mean annual chloride concentration exceeds 200 mg/l.
Acidification	Acid status classification not compliant with LAWA (2003).
Abstractions	Abstractions exceed one-third of mean low-water flow, 50 l/s, or 10% of mean flow.
Flow regulation	Structures with the following properties: • High vertical barriers exceeding 30 cm in height that are characterized by smooth surfaces are devoid of structures that promote river continuity and aquatic wildlife movement. • Backwater scenario in which at least 20% of the water body is almost stationary at the mean water level.
Morphological alterations	Water body morphology, transverse structures, backwater or other key structural parameters have been designated as class 6 or 7 along more than 70% of the length of the water body (LAWA 2000). If more than 30% of the length of the water body falls into this category, it is classified as possibly at risk of failing to meet the objectives.

Results

In the interests of obtaining a clear picture of the extent to which a given water body is at risk of failing to fulfil 'good status' requirements, Germany's surface water bodies were assigned to the following categories:

- Water bodies that are likely to achieve good status (not at risk of failing to meet the WFD objectives).
- Water bodies that may fail to achieve good status (possibly at risk of failing to meet the WFD objectives). This assessment is often attributable to a lack of sufficient data.
- Water bodies that are unlikely to achieve good status (at risk of failing to meet the WFD objectives).

According to the initial characterization undertaken in Germany in 2004, approximately 14% of the water bodies assessed are likely to meet the WFD objectives. 26% of the water bodies assessed are possibly at risk and approximately 60% of the water bodies assessed are at risk of failing to meet the WFD objectives.

Failure to meet the Directive objectives was generally attributable to physical alterations affecting the hydrology and/or geomorphology of a water body, as well as transverse structures such as weirs and sills that impede the upstream migration of fish and smaller aquatic organisms. Other significant causal factors were found to be nutrient input from diffuse sources, mainly agricultural activities, as well as other chemical pressures such as those resulting from wastewater treatment plants and precipitation drainage. The results of the characterization also showed that in most cases the (potential) failure of a water body to achieve good status is attributable to several factors acting simultaneously. Thus, for example, dam-regulated watercourses that are also subject to nutrient pressures regularly develop algal growth.

MONITORING BASED ASSESSMENT OF WATER BODY STATUS

Ecological status

Procedure and methods

Status assessments for water bodies in Germany are based on monitoring programmes. They have been designed and are being carried out on the basis of the results of the risk assessment in 2004. The monitoring programmes were elaborated for surface waters, groundwater and water-dependent protected areas. In Germany more than 8,500 surface monitoring sites were established (90% of which

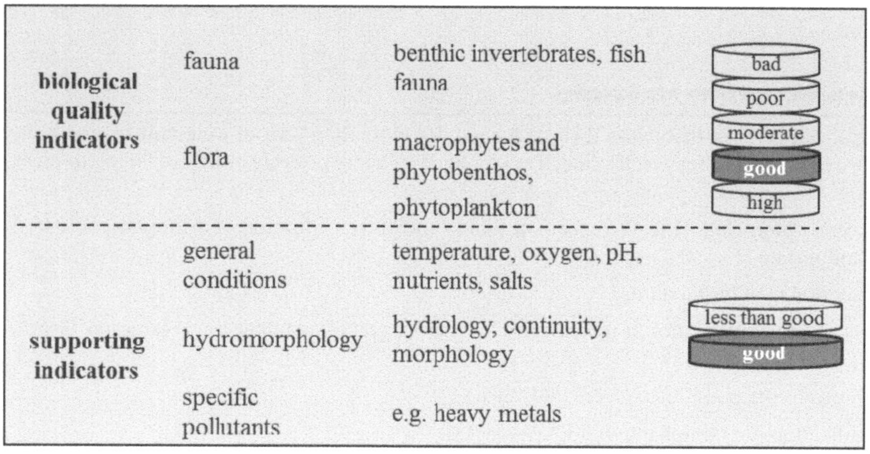

Figure 2 | Indicators and classification for assessment of the ecological status of water bodies according to the Water Framework Directive.

are located in rivers). Nine thousand sites were chosen to monitor groundwater bodies.

For surface water bodies, the ecological status is to be assessed in accordance with the biological elements fish fauna, benthic invertebrates and aquatic flora. Different assessment methods are used for each biological group (see e.g. Hering *et al.* 2003; AQEM Consortium 2004; Rolauffs *et al.* 2004; Meilinger *et al.* 2005). The assessment methods for the biological quality elements are based mainly on species taxonomic composition, and are adapted to reference conditions. Species composition and abundance under reference conditions are defined in accordance with the typology of rivers and lakes. For the typology of surface waters within Germany, this involved the definition and identification of reference conditions for a total of 24 different river types and 14 types of lakes.

The ecological status of water bodies in respect of biological quality elements is classified using five classes: 1: high; 2: good; 3: moderate; 4: poor; and 5: bad. From class 3 up to class 5 measures need to be taken to reach the objectives according to the WFD. The following aspects need to be assessed in conjunction with the biological elements: river basin specific pollutants, physico-chemical quality elements (e.g. temperature, oxygen, pH value and nutrient concentrations as descriptors of the general conditions) as well as hydromorphological features. These supporting elements are classified into 'good' and 'less than good' (Figure 2). The determining factor for the overall assessment is the worst assessment within a given group ('worst-case' or 'one out, all out' approach). The requirements for achieving 'good ecological status' are: (i) all biological elements must be rated 'good'; (ii) environmental quality standards (defined concentrations) of river basin specific pollutants must not be exceeded ('good'); and (iii) the results for the general conditions as well as the hydromorphology must fall within a range that allows for good ecosystematical functionality ('good').

Results

Of all the surface water bodies, 10% are classified as having a 'high' or 'good' ecological status at present. 90% of surface water bodies fall within the classes 'moderate' (30%), 'poor' (34%) and 'bad' (23%). A small percentage of surface water bodies (3%) has not yet been assessed and is thus classified as 'uncertain' (Figure 3).

In Germany, the main reasons for failure to achieve the WFD objectives in respect of ecological status are changes in hydromorphology including disrupted river continuity and high nutrient input introduced from wastewater treatment plants and agriculture.

Chemical status

For the assessment of the chemical status of surface water bodies, EU-quality standards for 33 priority substances in

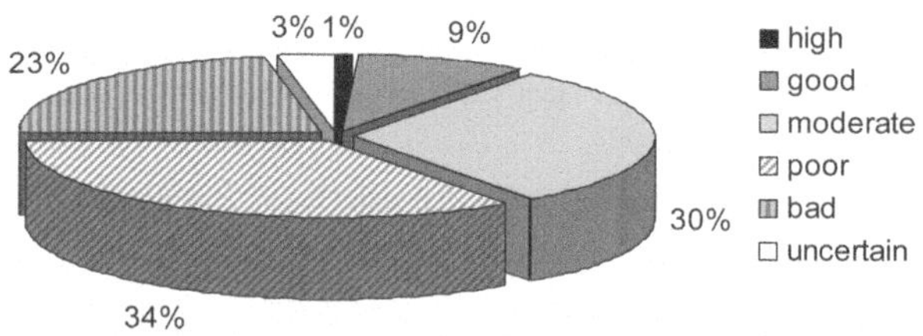

Figure 3 | Ecological status of surface water bodies in Germany. Data source: Portal WasserBLIcK (2010).

accordance with Annex X of WFD have to be taken into account. The list of priority substances includes pollutants that fall within the scope of Directive 2006/11/EC (European Parliament and Council 2006) as well as nitrates in accordance with Directive 91/676/EEC.

Of the 33 priority substances listed in Annex X of the WFD, 13 are classified as priority hazardous substances. They are subdivided into the groups heavy metals, pesticides, industrial pollutants and other pollutants. Ensuring that water bodies remain free of hazardous substances is one of the key goals of the European water protection effort.

The chemical status requirements were modified in the new subsidiary Directive 2008/105/EC that came into force in July 2010. Some states have already implemented the Directive. A comparison of the requirements of the new and old Directives shows that implementation of the former translates into significantly fewer water bodies with a 'good chemical status'. For example, under the old Directive 9% of all assessed water bodies in the Elbe river basin fail to achieve 'good chemical status', whereas under the new Directive the failure rate is 17%.

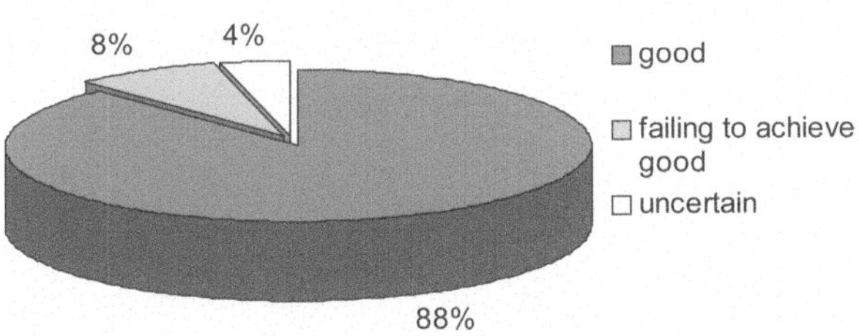

Figure 4 | Chemical status of surface water bodies in Germany. Data source: Portal WasserBLIcK (2010).

The chemical status of surface water bodies is classified into 'good status' and 'failing to achieve good status'. In sum, 88% of all surface water bodies exhibit 'good chemical status' according to present estimates (Figure 4). This figure will be less favourable following implementation of the new subsidiary directive on environmental quality standards (2008/105/EC) with its extended requirements for assessing chemical status.

A COMPARISON OF WATER BODY STATUS ASSESSMENT METHODS

The results obtained through monitoring based assessment for surface water bodies are, in the main, reasonably consistent with the risk assessment classification of German rivers on account of the large number of river water bodies involved (9,070 of the total of 9,900 surface water bodies in Germany). Comparison of the two sets of results yields useful insights.

Figure 5 shows the ecological and chemical status of the rivers as returned through risk assessment and monitoring based assessment respectively. For better clarity, the ecological and chemical statuses arrived at through monitoring based assessment are collated into the categories 'WFD objectives achieved' (classifications 'very good' and 'good'),

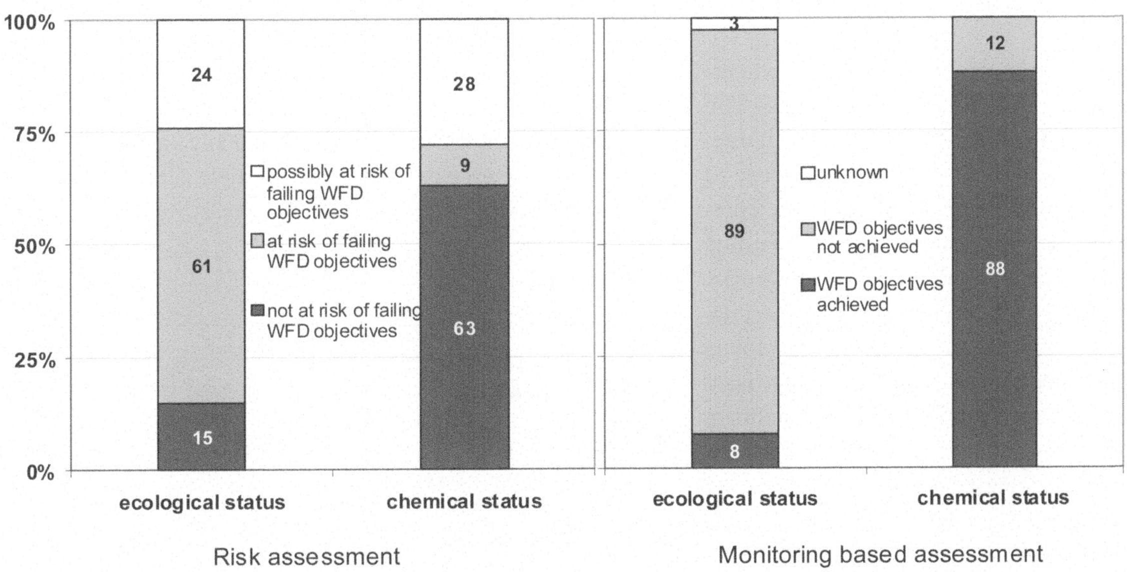

Figure 5 | Ecological and chemical river water body status based on risk assessment in 2004 (left) and the monitoring process in 2009 (right). Data source: Borchardt et al. (2005); Portal WasserBLIcK (2010).

'WFD objectives not achieved' (classifications 'moderate', 'poor' and 'bad'; i.e. failure to achieve 'good' status) and 'unknown'.

The results of the ecological water body status in the risk assessment show a relatively high number of water bodies being possibly at risk of failing to meet the WFD objectives. The main reasons for this were insufficient data and a lack of knowledge about pressures and impacts on the aquatic community (biological quality elements).

In order to obtain the missing information, operative monitoring was carried out for water bodies that are 'possibly at risk' or 'at risk' of failing to meet the WFD objectives. In addition, new assessment methods were developed with a focus on biological quality elements. In the comparison between the two assessments, water bodies 'possibly at risk' according to their ecological status were generally included in the 'WFD objectives not achieved' classification in the monitoring based assessment.

In contrast to the ecological status, a higher percentage is classified as being 'not at risk', i.e. 'WFD objectives are achieved' in respect of the chemical water body status. Water bodies that were originally considered to be 'possibly at risk of failing chemical WFD objectives' were generally moved to the 'WFD objectives achieved' category following monitoring based assessment.

The comparison of the two different assessment approaches reveals a generally good degree of consistency for a large proportion of the water bodies.

SUMMARY AND CONCLUSIONS

The assessment of the European water bodies has been carried out within the framework of the EU-WFD. The identification of pressures and impacts as part of the characterization process according to Art.5, WFD formed a useful basis for an initial estimation of water body status. The comparison of risk assessment and monitoring based results shows that risk assessment is a suitable tool for the initial evaluation of water bodies which also allows for the recognition of trends at an early stage. Methods that focus directly on existing pressures can be used to identify basic problems and provide explanations for failure to achieve the environmental objectives. Therefore, it is not necessary to wait for large-scale monitoring results before trends can be identified and effective action can be taken. However, it must be borne in mind that a risk assessment is no substitute for a comprehensive status assessment on account of several inherent uncertainties, in particular those involving status assessment on a water body scale. However, there are also many uncertainties in respect of water body assessment as well: the number of biological results realized for a water body may not be representative; the water body may not be amenable to assessment due to the absence of a reference water body or due to methodological uncertainties; the impact of management measures cannot be forecasted, and the prognostication of long-term changes in climatic patterns or the impact of natural disasters such as flooding or drought present significant difficulties. The use of biological indicators is common, but the varying resistance of indicators to the pollutants must be borne in mind. The exposure of biological indicators to contamination over a long period of time will render the indicators more resistant. However, the use of existing pressure and impact data indicators provides a helpful starting point for the identification of problems and development of effective management strategies. Subsequently, monitoring results can be used for backing up, confirming and as necessary modifying initial programmes of measures. In conclusion, we see risk and indicator based assessments as complementary elements in Integrated Water Resources Management which have to be linked in systematic and phased procedures.

Beyond the development of indicators to assess the status of water bodies and the efficiency of environmental improvement there is a need to develop indicators for water management purposes further. In Germany (as well as in the other European member states) different approaches have been adopted with respect to the status of the water bodies, the pressures and programmes of measures, the data and data quality, the management structures and prevailing socioeconomic conditions.

Current activities are being devoted to establish wider sets of indicators, e.g. to evaluate the improvements reached in a given river basin on the basis of highly aggregated data, to identify synergistic effects of multiple stressors and to assess the performance or effects of measures and of water management.

REFERENCES

AQEM Consortium 2004 Manual for the application of the AQEM system. http://www.aqem.de/.

Borchardt, D., Bosenius, U., Dörr, R.-D., Ewens, H.-P., Irmer, U., Jekel, H., Keppner, L., Mohaupt, V., Naumann, S., Rechenberg, B., Rechenberg, J., Richter, S., Richter, S., Rohrmoser, W., Stratenwerth, T., Willecke, J. & Wolter, R. 2005 Water Framework Directive – Summary of River Basin

District Analysis 2004 in Germany. Federal Ministry for the Environment, Nature Conservation and Nuclear Safety. Bonifatius, Paderborn.

CIS IMPRESS 2002a Analysis of Pressures and Impacts. WFD CIS Guidance Document 3, Directorate General Environment of the European Commission, Brussels.

CIS WATECO 2002b Economics and the Environment – The Implementation Challenge of the Water Framework Directive. WFD CIS Guidance Document 1, Directorate General Environment of the European Commission, Brussels.

European Parliament and Council 2000 Directive 2000/60/EC of the European Parliament and of the Council of 23 October 2000 establishing a framework for Community action in the field of water policy as amended by Decision 2455/2001/EC and Directives 2008/32/EC, 2008/105/EC and 2009/31/EC.

European Parliament and Council 2006 Directive 2006/11/EC of the European Parliament and of the Council of 15 February 2006 on pollution caused by certain dangerous substances discharged into the aquatic environment of the Community.

Hering, D., Buffagni, A., Moog, O., Sandin, L., Sommerhäuser, M., Stubauer, I., Feld, C., Johnson, R. K., Pinto, P., Skoulidikis, N., Verdonschot, P. F. M. & Zahradkova, S. 2003 The development of a system to assess the ecological quality of streams based on macroinvertebrates – design of the sampling programme within the AQEM project. *Hydrobiology* **88**, 345–361.

LAWA – Länderarbeitsgemeinschaft Wasser 2000 Gewässerstrukturgütekartierung in der Bundesrepublik Deutschland. Verfahren für kleine und mittelgroße Fließgewässer. Kulturbuch-Verlag, Schwerin.

LAWA – Länderarbeitsgemeinschaft Wasser 2003 Arbeitshilfe zur Umsetzung der EG- Wasserrahmenrichtlinie. Themenbezogenes Arbeitspapier Nr.4: Kriterien zur Erhebung von anthropogenen Belastungen und Beurteilung ihrer Auswirkungen zur termingerechten und aussagekräftigen Berichterstattung an die EU-Kommission.

Meilinger, P., Schneider, S. & Melzer, A. 2005 The Reference Index method for the macrophyte-based assessment of rivers – a contribution to the implementation of the European Water Framework Directive in Germany. *Int. Rev. Hydrobiol.* 322–342.

Portal WasserBLIcK: Bund-Länder- und Kommunikationsplattform der Bundesrepublik Deutschland zum Themenschwerpunkt Europäische Wasserrahmenrichtlinie. www.wasserblick.net

Richter, S. & Völker, J. 2010 Water Framework Directive – The way towards healthy waters. Results of the German River Basin Management Plans 2009. Published in July 2010 by the Federal Ministry for the Environment, Nature Conservation and Nuclear Safety.

Rolauffs, P., Stubauer, I., Zahradkova, S., Brabec, K. & Moog, O. 2004 Integration of the saprobic system into the European Union Water Framework Directive. *Hydrobiologia* **516**, 285–298.

First received 19 April 2012; accepted in revised form 24 July 2012

Attributiveness of a mass flow analysis model for integrated water resources assessment under data-scarce conditions

B. Helm, T. Terekhanova, J. Tränckner, M. Venohr and P. Krebs

ABSTRACT

Nutrients in river systems originate from multiple emission sources, follow various pathways, and are subject to processes of conversion and fate. One approach to tackle this complexity is to apply balance-oriented models. Although these models operate on a coarse temporal and spatial scale, they are capable of assessing the significance of the different emission sources and their results can be the basis for developing integrated water quality management schemes. In this paper we propose and apply a methodology to evaluate the attributiveness of such model results with regard to the modelled emission pathways. The MONERIS (MOdelling Nutrient Emissions in RIver Systems) model is set up, assuming plausible ranges of emission levels from four principal sources. The sensitivity of model performance is computed and related to the contribution from the pathways. The approach is applied for a case study in the upper Western Bug catchment (Ukraine). Coefficient of determination (R^2) is found insensitive against the model assumptions, at levels around 0.65 for nitrogen and 0.55 for phosphorous emissions. Relative mean absolute error is minimized around 0.2 for both nutrients, but with equifinal combinations of the varied emission pathways. Model performance is constrained by the ranges of the emission assumptions to a limited extent only.

Key words | mass flow analysis, model attributiveness, model performance, nutrient transport model, water quality management

B. Helm (corresponding author)
T. Terekhanova
J. Tränckner
P. Krebs
Institute of Urban Water Management,
TU Dresden,
01062 Dresden,
Germany
E-mail: *bjoern.helm@tu-dresden.de*

M. Venohr
Department of Shallow Lakes,
Institute of Freshwater Ecology and Inland Fisheries,
Müggelseedamm 310,
12587 Berlin,
Germany

INTRODUCTION

Water as a resource is affected in quality and quantity by multiple human activities. Among others, agriculture, industrial activities and human sanitation are major potential sources of nutrient emissions to surface waters and they demand an integrative perspective for their assessment.

Mass flow analysis (MFA) models categorize a group of tools aimed at the mass conservative routing of substance fluxes through different sub-systems. As a modelling paradigm in environmental sciences it was mainly propagated by Baccini & Bader (1995), and one of the earliest applications for river basins was presented with MONERIS (MOdelling Nutrient Emissions in RIver Systems) by Behrendt *et al.* (1999). Most tools operate statically, i.e. aim at annual fluxes and average storage capacities, and input–output relations are described with conceptual or empirical approaches.

MFA models are commonly applied for the source attribution of nutrient emission in river basins. They are especially promoted for data-scarce conditions, because the input information for the parameterization and calibration of process-based models is not available. Silgram *et al.* (2009) emphasizes that model performance depends more on data availability and driving boundary conditions than on model complexity. Process-oriented models are in most cases more expensive in data demand and calibration effort (Perrin *et al.* 2001) and are reported to tend stronger towards equifinality (Shepherd *et al.* 1999), i.e. with different model configurations same results are achieved.

Frequently MFA-oriented models have been propagated for the estimation of river nutrient loading over a large range of spatial scales from global (Bouwman *et al.* 2005), continental (Aritoli *et al.* 2008; Behrendt *et al.* 2005) to mesoscale rivers (Pieterse *et al.* 2003; Palmeri *et al.* 2005; Nikolaidis *et al.* 2009) and urban systems (Huang *et al.* 2007). In many cases these studies involve pathway-specific reduction potentials or scenarios. Palmeri *et al.* (2005)

doi: 10.2166/wst.2013.497

specify reduction potentials from agriculture and wastewater treatment plants (WWTPs); Nikolaidis *et al.* (2009) investigate the effect of measures in WWTP efficiency, detergent policy, agriculture, and erosion reduction. Artioli *et al.* (2008) rear-project the success of reduction policies implemented in the last 30 years.

In most applications model performance is evaluated, but only with focus on total nitrogen (TN) and phosphorous (TP) loads. Little regard is given to multi-objective evaluation, an approach that is established in application of stronger process-oriented models (Bekele & Nicklow 2007; Arabi *et al.* 2007). An approach to consider multi-objectivity for an agriculture–groundwater–river subsystem is proposed but not evaluated in Schilling *et al.* (2005). Hence a gap between model capabilities (estimation and routing of nutrients through different subsystems) and model evaluation can be assumed.

In this paper we propose a method to address uncertainty of emission estimates from the different pathways or sub-systems. Emission levels from these pathways are varied and model performance is used as evaluation criterion. Low variability in the emission levels of well-performing model realizations is considered as high attributiveness for the pathway.

The method is applied in a case study for the Western Bug catchment in the Ukraine with the MONERIS nutrient emission model.

STUDY AREA AND INPUT DATA

Western Bug catchment

The study area is located in the Western Bug basin. With a catchment area of almost 40,000 km², Western Bug is the largest tributary of the Vistula river (Figure 1, left). It has a transboundary basin within the Ukraine, Poland and, Belarus.

The river is one of the major sources of, but also threats to, Warsaw's drinking water supply (Niemirycz 1997) and accounts for 15% of the nutrient pollution in the Vistula (Forsius 2005; Zabokrytska *et al.* 2006). The discussion about the main sources of pollution is controversial. TACIS (2001) mentions urban and rural wastewater as a main polluter of the river. In contrast the United Nations Economic Commission for Europe (UNECE 2008) found that 84% of nitrogen (N) and 68% of phosphorous (P) pollution are caused by diffuse sources. According to Forsius (2005), the Lviv wastewater system is identified to be among the pollution hot spots in the Baltic Sea catchment.

The uppermost 2,500 km² of the basin were chosen as study area because all relevant potential sources of nutrient emissions are present, and data availability was best for the Kamianka Buska hydrological gauge and water quality sampling station at the basin outlet of the study area.

With a portion of 70%, arable land dominates land use (Figure 1, right). Agriculture mainly takes place on soils with loess parent material and in fluvial plains.

About 750,000 inhabitants live directly in the basin and the treated wastewater of an additional 200,000 person equivalents in the city of Lviv is transferred to the river system. Thus the effective population density is high (380 inh./km²) but unevenly distributed, with 700,000 people living in the urban agglomeration of Lviv (Figure 1, right).

As displayed in Figure 2 availability of data for the study region is limited and a strong decrease in measurement activities after the political changes in 1989–1991 is apparent. Uncertainties are calculated as 90% confidence intervals of mean annual concentrations and average absolute deviation of discharge in periods represented by each sampling event. They relate mainly to low and irregular

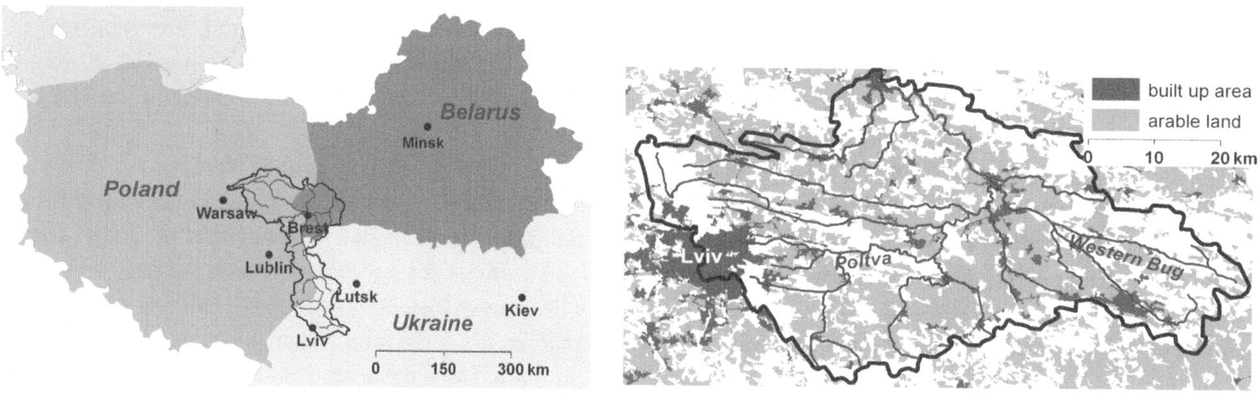

Figure 1 | Location of Western Bug basin and land use in the study area.

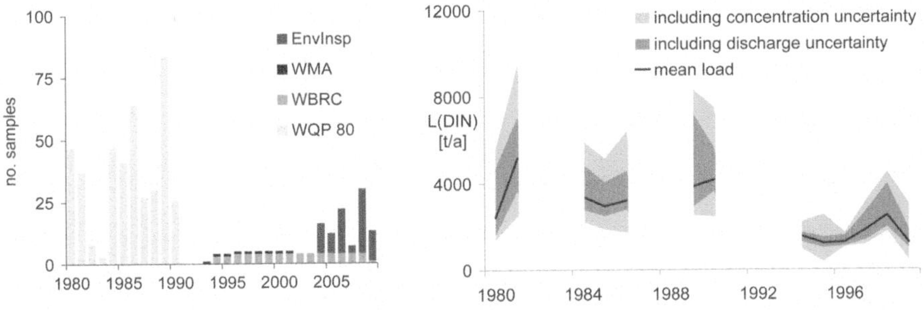

Figure 2 | Number of available water quality samples per year in the study area (1980–2008) and uncertainty in load estimation (1980–1998) for dissolved inorganic nitrogen (Helm 2010).

sampling frequency. Besides low sampling density, data are obtained from four different institutions and is not always consistent.

APPROACH

Modelling environment

The MONERIS model is applied as an established tool for the evaluation of nutrient emissions. The model considers atmospheric deposition, erosion, groundwater, point sources, surface runoff, tile drainage, and urban areas as sources of nutrient emissions. The system of equations is static and non-recursive. The model operates with annual time steps and on sub-basin level. Model input is defined by roughly 80 stationary and 30 transient state variables, and 190 model parameters, which are primarily not intended to be varied.

Data acquisition and preprocessing and model setup

Shuttle Radar Topography Mission (SRTM) data (Jarvis *et al.* 2008) is used as topographic reference for delineation of hydrological features, and resampled to 1,000 m resolution for erosion calculations. Land use was classified within the International Water Research Alliance Saxony (IWAS) project from LANDSAT-TM5 data (NASA 1989). A 1:200,000 scale soil map (Krupskyi 1967) was digitized, classified, regionalized, and parameterized in the project. Population data are available from the 1979, 1989 and 2001 censuses at settlement level; information on water and agricultural management practices at settlement and district level is from a digital data base (LvivStat 2009).

A detailed paper on data acquisition, pre-processing, and nutrient modelling is under preparation. Preliminary results and some more insight on the data situation are available in Terekhanova *et al.* (2010) and Helm (2010).

Variation of input assumptions

Emission and transport of nitrogen and phosphorous is modelled independently in MONERIS. Emission inputs from the subsystems assumed to be most relevant are varied within plausible ranges and permutatively combined. For each subsystem one variable for each nutrient species that proved sensitive to emission load in preliminary tests is included in the variation. Table 1 shows the assumed levels of the considered variables. The levels were defined to reflect an even distribution over the range and/or specific literature values as cited below.

WWTP efficiencies affect the input from point sources. All WWTPs in the study area are plants with secondary biological treatment. Lower estimates for the efficiencies are the ranges cited in Bouraoui *et al.* (2009); high efficiencies are according to the direct discharger inventory of Lviv State Environmental Inspectorate for the years 2000–2008.

Estimation of the retention of nutrients from pit latrine exfiltrate is extremely uncertain. Distance between pit and groundwater level, distance to water body and aquifer properties are cited as influencing factors (McCray *et al.* 2009). Heatwole & McCray (2007) mentions, as a result of a parameter sensitivity evaluation, that a removal rate

Table 1 | Levels of input assumptions in model variation

	Low			High
WWTP efficiency N [%]	20	40	60	75
WWTP efficiency P [%]	5	25	50	75
Retention of pit latrine N [%]	10	30	60	90
Retention of pit latrine P [%]	10	30	60	90
Fertilizer N surplus [kg/(ha·a)]	60	90	120	150
Soil P accumulation [kg/ha]	450	600	750	900
Top soil N content [g/kg]	2	3.5	5	8
Top soil P content [g/kg]	0.43	0.62	0.80	0.94

for nitrogen of 40% is exceeded in 99% of the cases. Corbett *et al.* (2002) assumes a removal of 50% phosphorous for sources with more than 50 m distance to a water body. Pieterse *et al.* (2003) found retention of 75% N and 60% P from pit latrine filtrate as best estimate for a Belgian lowland basin.

Fertilizer nitrogen surplus affects groundwater and tile drainage while phosphorous accumulation influences erosion and surface runoff pathways. The assumed ranges are derived from the variability of values for the entire Ukraine cited in FAO (2005) and statistics of fertilizer application in Lviv oblast (LvivStat 2009). Interannual variations are adopted from a data base provided with MONERIS that is calculated with OECD method (OECD 1999).

The nutrient contents in the top soil are regarded in the erosion pathway. The range of top soil N content is assumed as cited in Catt (2001) in which references of several authors are concluded for agricultural loess derivates. The range of phosphorous content is adopted from Blake *et al.* (2000). MONERIS applies a linear clay–phosphorous model to calculate the phosphorous content; the coefficient parameter was varied in order to scale the phosphorous content.

These four to the power of four variations result in 256 model configurations for N and P each.

Evaluation of model results

Two main aspects of model performance are pattern similarity and bias between modelled and observed data (Taylor 2001). Silgram & Schoumans (2004) propose a set of performance indicators for nutrient transport modelling with respect to the two aspects.

In this paper coefficient of determination (R^2) is used in the form of squared correlation coefficient as a measure of pattern similarity. According to Everitt (2002) in this form it can be interpreted as a measure of how good a predictor might be constructed from modelled values. In Equation (1) and Equation (2) observed loads are denoted as L_{obs}, modelled loads as L_{mod}, and mean loads as $\overline{L_{\text{mod/obs}}}$:

$$R^2 = \frac{\left(\sum_{i=1}^{n}\left(L_{\text{obs}(i)} - \overline{L_{\text{obs}}}\right)*\left(\text{L}_{\text{mod}(i)} - \overline{\text{L}_{\text{mod}}}\right)\right)^2}{\sum_{i=1}^{n}\left(L_{\text{obs}(i)} - \overline{L_{\text{obs}}}\right)^2 * \sum_{i=1}^{n}\left(\text{L}_{\text{mod}(i)} - \overline{\text{L}_{\text{mod}}}\right)^2} \tag{1}$$

Relative mean absolute error (rMAE) is chosen as a measure of bias; it is favoured in this study over root square error measures because it gives linear weight to small and big deviations:

$$\text{rMAE} = \frac{\sum_{i=1}^{n}\text{abs}\left(L_{\text{obs}(i)} - L_{\text{mod}(i)}\right)}{\sum_{i=1}^{n} L_{\text{obs}(i)}} \tag{2}$$

RESULTS AND DISCUSSION

Overall model performance

The diagraphs in Figure 3 show observed time series of riverine nutrient loads at the basin outlet versus the domain of 256 model runs. Load of dissolved inorganic nitrogen (DIN) and phosphate phosphorous (PO_4-P) are visualized. While DIN is a direct model output, PO_4-P was scaled from modelled TP, because TP observations were available for only four years. The scaling factor was determined as 0.805 from a linear regression with a coefficient of determination of 0.86 and at an *F*-tested significance of 0.001 confidence level.

Some characteristics of the observed time series are not covered by any of the model configurations. Low DIN loads in 1980 and low PO_4-P loads in 1980–1983 are not reproduced by the model. Instead of the observed decrease of DIN in 1985 and PO_4-P in 1987 relative to the preceding year an increase is modelled.

Sensitivity of model performance indicators

Visual interpretation of the diagraphs in Figure 3 implies that the altered model configurations mainly lead to a parallel shift of resulting time series. This is confirmed by the model performance indicators. rMAE (Figure 3, right), as a measure of distance between modelled and observed values, is sensitive to model configuration (expressed by the grey shade of the different realizations). R^2 (Figure 3, left), as a measure of course correlation, hardly changes over the range of model configurations, visualized by the uniform shade of the realizations.

The descriptive statistics of model performance indicators are presented in Table 2. For both indicators nitrogen is modelled slightly better, for both mean performance and best realizations. As already indicated by the shading of the graphs in Figure 3, ranges for correlation coefficient (0.081 for DIN and 0.089 for PO_4-P) are smaller than for rMAE (0.725 for DIN, 1.146 for PO_4-P). The latter indicates that the range of input assumptions is wider for phosphorous than for nitrogen.

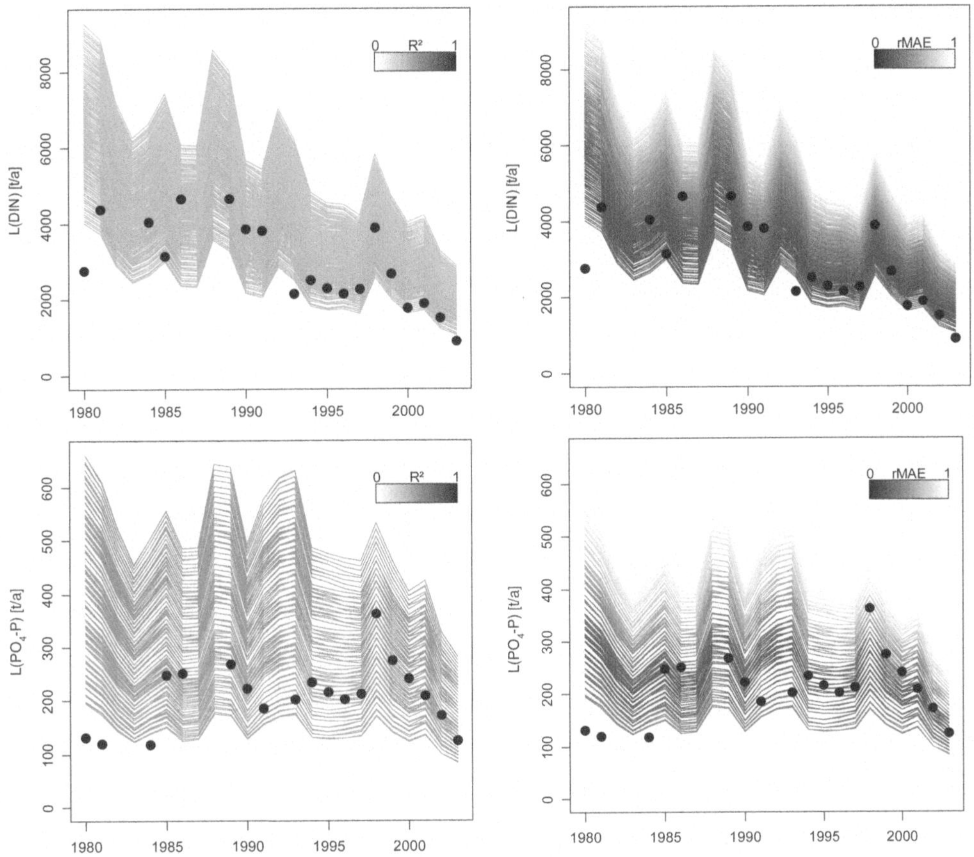

Figure 3 | Observed (black dots) and modelled (grey lines) loads of phosphate (lower figures) and dissolved inorganic nitrogen (upper figures) at basin outlet. Grey shades of modelled values indicate coefficients of determination (R^2, left) and relative mean absolute error (rMAE, right).

Table 2 | Descriptive statistical parameters of model performance indicators over all simulations

		Min	Q_{25}	Mean	Q_{75}	Max	Range
DIN	rMAE	0.205	0.267	0.424	0.547	0.930	0.725
	R^2	0.624	0.663	0.676	0.691	0.705	0.081
PO_4-P	rMAE	0.228	0.303	0.618	0.887	1.374	1.146
	R^2	0.501	0.506	0.523	0.535	0.590	0.089

Due to the insensitivity of R^2 to model configuration, it is discarded for further evaluations.

Figure 4 shows the relation between model performance and total emission from all pathways. In contrast to riverine loads, emission loads are evaluated as TN and TP. Good realizations are defined by rMAE <0.3, i.e. model runs with less than 30% deviation from observation. Best realizations correspond to the versions within lower 0.1 quantile of rMAE, i.e. the best 10% or 25 realizations. For TN emissions a clear Pareto-frontier can be determined, with a minimum at an emission level of 4,800 tonnes per year. Good realizations are observed for a range from 3,600 to 6,200 t/a; best realizations are found between 4,200 and 5,600 t/a of emission. For TP the course of the relation between emission level and rMAE is similar to TN with the global minimum at 540 t/a. The threshold for best realizations is slightly higher than for TN, which is in line with the descriptive statistics in Table 2. The range of emission is 420–640 t/a for good, and 480–540 t/a for best realizations.

Model performance for different emission pathways

In order to evaluate the attributiveness of MONERIS related to the different emission pathways, model performance is captured as dependent from emission level. The ranges of absolute emission levels of the pathways are shown in Table 3. Apparently under all assumptions WWTPs, groundwater and tile drainage are predominant TN pathways while WWTPs, urban systems and erosion dominate TP input.

In order to compare the emission levels of pathways despite their differing magnitudes, normalized values are displayed in Figure 5. Normalization is achieved by

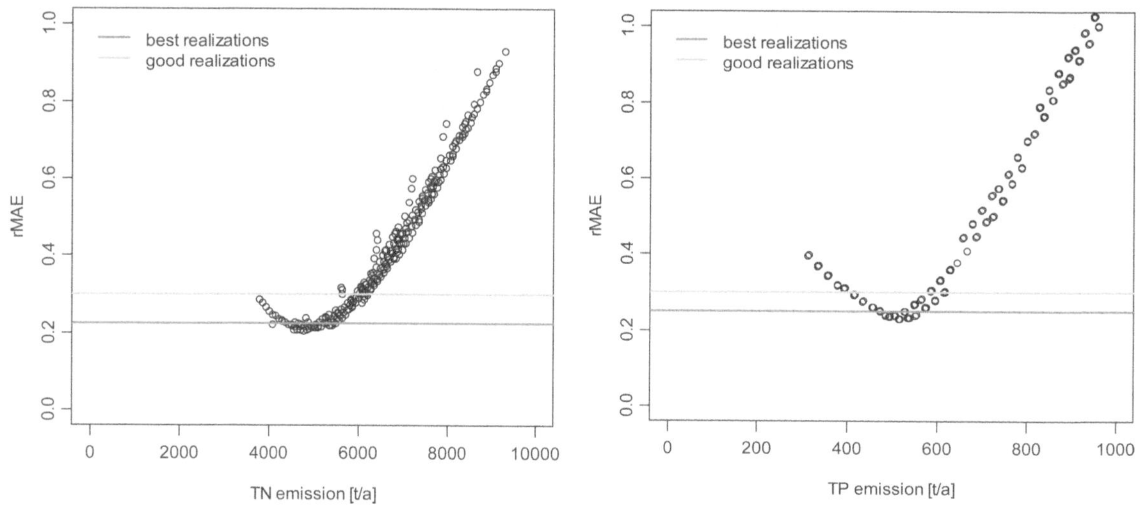

Figure 4 | Sensitivity of relative mean absolute error as a function of total nitrogen (left) and total phosphorous (right) emissions of all pathways.

Table 3 | Ranges of nutrient emissions by pathways

	WWTP	Urban systems	Groundwater	Tile drainage	Erosion
N [t/a]	1,026–3,340	153–806	960–1,600	1,390–3,030	100–386
P [t/a]	173–687	60–298	16–18	11	51–115

dividing absolute emissions by mean values for each pathway. The range in direction of the *x*-axis can be interpreted as model response to variation of input assumptions. Thus, urban systems respond slightly more strongly to the assumed variations than WWTPs, and while tile drainage varies by $+/-40\%$ for TN, it is insensitive for TP. Furthermore, the range and mean rMAE at the different path-specific emission levels are displayed. Lowest rMAEs for WWTPs are only found at the lower emission levels. In contrast, urban systems achieve low values over the whole range of emission levels. This tendency is confirmed by mean rMAE, which increases more strongly for WWTPs than for urban systems, i.e. average model performance decreases more with increasing WWTP emissions than with increasing urban system emissions.

Similar conclusions can be drawn for tile drainage, groundwater, and erosion. The variability of erosion is the largest among the three pathways for both nutrients, while groundwater and tile drainage exhibit a stronger effect on mean model performance. Low rMAE values are found over the whole range of the emission levels.

In general mean rMAEs in Figure 5 decrease more or less pronouncedly with lower emission levels. With regard to the concept of rMAE as a measure of bias that increases with over estimation as well as underestimation of the observed values, it could have been expected that minimum and/or mean rMAEs are lowest at an optimum level. This promotes the assumption that either the lower ranges of the input variables could have been further extended or other variables affecting the pathways should be additionally considered.

Attributiveness of emission pathways

So far evaluations focussed on the univariate sensitivity of river loads and emission levels against model performance. However, a synoptical perspective is required in order to capture the attributiveness of all pathways. Figure 6 and Figure 7 show the fractions of nutrient load from the different pathways for all, good, and best realizations within the model domain. The sub-settings of good and best realizations refer to the terms defined above, i.e. less than 30% deviation from observations, and belonging to the best 10% of all realizations, respectively.

The fractions of the different nitrogen pathways are displayed in Figure 6. For all subsets of the model domain main inputs are from WWTPs and tile drainage.

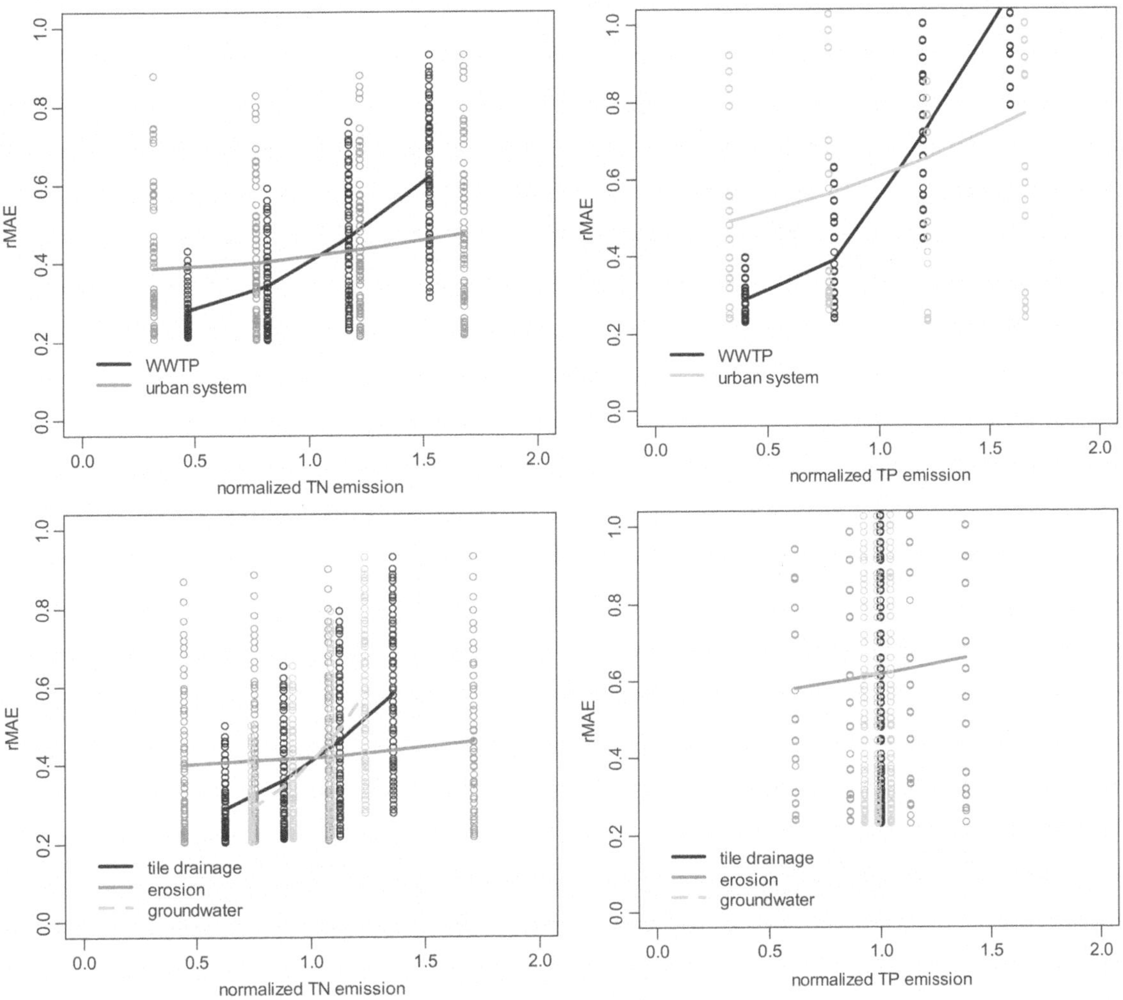

Figure 5 | Sensitivity of relative mean absolute error as a function of normalized total nitrogen (left) and normalized total phosphorous (right) emissions of all pathways. Dots indicate individual realizations while solid lines represent mean rMAEs of the different emission levels.

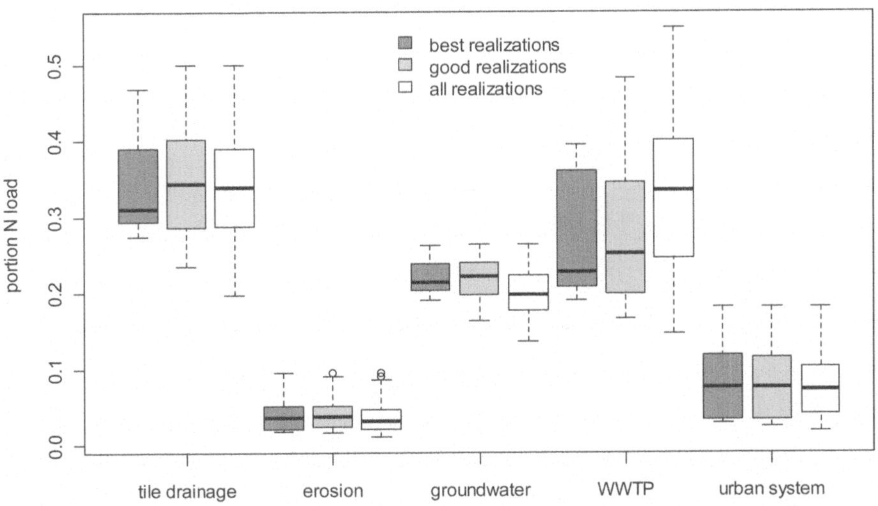

Figure 6 | Ranges of relative nitrogen loads from different pathway for all, good, and best realizations within the model domain.

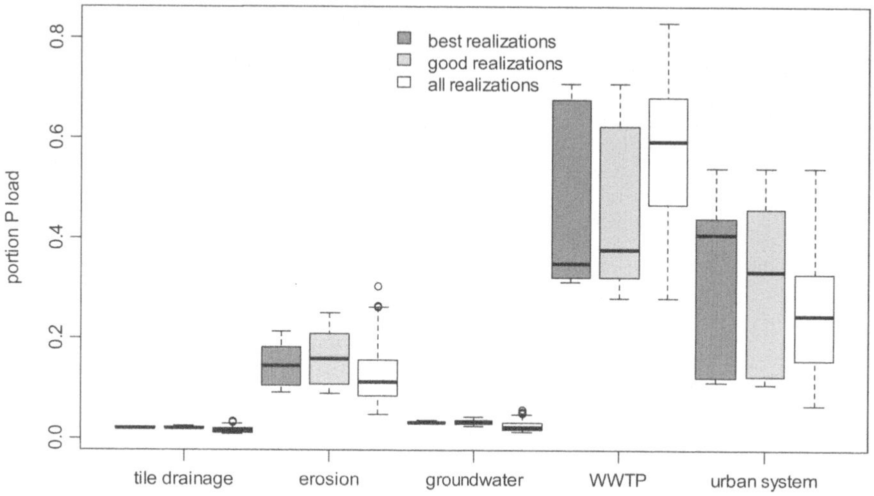

Figure 7 | Ranges of relative phosphorous loads from different pathway for all, good and best realizations within the model domain.

The range of the portion of WWTP emission in all realizations is 15–67%, with the central quartiles (visualized by the boxes) ranging from 25 to 40%, the spans of absolute values decrease for good (17–57%) and best (19–39%) realizations, while central quartiles slightly shift but do not decrease. A similar pattern can be observed for groundwater and tile drainage although the decrease of the span is less pronounced. The ranges of urban systems and erosion hardly alter between the different subsets, while the box-span is even slightly increasing with better realizations.

The portions of phosphorous pathways are shown in Figure 7. As already indicated in Table 3, emissions from WWTPs, urban systems and erosion are predominant.

Decrease in the span of absolute emission levels is again most distinct for the WWTPs, from 27 to 83% for all realizations to 32–71% for best realizations, but the span of the central quartiles increases at the same instance. Urban systems and erosion show similar patterns.

Based on model performance, only for a few of the evaluated pathways could a significant reduction of the wide range of possible input assumptions be achieved. The model can thus be considered attributive to the different emission pathways only to a limited extent.

CONCLUSIONS

In this study we introduced an approach to evaluate the attributiveness of a nutrient transport model with regard to input from different pathways. Model performance was used as indicator for attributiveness. While relative mean absolute error, as a measure of model bias, showed response to input data variation, correlation coefficient, as a measure of pattern similarity, was hardly affected. Due to overparameterization, equifinality is often cited as a problem of process-oriented models. With only one effective performance indicator for model evaluation, it is also found for the parameter configurations of the MONERIS model.

It has to be emphasized that the presented results do not cover the full ranges of uncertainty of the pathways' load estimation. The idea was rather to evaluate whether, inversely, model performance supports a reduction of input uncertainty. The conceptual approach applied is comparable to Monte-Carlo methods, assuming quasi-uniform distribution of the altered variables. The implementation environment of MONERIS restricts the use of established Monte-Carlo tools but such an extension could provide valuable additional information.

The ranges of contributions from the different subsystems are narrowed only to a small extend in good realizations when compared with the whole model domain. A calibration or optimization of model configuration that solely considers model performance of total loads would therefore lead to largely arbitrary quantification of emission sources.

In order to avoid this drawback, a set of quality-ensuring measures are suggested:

- Spatially distributed evaluation, which is already performed for the large-scale applications, may contribute to a reduction of uncertainty ranges, especially if subcatchments with a predominant pathway are included; at the same time the application of global or default assumptions, e.g. of WWTP treatment efficiencies, is limited.

- Multi-objective evaluation seems a promising approach wherever data from the different subsystems is available e.g. by considering nitrogen isotope signatures of the different pathways. For some systems this is more easily achievable than for others, i.e. WWTP effluent is monitored with less effort than subsurface nutrient fate.
- Extended input data acquisition in order to preliminarily reduce the ranges of the input assumptions.

Nevertheless the model proved largely robust in predicting the ranking of emission pathways and in separating dominant from minor inputs. It may thus serve as a first screening tool as proposed in the conceptual approach presented by Tränckner *et al.* (2010), but cannot be considered as a stand-alone evaluation tool under the given boundary conditions.

ACKNOWLEDGEMENTS

This work was financed by Federal Ministry of Education and Research (Bundesministerium für Bildung und Forschung, BMBF) (Grant Numbers 02WM1028/02WM1166) as part of the project International Water Research Alliance Saxony (IWAS). Data were provided by Lviv Statistical Department, Lviv Environmental Inspectorate, State Administration of Ecology and. Natural Resources, Volyn Hydrometeorological Service and Western Bug River Council.

REFERENCES

Arabi, M., Govindaraju, R. S., Engel, B. & Hantush, M. 2007 Multiobjective sensitivity analysis of sediment and nitrogen processes with a watershed model. *Water Resources Research* **43**, 11.

Artioli, Y., Friedrich, J., Gilbert, A. J., McQuatters-Gollop, A., Mee, L. D., Vermaat, J. E., Wulff, F., Humborg, C., Palmeri, L. & Pollehne, F. 2008 Nutrient budgets for European seas: A measure of the effectiveness of nutrient reduction policies. *Marine Pollution Bulletin* **56** (9), 1609–1617.

Baccini, P. & Bader, H.-P. 1995 *Regionaler Stoffhaushalt*. Spektrum Akademischer Verlag.

Behrendt, H., Huber, P., Kornmilch, M., Opitz, D., Schmoll, O., Scholz, G. & Uebe, R. 1999 Nährstoffbilanzierung der Flussgebiete Deutschlands, Umweltbundesamt.

Behrendt, H., van Gils, J., Schreiber, H. & Zessner, M. 2005 *Point and Diffuse Nutrient Emissions and Loads in the Transboundary Danube River Basin. – II. Long-term Changes*, Schweizerbart, 0945–3784.

Bekele, E. G. & Nicklow, J. W. 2007 Multi-objective automatic calibration of SWAT using NSGA-II. *Journal of Hydrology* **341** (3–4), 165–176.

Blake, L., Mercik, S., Koerschens, M., Moskal, S., Poulton, P., Goulding, K., Weigel, A. & Powlson, D. 2000 Phosphorus content in soil, uptake by plants and balance in three European long-term field experiments. *Nutrient Cycling in Agroecosystems* **56**, 263–275.

Bouraoui, F., Grizetti, B. & Aloe, A. 2009 Nutrient Discharge from Rivers to Seas for Year 2000 (JRC54459). Technical report, European Commission Joint Research Centre, European Commission.

Bouwman, A. F., Van Drecht, G., Knoop, J. M., Beusen, A. H. W. & Meinardi, C. R. 2005 Exploring changes in river nitrogen export to the world's oceans. *Global Biogeochemical Cycles* **19**.

Catt, J. A. 2001 The agricultural importance of loess. *Earth-Science Reviews* **54** (1–3), 213–229.

Corbett, D. R., Dillon, K., Burnett, W. & Schaefer, G. 2002 The spatial variability of nitrogen and phosphorus concentration in a sand aquifer influenced by onsite sewage treatment and disposal systems: a case study on St. George Island, Florida. *Environmental Pollution* **117** (2), 337–345.

Everitt, B. S. 2002 *The Cambridge Dictionary of Statistics* (entry for 'Coefficient of determination'), Cambridge University Press, Cambridge.

FAO (Food and Agriculture Organization) 2005 *Fertilizer Use by Crop in Ukraine*. FAO, Rome.

Forsius, K. (ed.) 2005 *Evaluation of Transboundary Pollution Loads*. Helsinki Commission, Helsinki.

Heatwole, K. K. & McCray, J. E. 2007 Modeling potential vadose-zone transport of nitrogen from onsite wastewater systems at the development scale. *Journal of Contaminant Hydrology* **91** (1–2), 184–201.

Helm, B. 2010 Mass flow analysis of nutrient emissions in Western Bug Basin. IWAS Ukraine final workshop, 15./16. 12. 2010, Dresden.

Huang, D.-B., Bader, H.-P., Scheidegger, R., Schertenleib, R. & Gujer, W. 2007 Confronting limitations: New solutions required for urban water management in Kunming City. *Journal of Environmental Management* **84** (1), 49–61.

Jarvis, A., Reuter, H., Nelson, A. & Guevara, E. 2008 Hole-filled seamless SRTM data V4, NASA Shuttle Radar Topographic Mission.

Krupskyi, M. 1967 Soil map of the Ukrainian SSR. Scale 1:200 000.

LvivStat 2009 Online Statistical Data Base of L'viv region. Main Statistical Office in L'viv region. Accessed: 05.09.2009.

McCray, J. E., Geza, M., Murray, K. E., Poeter, E. P. & Morgan, D. 2009 Modeling Onsite Wastewater Systems at the Watershed Scale: A User's Guide, Technical report, Water Environment Research Foundation, IWA Publishing.

NASA Landsat Program 1989 Landsat TM5 scene, LT51850251989228KIS00, USGS (U.S. Geological Survey), Sioux Falls.

Niemirycz, E. 1997 The Vistula River in Poland, in *International river water quality: pollution, restoration, 2nd International River Quality Symposium* (E. Niemirycz, Best Gerald Arthur & E. N. Teresa Bogacka, ed.), Gdansk, E, FN Spon, chapter, pp. 11–20.

Nikolaidis, N. P., Karageorgis, A. P., Kapsimalis, V., Drakopoulou, P., Skoulikidis, N., Behrendt, H. & Levkov, Z. 2009 Management of nutrient emissions of Axios River catchment: their effect in the coastal zone of Thermaikos Gulf, Greece. *Ecological Modelling* **220** (3), 383–396.

OECD 1999 *Environmental Indicators for Agriculture. Volume 1 Concepts and Framework*, Organisation for Economic Co-operation and Development, Paris.

Palmeri, L., Bendoricchio, G. & Artioli, Y. 2005 Modelling nutrient emissions from river systems and loads to the coastal zone: Po River case study, Italy. *Ecological Modelling* **184** (1), 37–53.

Perrin, C., Michel, C. & Andréassian, V. 2001 Does a large number of parameters enhance model performance? Comparative assessment of common catchment model structures on 429 catchments. *Journal of Hydrology* **242** (3–4), 275–301.

Pieterse, N. M., Bleuten, W. & Jørgensen, S. E. 2003 Contribution of point sources and diffuse sources to nitrogen and phosphorus loads in lowland river tributaries. *Journal of Hydrology* **271** (1–4), 213–225.

Schilling, C., Behrendt, H., Blaschke, A., Danielescu, S., Dimova, G., Gabriel, O., Heinecke, U., Kovacs, A., Lampert, C., Postolache, C., Schreiber, H., Strauss, P. & Zessner, M. 2005 Lessons learned from investigations on case study level for modelling of nutrient emissions in the Danube basin. *Water Science and Technology* **51** (11), 83–91.

Shepherd, B., Harper, D. & Millington, A. 1999 Modelling catchment-scale nutrient transport to watercourses in the U.K. *Hydrobiologia* **395–396**, 227–238.

Silgram, M. & Schoumans, O. 2004 Modelling Approaches: Model Parameterisation, Calibration and Performance Assessment Methods in the EUROHARP Project, Technical report, NIVA, Oslo, Norway.

Silgram, M., Schoumans, O. F., Walvoort, D. J. J., Anthony, S. G., Groenendijk, P., Stromqvist, J., Bouraoui, F., Arheimer, B., Kapetanaki, M., Lo Porto, A. & Martensson, K. 2009 Subannual models for catchment management: evaluating model performance on three European catchments. *Journal of Environmental Monitoring* **11**, 526–539.

TACIS 2001 *Transboundary Water Quality Monitoring and Assessment: Bug and Latorica/Uzh*, Ministry of Ecology and Natural Resources of Ukraine, Kyiv.

Taylor, K. E. 2001 Summarizing multiple aspects of model performance in a single diagram. *Journal of Geophysical Research* **106** (D7), 7183–7192.

Terekhanova, T., Helm, B. & Tränckner, J. 2010 Modeling Nutrient Balance for Western Bug. In: *6th Study Conference on BALTEX – Conference Proceedings* (M. Reckermann & H.-J. Isemer, eds.), 14–18 June 2010. Międzyzdroje, International BALTEX Secretariat.

Tränckner, J., Blumensaat, F., Helm, B., Terekhanova, T. & Krebs, P. 2010 A systematic approach for model-based integrated water resources management, In: *Integrated Water Resources Management Karlsruhe 2010 : IWRM, International Conference, 24–25 November 2010 conference proceedings* (Hartwig Steusloff, ed.), KIT Scientific Publishing, Karlsruhe.

UNECE 2008 *Our Waters: Joining Hands Across Borders – First Assessment of Transboundary Rivers, Lakes and Groundwater*, United Nations Economic Commission for Europe.

Zabokrytska, M., Khilchevskiy, V. & Manchenko, A. 2006 Hydroecological status of Zakhidnyj' Buh Basin in the territory of the Ukraine, NIKA Zentr.

First received 14 March 2012; accepted in revised form 11 July 2012

Theme V

Information and decision support systems

IWRM decision support with material flow analysis: consideration of urban system input

T. A. Terekhanova, B. Helm, J. Traenckner and P. Krebs

ABSTRACT

A review of material flow analysis (MFA) tools, comparison of case studies and analysis of approximately 20 MFA tools (static, semi-empirical models) are performed. The evaluation of the quantification procedures revealed several deficits in the approaches. The following principal complications for a reliable quantification of inputs from the urban water system are identified: (1) frequently insufficient data for urban system model validation (e.g. combined sewer overflow (CSO) discharges); (2) the necessity for additionally quantifying diffuse sources in order to verify modelling results at basin scale, where both input pathways occur, and (3) the contradictions arising when describing the highly dynamic urban system with the help of static MFA models. However, a wise selection of appropriate calculation procedures with regard to the concrete systems characteristics and available data can minimize the model deviations significantly. Criteria and suggestions for designing adapted quantification procedures are given.

Key words | basin modelling, IWRM, MFA, nutrients, urban system

T. A. Terekhanova (corresponding author)
B. Helm
J. Traenckner
P. Krebs
Institute for Urban Water Management,
Technische Universität Dresden,
Helmholtzstr. 10,
01069 Dresden,
Germany
E-mail: *Tatyana.Terekhanova@mailbox.tu-dresden.de*

INTRODUCTION

River basin modelling

According to its key principle of 'Natural System and Human System Integration', integrated water resources management (IWRM) takes the three often conflicting dimensions of ecology, economy and equity (Grambow 2007) into account. This conflict is especially prevalent in water quality problems of water resources which are subject to various activities and usages. Historically, the expectation has been that models can predict the response of a system to a range of environmental or anthropogenic drivers (Dunn *et al.* 2008). Existing modelling techniques result in quite a wide spectrum of quantification tools (Letcher *et al.* 1999; Silgram & Schoumans 2003; Quinn 2004; Anthony *et al.* 2009).

At a certain step of an IWRM process it is important to choose an appropriate tool with regard to its complexity and modelling effort requirements. Tränckner *et al.* (2010) therefore proposes a systematic screening procedure to reduce spatial and causal complexity as a prerequisite to apply dynamic process-oriented models for considering the most relevant deficits and pressures. The primary step of basin system analysis compiles the necessary knowledge of pressures on the system, their quantities and their sources, which are essential for decision support towards water quality management on a river basin scale or catchment planning (Anthony *et al.* 2009).

A model approach based on the methodology of material flow analysis (MFA) can be a useful framework for the screening procedure. MFA is a multidisciplinary systematic assessment of the flows and stocks of materials within a system defined in space and time (Brunner & Rechberger 2004). In water quality studies it allows us to identify the causes of surface and groundwater pollution and to propose the most promising preventive means to avoid them (Baccini & Brunner 1991). In contrast to dynamic modelling, lower complexity of processes representation and consequently lower input data requirements prevail (Brunner & Rechberger 2004).

There are several examples of MFA tool applications for the apportionment of pollution loads sources in river basins (see references later in the text). These are balancing tools, quantifying the matter loads on basin scale, their sources/sinks and pathways to the basin outlet, where water quality is of special consideration.

doi: 10.2166/wst.2013.470

Emissions assignment in a river basin: role of urban system

Sources assignment is based on the results of basin modelling (deWitt 2001) or statistical estimations (Van Drecht *et al.* 2009). The largest uncertainty in these model predictions was considered to be associated with the partitioning of riverine loads between sources, due to uncertain estimates of point source input and the net effect of in-river retention processes (Silgram *et al.* 2008).

In terms of local and acute impacts on the water body urban system emissions can be exceptionally important. Furthermore they can more easily be targeted for control and management compared with nonpoint sources scattered throughout a watershed (Puckett 1995). Van Drecht *et al.* (2009) estimates that urban emissions generally contribute relatively little to total stream loads compared with nonpoint sources (Figure 1(a)). Although, our own compilation of different published studies shows a broad range of estimates of nutrient load originating from point sources (Figure 2).

There is also no clear impact of river basin scale on the relevance of point sources impact. While the data of Van Drecht *et al.* (2009) suggest a decreasing impact with basin size (Figure 1(b)), the compilation in Figure 2 gives an inverse tendency. One essential reason for the observed diffuse pattern of point sources impact is the population density in the investigated areas as demonstrated by Garnier *et al.* (2002).

However, even for the same river basin, the assessed ratio of point sources input deviates and is strongly dependent on the model basis used (Figure 3). These uncertainties in the

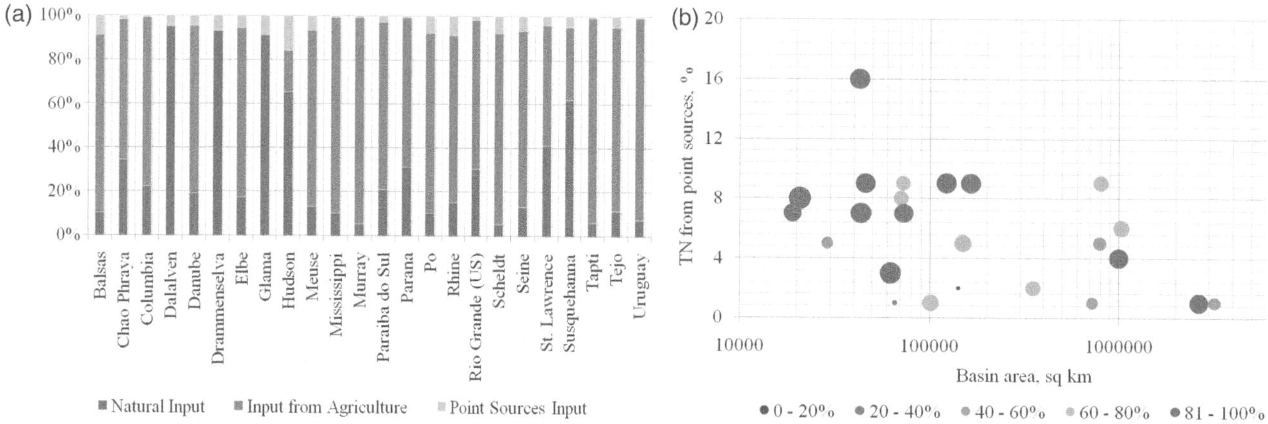

Figure 1 | (a) Nitrogen input sources partitioning; (b) nitrogen input from point sources with regard to basin area and fraction (%) of urban population in river basin (red circles) (adapted from Van Drecht *et al.* 2009). The online version of this figure, available at http://www.iwaponline.com/wst/toc.htm, is in full colour.

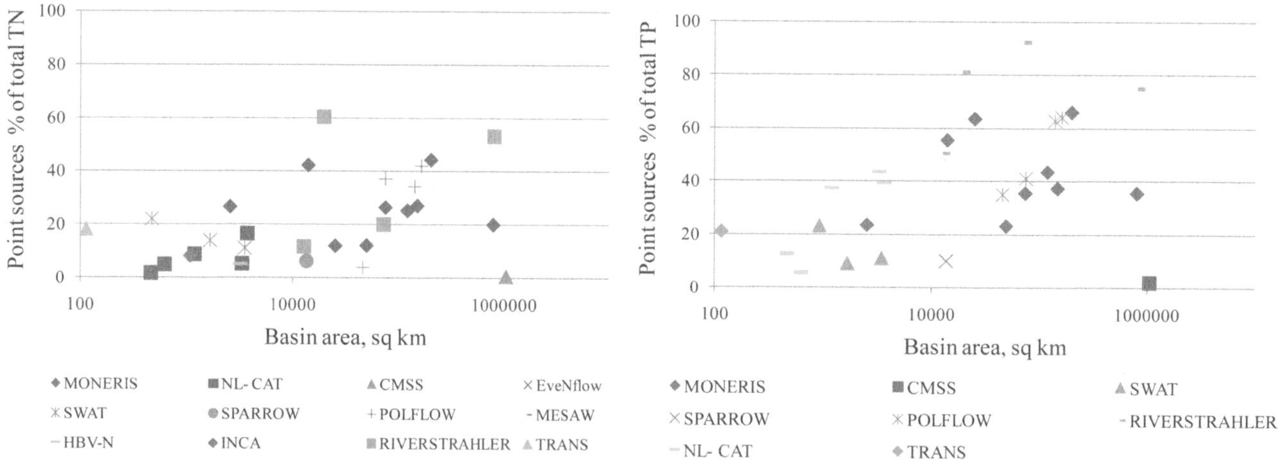

Figure 2 | Point sources input modelled with different watershed tools (from published modelling results: references can be requested from the author). The online version of this figure, available at http://www.iwaponline.com/wst/toc.htm, is in full colour.

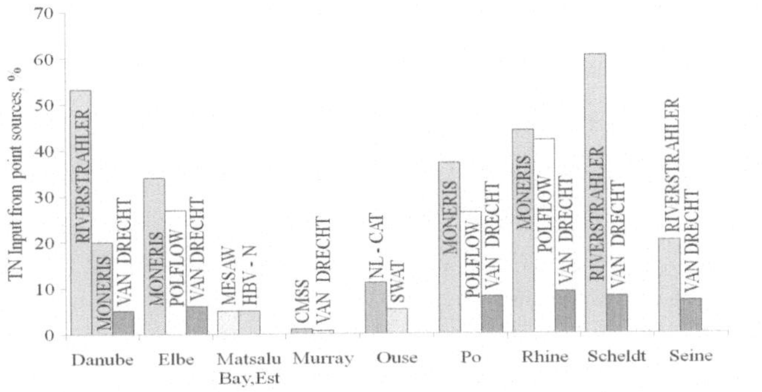

Figure 3 | TN input from point sources in river basins estimated with different tools (deWitt 2001; Garnier *et al.* 2002; Van Drecht *et al.* 2009; other references can be requested from the author).

estimates of point sources input are potentially originating from the diverse conceptual definition of point sources, perceived in different models or in model assumptions and approaches. Generally, point sources emissions consider the loads entering the river system from wastewater treatment plants (WWTPs) as the overall estimation of an urban system input, with the exception of some approaches.

Therefore, the objectives of this work are: (i) to set up a database of water quality (nutrients) models on river basin scale and their applications; (ii) to assess their compliance with MFA methodology; (iii) to select tools that consider emissions from the urban system; (iv) to analyse the quantification procedure of urban system emissions used in the selected tools; and (v) to recommend a model choice procedure with regard to appropriate quantification of emissions from urban systems.

METHODS

Literature survey

Based on a literature review, models were indentified that can quantify N and P nutrient emissions or/and loads on a catchment scale. In parallel a database of available tools for water quality modelling based on emissions quantification has been compiled.

Selection of models for analysis of urban system input quantification

Catchment scale modelling

With regard to MFA (Brunner & Rechberger 2004) the system boundaries have to be defined first. A river basin represents a naturally defined system bordered by watershed and catchment outlet. For analysis of urban input quantification, only the tools performing modelling of water quality on a river basin scale have been considered.

Compliance with MFA methodology

The tools based on matter conservation were considered to be in compliance with MFA. Taking into account the representation of conventional processes in MFA as transformation coefficients (Brunner & Rechberger 2004), only tools with static calculation approaches were chosen for further analysis. Automatically, the resolution of the models is then limited to annual or sub-annual time-steps. Moreover all selected tools had to be able to describe total emissions of nutrients and their sources assignment.

Urban system

Afterwards the models were assessed based on the criterion, whether they do include the emission from an urban system. The selection of tools for analysis was performed based on consideration of emissions from the urban system due to defined boundary conditions, i.e. including at least one of the below-mentioned compartments of the urban system emitting into surface water.

Urban system compartments

Urban systems are identified as sub-systems of a river basin and its boundaries are delineated by infrastructure distribution. The compartments of this sub-system include the structures visualized in Figure 4. In terms of MFA these compartments represent processes (Brunner & Rechberger 2004) where materials such as water and eventually nutrients are introduced. In an urban system we distinguish three types of flows: wastewater flow (F), stormwater flow from sealed areas (S) and treated water flow (E, with reduced material loads). Other flows, such as evaporated/transpired or infiltrated into aquifer are not considered. As nutrient emission pathways into receiving water, several processes have been chosen (Figure 4).

Modelling tool documentation

Availability of tool documentation was chosen as a must have. During the literature survey it was noted that some

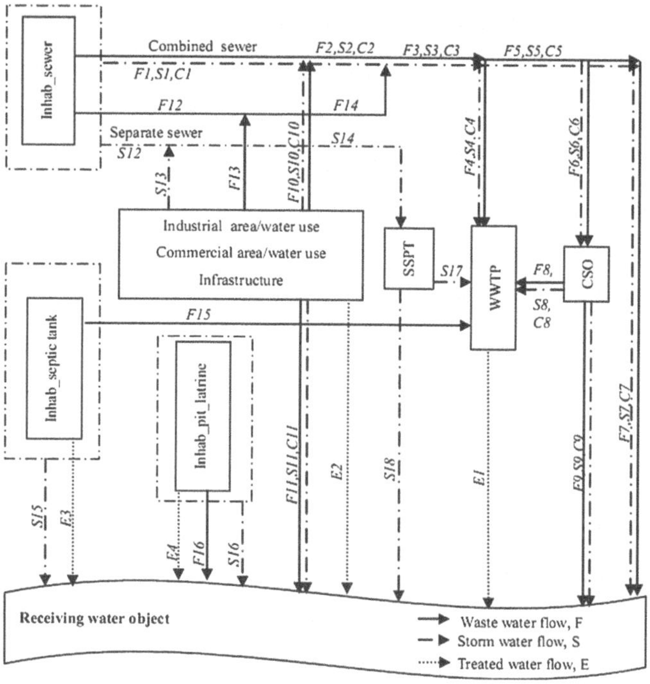

Chosen Emission pathways:

- *C7* Flows from combined sewer system diverted directly into receiving water object
- *C9* Overflow volumes from CSO structure into RWO
- *C11* Direct mixed water discharge from non-residential users/areas into RWO
- *S15* Flows (F,S) from residential areas with septic tanks
- *S,F 16* Direct (criteria: distance from RWO) discharge (F,S) from residential areas with pit latrines or similar
- *S18* Discharge into RWO from SSPT
- *E1* Effluent from municipal WWTP
- *E2* Effluent from industrial WWTP
- *E3* Effluent from septic tank
- *E4* Effluent from pit latrine

Comment: Emission pathways to groundwater, such as infiltrated rain water, leaching from sewer systems, pit latrines and septic tanks were not considered in this study.
SSPT is for separate sewer pre-treatment;
CSO is for combined sewer overflow;
WWTP is for waste water treatment plant.

Figure 4 | Urban drainage compartments.

models lacked documentation from the developer. In such cases model assumptions, boundary conditions and calculation approaches remained unclear.

Analysis of quantification procedures

To evaluate the capability of the tools to assess urban system emissions appropriately, a ranking procedure has been applied. Depending on the way models include urban system compartments, points ranging from 0 to 3 were given:

- 0 emissions from specified compartments are not considered
- 1 emissions from specified compartments are involved as pure input data
- 2 emissions can be assessed based on input data and a simple transport/transformation model
- 3 emissions are calculated case specific based on input data and additional systems information using more advanced modelling approaches.

The emissions from inhabitants not connected to the wastewater system (E4, F7) and combined sewer overflow (CSO) are considered separately. Their quantification is usually complicated due to scarce monitoring data to validate a computational approach. Finally, the general legitimacy to apply a computational approach has been discussed with regard to boundary conditions introduced in the models.

RESULTS AND DISCUSSION

Analysis of MFA tools

Database and selection of tools

The collected database includes 65 tools for water quality modelling. Among them 42 are able to represent water quality on a basin scale. The compliance with MFA methodology (static models) was found in 23 models. Selection based on inclusion of both point and non-point sources of nutrients results in 18 tools.

Sources apportioning models can be divided into two types: export coefficient models and discharge multiplied by concentration of matter ($C \times D$) approach (Grimvall & Stalnacke 1996). Although some selected models have mixed approaches it was decided to split them into these two groups for further analysis. Open tools such as SIMBOX (Schaffner & Scheidegger 2006) and WEAP (2011) are not regarded, as the elements of water balance and loads can be set up by the user.

Urban system consideration in export coefficient models

Although these models are based on simple empirical relationships between nutrient generation and catchment characteristics such as climate, topography and land use, the representation of emission loads from the urban system is differing in degree of comprehensiveness. In pure export coefficient models e.g. CMSS (2011), MESAW (Grimvall & Stalnacke 1996), Smith *et al.* (2005), SPARROW (2011), WATERSN (Castro *et al.* 2003), WBLMER (Valiela *et al.* 1997) the total load from a catchment is calculated as the sum of specific sources (area, population and direct discharge) multiplied by specified nutrient delivery ratios. The models TEOTIL (Barkved *et al.* 2008), WMM (2011), PLOAD (2011) include additionally the technique of the $C \times D$-approach to quantify the pollution loads of stormwater runoff from sealed areas (Figure 5). Although in terms of urban system quantification, the model STOFFBILANZ (2011) is more comprehensive than others, it is included in this group because it operates with source-specific export coefficients.

The calculation approaches of effluents from WWTP (E1, E2), septic tanks (E3) and pit latrines (E4), which are designated as point sources in these models, are based on number of inhabitants and the delivery ratio. The efficiency of wastewater treatment facility (or delivery ratio to a water body) is considered in the tools WATERSN, WBLMER, and WMM. The most simplified way of representing point source emission is the use of official effluent discharge data (Figure 5).

STOFFBILANZ is the only tool that considers urban drainage infrastructure and hence allows us to estimate the magnitude of stormwater runoff impacts. But the presented calculation of the load from paved areas diverted into the storm sewer system is valid only for urban areas with a separate sewer system. Without consideration of storm runoff partitioning between drainage systems this approach cannot be applied to an area with varying drainage infrastructure.

Among the presented EC-models a CSO is mentioned only in the WMM tool in terms of input data, i.e. overflow volumes and event mean concentrations should be evaluated based on regional monitoring data, which hardly exist. Inhabitants not connected to the sewer system are accounted for only in STOFFBILANZ, where partitioning of population connection to sewer is considered.

Application legitimacy

In terms of input data requirements and flexibility the EC-approach is very attractive. The models of this type can be run in conditions with data scarcity. The export coefficient values are provided differently: as look up values provided with a model (e.g. CMSS), parameterized through model calibration (MESAW), estimated by regression (SPARROW) or introduced as default values based on data from empirical studies (in Smith *et al.* (2005) or from references, e.g. WBLMER). Default values of export coefficients should be assigned with special care, because they may significantly vary for the same land use categories even within one river basin (Grimvall & Stalnacke 1996).

Therefore EC-models can be applied for primary estimations of two main matter loads from the urban system which have a significant impact on receiving waters: sewage and stormwater runoff. Emissions accounting from the urban system is strongly determined by water services infrastructure, its performance and volumes of water flows. Hence, without case specific parameterization, the application of EC-models may result in insufficient estimation of urban system contribution. If the different urban system sources are included insufficient detail, the export coefficients can be assigned to the desired emissions pathway.

Name	type	E1	E2	E3	E4	C7	C9	C11	S11	S15	S16	S18	R
MESAW	EC	1	1	0	0	0	0	0	1	1	1	1	0.18
CMSS	EC	1	1	0	0	0	0	0	2	2	2	2	0.3
WATERSN	EC	3	0	3	0	0	0	0	1	1	1	1	0.3
SPARROW	EC	1	1	1	0	0	0	0	2	2	2	2	0.33
Smith (2005)	EC	2	1	1	0	0	0	0	2	2	2	2	0.36
WBLMER	EC	0	0	3	0	0	0	0	3	3	3	3	0.45
PLOAD	EC/CxD	1	1	0	0	0	0	1	3	3	3	3	0.45
TEOTIL	EC/CxD	1	1	1	1	0	0	0	3	3	3	3	0.48
WMM	EC/CxD	1	1	0	3	0	1	0	3	3	3	3	0.54
STOFFBILANZ	EC**	1	1	0	3	3	0	0	3	3	3	3	0.61

Name	type	E1	E2	E3	E4	C7	C9	C11	S11	S15	S16	S18	R
POLFLOW	CxD	1	1	1	1	1	0	1	0	0	0	0	0.18
GISPLM	CxD	1	1	0	0	0	0	0	2	2	2	2	0.30
MOBINEG	CxD	1	1	1	1	1	1	1	2	2	2	2	0.45
HARP-appr	CxD	3	1	0	1	1	1	1	2	2	2	2	0.48
STEPL	CxD	1	1	3	0	3	0	1	2	2	2	2	0.52
MONERIS	CxD	1	1	3	3	3	3	1	3	3	3	3	0.82
WEAP	CxD	-	-	-	-	-	-	-	-	-	-	-	-
SIMBOX	CxD	-	-	-	-	-	-	-	-	-	-	-	-

0 not considered | 2 simplified calculation
1 input data in loads | 3 comprehensive calculation

Figure 5 | Urban emission representations in export coefficient models (left) and in $C \times D$-models. *R* is for relevance of the tool for quantification of urban system emissions.

Urban system consideration in C × D-models

The computational approach of load models is based on two steps. At first, elements of water balance and water flows between subsystems are calculated. Then, nutrient load is calculated via multiplying matter concentrations or specific loads in sub-systems with identified water flow volumes. Therefore, the definition of the emission pathways is strongly determined by the conceptual model implemented in a tool.

The complexity of quantification procedures in this group is extremely heterogeneous (Figure 5). The POLFLOW (deWitt 2001), STEPL (2011) and GISPLM (2011) tools quantify surface runoff loads by multiplication of water flow volume with constant specific concentrations, which makes these models similar to the EC-group. Additionally GISPLM includes load reduction factors determined by receiving water proximity.

A more sophisticated approach for surface runoff load quantification is presented in the tool MONERIS (2011). It allows separate accounting for loads from impervious areas connected (S11, S18) or not (S15, S16) to stormwater sewers and for loads of direct discharges from combined sewers (C7, C11). This approach requires significantly more input data and allows more detailed quantification of the stormwater loads from paved urban areas, than comparable numbers of described emission pathways in the quantification approaches of MOBINEG (2011) and HARP-Guidelines (2000). The differentiation of sewer systems is an advantage of the MONERIS tool, which allows the quantification of CSO volumes based indeed on German design standards (ATV-128) and adopted for German climatic conditions (MONERIS 2011).

Although C × D-tools operate with water flows, the quantification of point sources is quite simplified, e.g. as input data (Figure 5). This may be caused by parameterization of water intake and the assumption of the equality between water intake and wastewater discharge. This assumption reduces the number of input data, but the estimation of the matter flux entering WWTPs may deviate from the load introduced after water use. However, wastewater loss is typically in the order of a few % and thus negligible in a balance model.

Application legitimacy

The C × D-approach engages a strong conceptual background for loads quantification on basin scale. The application of these models should be performed with special care to data availability for the studied river basin. Moreover, the approaches (such as MONERIS, STEPL) include regional-specific features in their input data and quantification procedures. Therefore, their application to different natural and infrastructural conditions demands good modelling experience for parameter adaptation to local conditions.

CONCLUSIONS AND OUTLOOK

Analysis of 18 MFA tools (static, semi-empirical models) has shown that, regarding the crucial importance of urban systems for water quality, the modelling approaches for emissions from those systems have to be improved. Independently of model type the basic loads, i.e. load from sewage and runoff from paved areas, are introduced in all models. But the consideration of other urban emissions varies significantly. This concerns the deficits in the conceptual representation of urban systems as well as the quantification procedures involved in a tool.

Typical constraints for an appropriate quantification of urban system sources are:

- coarse definition of urban system boundaries and compartments;
- disagreement in type of pollution source;
- inadequate quantification approaches for CSOs and stormwater runoff;
- poor consideration of WWTP efficiency;
- high requirements on input data, instead of a computational approach;
- doubtful transferability of reference pollutant concentrations/loads to other systems.

Summarizing, no quantification approach will be able to overcome the above mentioned deficits completely. However a wise selection of appropriate calculation procedures with regard to the concrete systems characteristics and available data can minimize the model uncertainty significantly.

Based on the results of this study for the IWRM decision making, we propose to use:

- Export coefficient models, when available data are scarce, or the impact of urban system emissions is a priori known to be less relevant.
- More comprehensive quantification of urban emission pathways with C × D-models, applied to areas with greater urban pressures. However, data availability must be satisfactory.
- In the basins with higher population density and heterogeneous urban infrastructure, an explicit urban MFA tool modelling results in the basin model as boundary conditions.

REFERENCES

Anthony, S., Silgram, M., Collins, A. & Fawcett, L. 2009 Modelling nitrate river water quality for policy support. *International Journal of River Basin Management* **7** (3), 259–275.

Baccini, P. & Brunner, P. 1991 *Metabolism of the Anthroposphere*. Springer, New York.

Barkved, L., Fazi, S. & LoPorto, A. 2008 STRIVER: Strategy and methodology for improved IWRM D7.1 Part 2.

Brunner, P. H. & Rechberger, H. 2004 *Practical Handbook of Material Flow Analysis*. Lewis Publishers, Boca Raton, Florida.

Castro, M., Driscoll, C., Jordan, T., Reay, W. & Boynton, W. 2003 Sources of nitrogen to estuaries in the United States. *Estuaries* **26**, 803–814.

CMSS 2011 Available from: www.toolkit.net.au/wincmss.

deWitt, M. 2001 Nutrient fluxes at the river basin scale. I: the PolFlow model. *Hydrological Processes* **15**, 743–759.

Dunn, S., Freer, J., Weler, M., Kirkby, M., Seibert, J., Quinn, P., Lischeid, G., Tetzlaff, D. & Soulsby, C. 2008 Conceptualization in catchment modelling: simply learning? *Hydrological Processes* **22**, 2389–2393.

Garnier, J., Billen, G., Hannon, E., Fonbonne, S., Videninac, Y. & Souliec, M. 2002 Modelling the transfer and retention of nutrients in the drainage network of the danube river. *Estuarine, Coastal and Shelf Science* **54**, 285–308.

GISPLM 2011 User Manual. Available from: http://www.wwwalker.net/gisplm/.

Grambow, M. 2007 Wassermanagement, Vieweg + Teubner.

Grimvall, A. & Stalnacke, P. 1996 Statistical methods for source apportionment of riverine loads of pollutants. *Environmentrics* **7**, 201–213.

HARP-Guidelines 2000 Harmonised Quantification and Reporting Procedures for Nutrients. Available from: http://www.klif.no/publikasjoner/vann/1759/ta1759.pdf.

Letcher, R., Jakeman, A., Merritt, W., McKee, L., Eyre, B. & Baginska, B. 1999 *Review of Techniques to Estimate Catchment Export*. NSW Environment Protection Authority, Sydney.

MOBINEG 2011 Available from: http://www.saarland.de/dokumente/thema_wasser/Abschlussbericht_kubiniok_2005.pdf.

MONERIS 2011 User Manual. Available from: http://moneris.igb-berlin.de/.

PLOAD 2011 User Manual. Available from: http://water.epa.gov/scitech/datait/models/basins.

Puckett, L. 1995 Identifying the major sources of nutrient water pollution. *Environmental Science and Technology* **29** (9), 408A–415A.

Quinn, P. 2004 Scale appropriate modelling: representing cause-and-effect relationships in nitrate pollution at the catchment scale for the purpose of catchment scale planning. *Journal of Hydrology* **291**, 197–217.

Schaffner, M. & Scheidegger, R. 2006 Using a material flow analysis model to trace and quantify pollution sources in river basins of developing countries. In: *Proceedings of International Conference on Management of Water, Wastewater and Environment: Challenges for the Developing Countries*, Kathmandu 13–15 Sept., http://www.eawag.ch/forschung/sandec/publikationen/sesp/dl/MFA_rice_NEAconf_2006.pdf.

Silgram, M. & Schoumans, O. (eds) 2003 Review and literature evaluation of quantification tools for the assessment of nutrient losses at catchment scale. EUROHARP report 1-2003. Oslo, Norwegian Institute for Water Research (NIVA).

Silgram, M., Anthony, S., Fawcett, L. & Stromqvist, J. 2008 Evaluating catchment-scale models for diffuse pollution policy support: some results from the EUROHARP Project. *Environmental Science and Policy* **11**.

Smith, R., Jordan, C. & Annett, J. 2005 A phosphorus budget for Northern Ireland: inputs to inland and coastal waters. *Journal of Hydrology* **304** (1–4), 193–202.

SPARROW 2011 User Manual. Available from: http://pubs.usgs.gov/tm/2006/tm6b3/PDF/tm6b3_part1a.pdf.

STEPL 2011 User Manual. Available from: https://wiki.epa.gov/watershed2/index.php.

STOFFBILANZ 2011 User Manual. Available from: http://galf-dresden.de.

Tränckner, J., Blumensaat, F., Helm, B., Terekhanova, T. & Krebs, P. 2010 A systematic approach for model-based integrated water resources management. *IWRM Karlsruhe 2010*, Germany.

Valiela, I., Collins, G., Kremer, J., Lajtha, K., Geist, M., Seely, M., Brawley, J. & Sham, C. 1997 Nitrogen loading from coastal watersheds to receiving estuaries: new method and application. *Ecological Applications* **7** (2), 358–380.

Van Drecht, G., Bouwman, A., Harrison, J. & Knoop, J. 2009 Global nitrogen and phosphate in urban wastewater for the period 1970 to 2050. *Global Biogeochemical Cycles* **23**, GB0A03.

WEAP 2011 User Manual. Available from: http://www.weap21.org/WebHelp/index.html.

WMM 2011 User Manual. Available from: http://www.rougeriver.com/proddata/wmmmanul.pdf.

First received 26 January 2012; accepted in revised form 28 June 2012

A decision support procedure for integrative management of dammed raw water reservoirs

I. Slavik, W. Uhl, B. Skibinski, S. Rolinski, T. Petzoldt, J. Benndorf, N. Scheifhacken, L. Paul, M. Funke, H. Lohr, J. Völker and D. Borchardt

ABSTRACT

Dammed drinking water reservoirs with their catchment areas and the downstream rivers are dynamic systems that change permanently under the influence of many factors. Their multifunctional use for drinking water supply, flood control, energy production, nature conservation and recreation as well as ecological constraints for the rivers downstream requires an integrative management considering and balancing between different requirements. Thus, an optimal reservoir management has to take into account scenarios of external influences and must be based on predictions of prospective raw water qualities. Furthermore, the impacts of short- and long-term changes of the raw water quality on drinking water treatment have to be considered. The problem is very complex and cannot be solved intuitively but requires the application of hydrological, ecological and process models. This approach was followed in the work presented here, as a tool to predict and evaluate the impacts of different reservoir management strategies in an integrative way is currently not available. The developed decision support procedure (DSP) allows for the estimation of the effects of different hydrological and water quantity management scenarios on raw water quality, water processing costs and ecology in the downstream river. Extreme hydrological events or changing boundary conditions (e.g. climate change) are taken into account.

Key words | decision support procedure (DSP), drinking water, integrative modelling, management of reservoirs, water quality

I. Slavik (corresponding author)
W. Uhl
B. Skibinski
Technische Universität Dresden,
Institute of Urban Water Management,
Chair of Water Supply Engineering, 01062 Dresden,
Germany
E-mail: *irene.slavik@tu-dresden.de*

S. Rolinski
Potsdam Institute for Climate Impact Research (PIK), PO Box 60 12 03, 14412 Potsdam,
Germany

T. Petzoldt
J. Benndorf†
N. Scheifhacken
Technische Universität Dresden,
Institute of Limnology, 01062 Dresden,
Germany

L. Paul
Technische Universität Dresden,
Neunzehnhain Ecological Station,
Neunzehnhainer Straße 14, 09514 Lengefeld,
Germany

M. Funke
H. Lohr
SYDRO Consult GmbH,
Mathildenplatz 8, 64283 Darmstadt,
Germany

J. Völker
D. Borchardt
Helmholtz Centre for Environmental Research-UFZ,
Aquatic Ecosystems Analysis and Management,
39114 Magdeburg,
Germany

NOMENCLATURE

A_e, km^2	Catchment area	Q_{in}, L s^{-1}; m^3 s^{-1}	Inflow
DOC, mg L^{-1}	Dissolved organic carbon	Q_m, L s^{-1}; m^3 s^{-1}	Average flow rate
DSP	Decision support procedure	Q_{max}, L s^{-1}; m^3 s^{-1}	Maximum flow rate
DW	Drinking water	Q_{out}, L s^{-1}; m^3 s^{-1}	Outflow
DWT	Drinking water treatment	TOC, mg L^{-1}	Total organic carbon
P	Phosphorus	WFD	Water Framework Directive

† Deceased

doi: 10.2166/ws.2013.032

INTRODUCTION

The management of dammed drinking water reservoirs is complex and many aspects partially counteracting have to be considered. For instance, there is a conflict between reliable supply of raw water in adequate quantity and quality and sufficient flood control (Sieber 2003). For optimum flood protection an almost empty dam is required which brings lower water quality along. To maintain high reservoir water quality due to the natural self-cleansing ability of the water body, the water level should be maintained as high and constant as possible. If drinking water supply has priority in reservoir management, a sufficient volume for drinking water production has to be primarily guaranteed. The flood retention is also limited with respect to water quality aspects. Consequently, a compromise and an integrative strategy are needed to cover these aspects in reservoir management. Another conflict concerns the amount and quality of compensation water released, which should be minimal in order to save high quality raw water, but there are also requirements and objectives of the EU Water Framework Directive (WFD) (European Commission 2000) which have to be fulfilled. In this context also the impact of changes in compensation water quality on the value and type of river ecosystem services should be considered (Costanza *et al.* 1997; Strayer & Dudgeon 2010). According to the WFD, most of the rhithral and potamal rivers beneath dams will not achieve a 'good ecological status' or a 'good ecological potential', respectively, until 2015 (Richter & Völker 2010). In spite of good water quality in the water column with high ecological potential, interacting processes between surface water, riverbed (morphology) and hyporheic interstitial can be interfered by fluctuating reservoir discharges. Hydraulic conditions and flow regime can be seen as a key variable which shapes the many fundamental ecological characteristics in rivers and streams (Richter *et al.* 1996). A possible consequence of unnatural hydraulic conditions is a functional loss of the hyporheic interstitial as a fundamental habitat for aquatic communities (e.g. Saenger & Zanke 2006; Borchardt & Pusch 2008).

Furthermore, raw water quality changes inevitably have an effect on drinking water treatment efficiency and costs. Increasing inputs of organic and particulate matter into the raw waters require a higher degree of treatment plant efficiency, mostly requiring changes to the whole treatment technology and supply system operation. Increasing concentrations of, e.g. humic substances deteriorate the coagulation of water contaminants. The disinfection by-product formation potential and the microbial contamination within the supply system will increase with decreasing treatment effectiveness. As a consequence of increasing concentrations of particles, algae and organic matter and temporary changes in iron and aluminium concentration within the raw water, more chemicals and volumes of backwash water are needed, higher amounts of sludge are produced and filter breakthrough will take place earlier. This will result in rising costs of operation and waste disposal.

Reservoir specific water management plans have to take into account the following aspects:

- Long (e.g. climate change) and short-term extreme hydrological events (floods, droughts).
- Structure and management of the drainage basin.
- Water demand and treatment technology installed at the water works.
- Ecological functionality of the lower reaches.

In contrast to these complex processes and partly counteracting interests, reservoir management is still mainly based on experience only. However, an increasing interest of water authorities can be recognised to support their decision by scientifically proven algorithms and models for future management. Therefore, tools integrally considering the different aspects are urgently needed. However, the evolution of reservoir management strategies that harmonise all demands is very difficult.

Consequently, the main and innovative objective of the multi-disciplinary study presented here was to develop a decision support procedure (DSP) for an integrative management of multipurpose drinking water reservoirs considering all the aspects mentioned above. For the DSP, a modular structure and the development and linking of models describing processes and interrelations in the subsystems were intended:

- water quantity and quality
- drinking water treatment
- downstream river water quality.

The linking of models should include all aspects of the physical, chemical or biological structure of the reservoir and consider the effects of meteorological or hydrological extreme events on:

- the raw water quality,
- the drinking water treatment, and
- the ecological state of the downstream river.

to establish a consistent framework.

Considering targets and limits defined by the reservoir operator and based on observed or simulated long-term run-off series, water quantity management rules should be derived and optimised. The aim was to predict their influences on the reservoir's physical structure and water quality depending on varying release depths of raw water and compensation water. Finally, the impacts of the resulting water quality on costs of drinking water treatment and on the ecology of the downstream river should be simulated. Moreover, it was intended to evaluate the results on the basis of the targets and limits defined. If the results do not fulfil the targets, the procedure can be repeated with a modified scenario. Thus, users of the DSP will be enabled to simulate and evaluate the complex consequences of different management strategies and flood protection demands on:

- reservoir water quantity and quality
- drinking water treatment
- river water quality.

MATERIALS AND METHODS

The reservoir system investigated comprised the reservoirs Klingenberg and Lehnmühle including their tributaries and pipelines (Figure 1). Both are situated in the South-eastern low mountain range of Saxony (Germany).

To optimise the operation rules for quantity related reservoir management, objective functions describing optimal management were set up. In this case, the guarantee of drinking water supply, dam stability and the protection of downstream sites have been the main objectives. Both flood protection and low-flow requirements downstream are achieved by keeping maximum and minimum tolerable discharges, respectively. By setting up the reservoir system

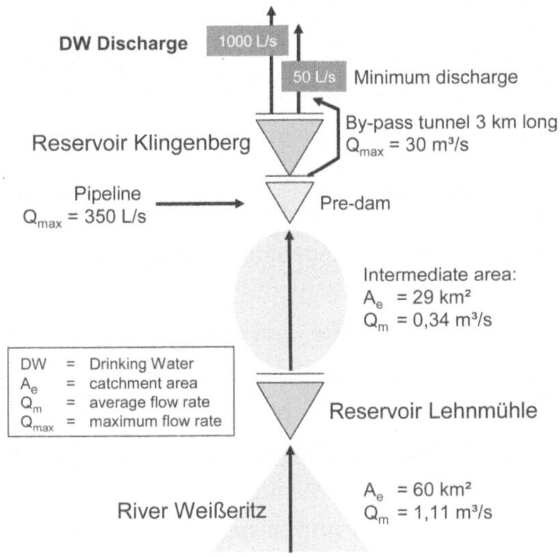

Figure 1 | Reservoir system investigated.

(Figure 1), the objective functions describing outflows from the reservoir are functions of inflows, season and the volume of all reservoirs in the system. In order to generate inflow as a function of time during the year, data from past precipitation-drainage-events and measured inflows to the reservoir were subjected to pattern recognition. Here, 100 scenarios were generated stochastically covering a period of 100 years, each describing typical inflow events. Thereby, all components were available to apply quantity optimisation using evolutionary strategy. As a parameter for the evolutionary strategy, the operating rules as relationships between discharges and selected state variables are considered, for example, in the simplest case, a relationship between discharge and reservoir volume. The rules themselves are described as 'if-then-functions'.

The integration of water quantity and quality management was performed using the software TALSIM developed by SYDRO Consult GmbH (Lohr 2001) but extended now to include several water quality criteria and allow the adoption of quantity based operation strategies according to the quality-related criteria (Rolinski *et al.* 2007).

To estimate the effects of different management strategies on water quality (at optimum quantity-management), the coupled physical–ecological model SALMO-HR was used for deterministic predictions. The model SALMO-HR is a vertically resolved one-dimensional hydrophysical–ecological model for lakes and reservoirs (Petzoldt *et al.* 2005; Rolinski *et al.* 2005; Rolinski *et al.* 2008). Its equations and parameters are defined in a generalised manner, so that site-specific calibration is restricted to waterbody specific properties such as planktivorous fish density, light extinction and sediment-P-release. The model consists of the ecological sub-model SALMO-1D developed at Technische Universität Dresden (Benndorf & Recknagel 1982; Petzoldt *et al.* 2005) and the hydrophysical k-ε-model LAKE of the Institute IAMARIS eV. (Baumert *et al.* 2005). The current final version of SALMO-HR simulates the seasonal development of temperature, stratification and turbulence (physical components) as well as the concentrations of phosphorus, nitrogen, phytoplankton (three groups), zooplankton, oxygen, DOC (including humic substances) and suspended matter (four particle classes) (Scheifhacken & Paul, unpublished data).

Extreme hydrological events, e.g. extreme floods or severe low water levels can substantially deteriorate the water quality and may even result in a malfunction of drinking water treatment (Paul & Scheifhacken 2009). Therefore, inflow concentrations and in-lake distributions of pollutants were investigated in order to develop dam-specific as well as general standards to complete the optimisation rules for the management of drinking water reservoirs in both water quantity and quality aspects. Special attention was paid to turbidity caused by floods and the characteristics of suspended particulate matter (Scheifhacken *et al.* 2010). Besides general water quality data, flux, concentration, size distribution and settling velocity of particles as well as fractions of dissolved natural organic matter were determined. The dataset was completed with historical data provided by the reservoir administration. Empiric models were used to describe the development of turbidity in the reservoirs depending on particle inflow and sedimentation. Following guidelines given in ATT (2009), a procedure was developed and implemented that allows for the determination of the minimum storage capacity and minimum hypolimnion volume of stratified reservoirs at the beginning of the summer stratification, which are needed to ensure basic raw water quality and quantity requirements with a user-defined statistical probability (Scheifhacken & Paul, unpublished data).

To model the impact of management strategies on river water quality downstream of the reservoir, the oxygen and nutrients concentrations in the water and the hyporheic interstitial were measured with probes and a data logger and were recorded automatically to investigate the influence of reservoir discharge on benthic invertebrates. Furthermore, oxygen, temperature, pH, conductivity, nitrogen, phosphorus, DOC and pressure variability were studied at three reference sampling sites in the catchment area downstream of the reservoirs. The interactions between surface and pore water in the sediment were modelled based on the temperature variability, the functionality was shown due to decelerated temperature exchange between both: vertical infiltration rates, reaction potential, detrital drive and mass transfer between the flowing water and the hyporheic interstitial dependent on discharge. Results of the water quality characteristics were compared with thresholds of chemical and physico-chemical variables given by LAWA (2007) to estimate the ecological state of water bodies according to WFD. Oxygen concentrations in the surface water were modelled according to different scenarios to predict water quality for specific management strategies. Benthic invertebrates were sampled in spring 2007 and 2008 according to standardised methods (Haase *et al.* 2004; Hering *et al.* 2004, Sundermann *et al.* 2008). Biological attributes of benthic invertebrates (e.g. taxonomic or functional composition) were selected to identify possible effects of unnatural inflow to the aquatic community.

To assess the impact of water quality on drinking water treatment, coagulation filtration experiments were planned according to factorial design and carried out at pilot scale. Treatment performance and costs were described as a function of organic load and turbidity using an empirical model. Required information had to be determined with maximum accuracy from as few as possible experiments. Using the method of factorial design, simultaneous variation and investigation of all variables within a single experimental plan was possible (Montgomery 2005). Thus, distinction of significant from insignificant variables as well as identification of relationships and optimum values could be achieved. Treatment performance was measured as filter run time until breakthrough, head loss, coagulant demand, acid/base demand for pH-adjustment, backwash water demand and sludge produced. The method of factorial design was applied to quantitatively describe the relationship between the investigated process parameters (raw water quality parameters, treatment parameters) and the dependent variable (treatment capacity). This approach was chosen because of the high complexity of mechanistic models describing treatment processes and the advantage of not being plant specific or requiring considerable model calibration and validation as mechanistic models do. The use of the method of factorial design was meant to develop regression models in the form of regression equations. This and the application of an optimisation algorithm finally resulted in the determination of optimum treatment parameters (pH, coagulant dosage and filter velocity) for each combination of organic load and turbidity. Then, with these optimised treatment parameters, specific treatment costs can be determined. As a result, the expected treatment costs for optimised treatment of each expected raw water quality can be calculated and plotted in three-dimensional diagrams (Slavik *et al.* 2010).

RESULTS

The DSP developed consists of a number of modules. The modules include model based mathematical descriptions and stochastic analysis of the processes and interrelations of the subsystems of a reservoir system (Figure 2). The data interchange among these modules offers the possibility to predict the consequences of different management strategies (different intake levels and amounts of raw water withdrawal and discharge to the lower reaches) in interaction with external factors (catchment area management, hydrological and climatic factors) on the particular subsystems (Figure 2). Finally, the optimal reservoir management strategy for current or prospective variable constellations can be derived and validated with respect to defined target values and limits.

The optimised flow of information and data between the particular modules are mandatory to apply the DSP (Figure 3). The starting point is the choice of a management strategy and the definition of ranges of acceptance for quality and treatment cost indicators (Figure 3). Different scenarios of raw water abstraction and discharge to the downstream reaches can be applied. Each management scenario is characterised by certain preferences of release levels for raw water and water to the downstream reaches, respectively. For example, raw water is taken from either the deepest (scenario 1) or the highest (scenario 2) possible hypolimnetic outlet structure. Alternatively, epilimnetic water will be abstracted to the downstream reaches and raw water is taken from a medium hypolimnetic outlet (scenario 3). For these three scenarios, the development of water quality parameters and their effects on drinking water treatment costs and the ecological state of the river downstream of the reservoir can be described (Figures 3 and 4).

Then, a mere quantity optimisation for the chosen management strategy follows within the module 'Simulation water quantity'. To optimise quantity, objective functions were set up as described before to guarantee drinking water supply, dam stability and the protection of downstream sites. Using these objective functions and the stochastically generated inflow loads, operation rules can be defined to optimise quantity management for the reservoir system investigated. On the basis of these operation rules, relationships between the inflow to the reservoir, the discharges

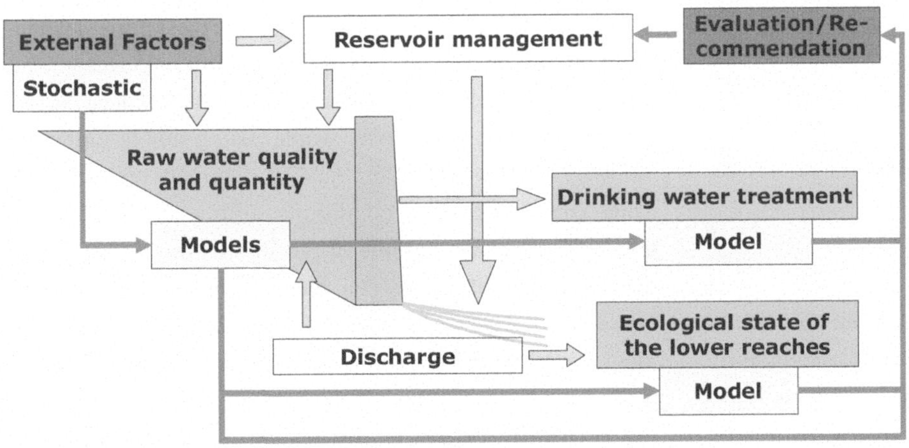

Figure 2 | Subsystem interactions and data flows considered in the decision support procedure.

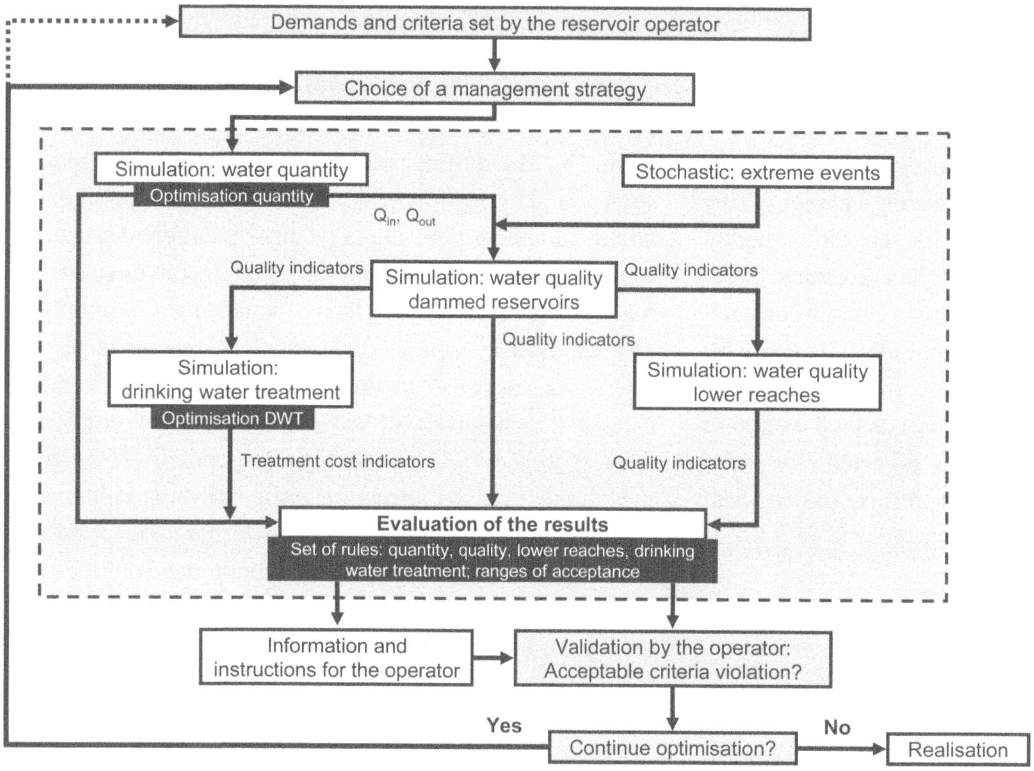

Figure 3 | Flow of information and data between the modules of the decision support procedure, DWT = drinking water treatment, Q_{in} = inflow, Q_{out} = outflow.

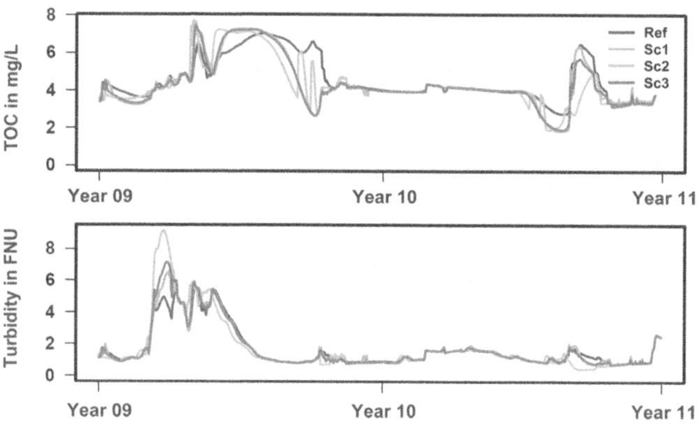

Figure 4 | Development of the TOC-concentration and the turbidity in the raw water reservoir during two generated years for the scenarios 1 to 3 (Sc1, Sc2, Sc3, raw water abstraction at the deepest, highest or medium hypolimnetic outlet, respectively) compared with a mere quantity optimised reservoir management strategy (Ref).

(drinking water treatment, rivers downstream) and the resulting storage level can be described. The results of the quantity related calculations (Q_{in}, Q_{out}, storage level) serve as input data for the simulations of dammed reservoir water quality and directly enter the all-up evaluation.

Within the quality-related optimisation, investigation results of inflow and in-lake distribution of pollutants during extreme hydrological events (e.g. extreme floods or severe draw-downs of the storage level) are also taken into account. Using stochastic simulations, the minimum storage and the minimum hypolimnion volume are calculated under consideration of the water quantity management rules determined before, which are necessary to ensure raw water withdrawal from hypolimnetic layers during summer stagnation with a user defined probability. Additional rules for the quantity management at extreme situations are derived.

The subsequent reservoir water quality simulation in the module 'Simulation: water quality dammed reservoirs' is using these boundary conditions and gives information under which constellations of the inflow, the reservoir storage and the discharge etc. quality-related problems can occur. Moreover, scenario based modelling shows how quality problems can be prevented by changing the height of raw water intake for drinking water production and/or discharge to the river, respectively. The stochastically generated time series from the water quantity management model were used as hydrological input data and complemented with meteorological records and empirically derived import concentrations (nutrients, humic substances, particles). Subsequently, indicators for the quality of raw water and ecological impacts on the downstream reaches of the discharge were determined using these simulation results. The comparison of different variants of operational reservoir management on the basis of simulated scenarios allows objectifying management decisions, which guarantee a best possible raw water quality for drinking water production and good ecological conditions downstream. The results of these simulations do not only contribute to the all-up evaluation but are also used as input data for the calculations within the modules 'Simulation: drinking water treatment' and 'Simulation: water quality lower reaches'.

The quality simulation of the lower reaches within the module 'Simulation: water quality lower reaches' shows under which constellation of the discharge quality-related problems occur downstream and whether, and if necessary how, these can be reduced by changes in discharge. By using a model describing transport and transformation processes, especially the oxygen conditions (River Water Quality Model No. 1; Reichert *et al.* 2001) in the flowing water and the hyporheic interstitial dependent on different reservoir management strategies, river water quality can be predicted. As a result, these simulations provide indicators for the quality of raw water being discharged to the lower reaches. According to the different defined scenarios of discharge, model based results of water quality variables (e.g. oxygen, phosphorus, nitrogen) do not exceed thresholds in surface water by LAWA (2007), but long-lasting minimum discharges to the lower reaches have an impact on the oxygen content of the deeper sediment layers. The benthic invertebrate community is not influenced by reservoir discharge with regard to assessment methods used to classify ecological status by WFD. However, significant changes in biocoenotic composition of invertebrates were identified in consideration of selected biological attributes adapted to functionality of hyporheic interstitial (Oligochaeta, Diptera) and discharge (Rheoindex, active filter feeders).

To investigate impacts of different raw water qualities on drinking water treatment within the module 'Simulation: drinking water treatment', a model was developed to optimise drinking water treatment which takes into account short-, mid- and long-term factors impacting raw water quality (Slavik & Uhl 2007; Slavik & Uhl 2009; Slavik *et al.* 2010). To assess the impact of water quality resulting from reservoir management on drinking water treatment, at first analyses of historical quality data and treatment performance during extreme events were carried out. Furthermore, an extensive series of experiments using pilot-scale filter plants was run, planned and evaluated by factorial design. Input parameters for simulation of drinking water treatment are TOC and turbidity, obtained from the water quality module. Based on the results of data analysis and pilot investigations and the model developed herewith, optimum treatment parameters (filter velocity, coagulant dosage, pH-value) were determined for minimum costs per m^3 drinking water produced. To evaluate drinking water treatment within the framework of the DSP, treatment cost indicators can be set by the user.

Finally, an overall evaluation of the simulation results can be done. After running the combined models, the user of the DSP will have to decide whether a possible violation of the chosen criteria is acceptable or not. Furthermore, the decision on continuing the optimisation has to be made. An optimisation and iteration cycle will be finished on the decision of the user if a management strategy can be found which ideally fulfils the quality and quantity-related objectives as well as the demands on drinking water treatment efficiency or rather will show least possible deviation. Moreover, the user will get information and instructions for the operation and management of a raw water reservoir. Otherwise, a new management strategy has to be chosen to start the optimisation and iteration cycle again.

For the three abstraction scenarios described above, the simulation results with respect to raw water quality and drinking water treatment costs are exemplarily shown (Figure 4). Within the water quality model, the seasonal dynamics of the reservoir water can be reproduced reasonably well, especially in terms of particulate substances, physical variables (e.g. temperature) and phytoplankton (abundances, biomass). As this seasonal dynamic is decisive for raw water quality with regard to drinking water treatment, results for e.g. the TOC-concentration are derived on a daily time step. As result, the development of raw water quality parameters in the reservoir can be simulated, exemplarily shown in Figure 4 for the TOC-concentration and the turbidity for two generated years. Following this, specific costs (in Euro per m^3) of drinking water treatment by coagulation–filtration can be calculated on the basis of the simulated TOC-concentration and turbidity when the model for drinking water treatment simulations is applied. The effects of the abstraction level and the resulting development of the water quality parameters on drinking water treatment costs are shown in Figure 5.

SUMMARY

The entire system of a raw water reservoir for drinking water production with its catchment area, the river downstream and the drinking water treatment plants is very complex and dynamic. With increasing demands and external pressures such as climate change, future management of drinking water reservoirs should not only be based on water management plans and the operator's experience any more. Here, we introduce a newly developed integrative approach to manage raw water reservoirs by applying a DSP. All aspects of water quality and quantity, flood protection, cost efficient drinking water treatment as well as guarantee of an acceptable ecological state of downstream river sections are implemented. Furthermore, the impact of extreme hydrological events and long-term changes (e.g. due to climate change) are included. This was done on the basis of linking and partly coupling of models describing the processes and interrelations within a drinking water reservoir system. Within the modules water quantity and raw water quality, already existing software tools were developed further with respect

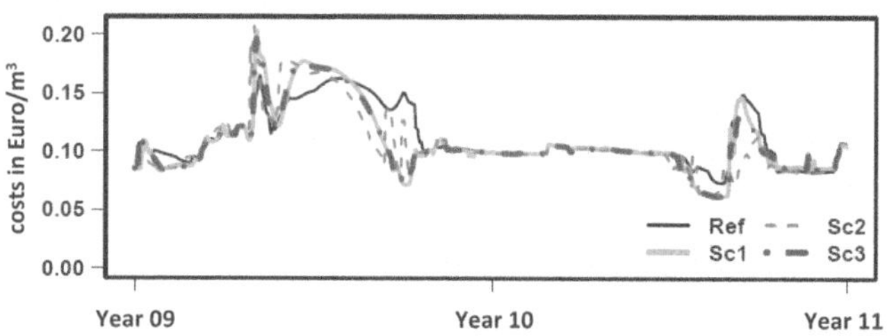

Figure 5 | Development of specific costs (in Euro per m^3) of drinking water treatment by coagulation–filtration during two generated years for the scenarios 1 to 3 (Sc1, Sc2, Sc3 see Figure 4) compared with a mere quantity optimised management strategy (Ref).

to the intended purposes of this project and interlinked. The results of the water quantity and quality simulations serve as data for the calculations to optimise treatment efficiency for drinking water production as well as to describe transport and transformation processes in the lower reaches. By combining all simulation results and by setting up demands and criteria for raw water reservoir operation, users of this DSP will be able to simulate and to evaluate the complex consequences of different management strategies and flood protection demands on reservoir water quantity and quality, drinking water treatment and river water quality.

According to ecological health and environmental objectives by WFD of lower reaches, there is a need to improve a more dynamic discharge and flow regime out of dams. Hydraulic conditions close to natural ones ensure ecological functionality with regard to aquatic communities, morphological alterations and hyporheic interstitial.

ACKNOWLEDGEMENTS

The authors would like to thank the Federal Ministry of Education and Research (BMBF) for funding under grant no. 02WT0719. The DREWAG NETZ GmbH, the Landestalsperrenverwaltung (LTV) and the Wasserversorgung Weißeritzgruppe are gratefully thanked for supporting the project.

REFERENCES

ATT 2009 *Integrale Bewirtschaftung von Trinkwassertalsperren gemäß DIN 19700. Arbeitsgemeinschaft Trinkwassertalsperren e.V., ATT-Schriftenreihe Band 7.* Oldenbourg Industrieverlag, München 2009.

Baumert, H. Z., Benndorf, J., Bigalke, K., Goldmann, D., Nöhren, I., Petzoldt, Y., Post, J. & Rolinski, S. 2005 Das hydrophysikalisch ökologische Talsperren- und Seenmodel SALMO-HR – Modelldokumentation und Leitfaden für den Anwender. Dresden.

Benndorf, J. & Recknagel, F. 1982 Problems of application of the ecological model SALMO to lakes and reservoirs having various trophic states. *Ecological Modelling* **17**, 129–145.

Borchardt, D. & Pusch, M. (eds) 2008 An integrative, interdisciplinary research approach for the identification of patterns, processes and bottleneck functions of the hyporheic zone of running waters. *Fundamental and*

Applied Limnology. Special Issue Advances in Limnology **61**, 1–7.

Costanza, R., d'Arge, R., de Groot, R., Farber, S., Grasso, M., Hannon, B., Limburg, K., Naeem, S., O'Neill, R. V., Paruelo, J., Raskin, R. G., Sutton, P. & van den Belt, M. 1997 The value of the world's ecosystem services and natural capital. *Nature* **387**, 253–260.

European Commission 2000 Official Journal L 327, 22.12.2000, p. 1; Directive 2000/60/EC (EU Water Framework Directive).

Haase, P., Lohse, S., Pauls, S., Schindehütte, K., Sundermann, A., Rolauffs, P. & Hering, D. 2004 Assessing streams in Germany with benthic invertebrates: development of a practical standardised protocol for macroinvertebrate sampling and sorting. *Limnologica* **34** (4), 349–365.

Hering, D., Meier, C., Rawer-Jost, C., Feld, C. K., Biss, R., Zenker, A., Sundermann, A., Lohse, S. & Böhmer, J. 2004 Assessing streams in Germany with benthic invertebrates: selection of candidate metrics. *Limnologica* **34** (4), 398–415.

LAWA – Länderarbeitsgemeinschaft Wasser AO 2007 Rahmenkonzeption Monitoring. Teil B – Bewertungsgrundlagen und Monitoring. Arbeitspapier II Hintergrund- und Orientierungswerte für physikalisch-chemische Komponenten. Stand 07.03.2007.

Lohr, H. 2001 *Simulation, Bewertung und Optimierung von Betriebsregeln für wasserwirtschaftliche Speichersysteme.* Mitteilungen des Instituts für Wasserbau und Wasserwirtschaft Heft 118, Technische Universität Darmstadt, Darmstadt.

Montgomery, D. C. 2005 *Design and Analysis of Experiments*, 6th Edition 2005. John Wiley, New York.

Paul, L. & Scheifhacken, N. 2009 *Einfluss von Wassermengenbewirtschaftung auf die vertikale Verteilung allochthoner Trübstoffe in Trinkwassertalsperren nach Hochwässern.* DGL Tagungsband 2009, Oldenburg. Eigenverlag der DGL, S. pp. 144–148, Hardegsen 2010.

Petzoldt, T., Rolinski, S., Rinke, K., König, M., Baumert, H. Z. & Benndorf, J. 2005 SALMO: Die ökologische Komponente des gekoppelten Modells. *Wasserwirtschaft* **95** (5), 28–33.

Reichert, P., Borchardt, D., Henze, M., Rauch, W., Shanahan, P., Somlyody, L. & Vanrolleghem, P. A. 2001 *River Water Quality Model No.1, Scientific & Technical Report No. 12.* International Water Association, London. ISBN: 9781900222822, 144 pp.

Richter, B. D., Baumgartner, J. V., Powell, J. & Braun, D. P. 1996 *A method for Assessing Hydrologic Alteration within Ecosystems. Conservation Biology* 10 (4), pp. 1163–1174.

Richter, S. & Völker, J. 2010 *Water Framework Directive – The Way Towards Healthy Waters. Results of the German River Basin Management Plans 2009.* Published in July 2010 from the Federal Ministry for the Environment, Nature Conservation and Nuclear Safety, Berlin, 75 pp.

Rolinski, S., Petzoldt, T., Baumert, H. Z., Bigalke, K., Horn, H. & Benndorf, J. 2005 Das physikalisch-ökologisch gekoppelte Talsperrenmodell. *Wasserwirtschaft* **95** (5), 34–38.

Rolinski, S., Petzoldt, T. & Benndorf, J. 2008 Nutrients, Weather, Food Web: What do we need to model Trophic Change? In: *Proceedings of the Workshop 'Perspectives of lake modeling towards predicting reaction to trophic change'*, Kompetenzzentrum Wasser Berlin Publication Series, 9, pp. 30–37, ISBN 978-3-9811684-3-3.

Rolinski, S., Petzoldt, T., Hillert, K., Paul, L. & Benndorf, J. 2007 Linking water quality to water quantity: Integration of disturbance variables into the management strategies of drinking water reservoirs, In: *American Geophysical Union Fall Meeting*, abstract #B51B-0378, San Francisco, USA.

Saenger, N. & Zanke, U. C. E. 2006 A depth -oriented view of hydraulic exchange patterns between surface water and the hyporheic zone: analysis of field experiments. *Fundamental and Applied Limnology. Special Issue Advances in Limnology* **61**, 9–27.

Scheifhacken, N., Horn, H. & Paul, L. 2010 Comparing *in situ* particle monitoring to microscopic counts of plankton in a drinking water reservoir. *Water Research* **44**, 3496–3510.

Sieber, H.-U. 2003 Talsperren als multifunktionale Anlagen. *WWT* **12**, 17–21.

Slavik, I. & Uhl, W. 2007 Optimizing water treatment to respond to quality changes due to reservoir management and climate change, In: *Proc. Am. Water Works Ass. 'Water Quality Technology Conference'*, Charlotte, USA.

Slavik, I. & Uhl, W. 2009 Analysing water quality changes due to reservoir management and climate change for optimization of drinking water treatment. *Water Science and Technology - Water Supply* **9** (1), 99–105.

Slavik, I., Skibinski, B., Watzlawik, S. & Uhl, W. 2010 Beschreibung von Einflüssen der Wasserqualität auf den Aufwand für die Trinkwasseraufbereitung – ein statistisches Modell. In: *Proc. Wasser 2010 – Jahrestagung der Wasserchemischen Gesellschaft*, Bayreuth, 10–12. May, 2010.

Strayer, D. L. & Dudgeon, D. 2010 Freshwater biodiversity conservation: recent progress and future challenges. *Journal of the North American Benthological Society* **29** (1), 344–358.

Sundermann, A., Pauls, S. U., Clarke, R. T. & Haase, P. 2008 Within-stream variability of benthic invertebrate samples and EU Water Framework Directive assessment results. *Fundamental and Applied Limnology (Archiv fuer Hydrobiologie)* **173**, 21–34.

First received 22 June 2012; accepted in revised form 24 September 2012

Estimating the recreational carrying capacity of a lowland river section

Stefan Lorenz and Martin T. Pusch

ABSTRACT

Recreational boating represents a major human use of inland waters in many regions. However, boating tourism may affect the ecological integrity of surface waters in multiple ways. In particular, surface waves produced by boating may disturb freshwater invertebrates, such as interrupting the filtration activity of benthic mussels. As mussels may significantly contribute to self-purification, disturbance may have crucial impacts on water quality, and thus on water tourism. In this paper we calculate the carrying capacity of a river section for sustainable boating tourism based on the preservation of water quality. This approach is complemented by spatial and social approaches for carrying capacity estimates. The ecological carrying capacity significantly decreases with lower water levels during summer. Hence, the analysis of variables that influence the river's carrying capacity allows the formation of recommendations for management measures that integrate social, touristic and ecological aspects.

Key words | boating tourism, mussels, self-purification, Spree river, sustainability, water management

Stefan Lorenz (corresponding author)
Martin T. Pusch
Leibniz-Institute of Freshwater Ecology and Inland Fisheries, IGB,
Müggelseedamm 301,
12587 Berlin,
Germany

Stefan Lorenz
Institute of Biology,
Freie Universität Berlin,
Takustraße 3,
14195 Berlin,
Germany
E-mail: *stefan.lorenz@igb-berlin.de*

INTRODUCTION

Rivers and lakes constitute some of the most valuable natural resources for tourism, as they provide opportunities for multiple touristic activities, like swimming, fishing, sailing or boating (Postel & Carpenter 1997). The multiple ways to use the natural resources of lakes and rivers allow the establishment of a broad array of secondary business segments, such as boat rentals, camping grounds, restaurants and hotels (e.g. CBSPOR 2011).

Growing urban development of the shores of inland waters is expected to produce increasing detrimental effects on surface water systems, such as pollution, physical alteration of shoreline areas, and disturbance of aquatic flora and fauna (Gabel *et al.* 2008). To avoid devaluing major lake resources, the detrimental effects of water-based tourism must be mitigated and regulated. Hence, the carrying capacity of the surface water body under use must be estimated to ensure the sustainable operation of water-related tourism businesses.

The carrying capacity concept was first developed in the field of population ecology and wildlife management, where it was defined as the number of individuals of one species that can be maintained within a given habitat area (Wagar 1964). This concept was also applied to the management of visitor numbers to national parks and wild lands. By applying the concept of carrying capacity to humans, two additional aspects emerged. First, visitors affect the environmental resources of the area they use in a more variable way (as previously hypothesised in wildlife management), and second, visitors affect their own perception of the environment at the same time. Furthermore, crowding or social conflicts also influence a visitor's personal experience and recreational effect (Wagar 1964).

Previous efforts to estimate the carrying capacity of lakes for water-based tourism accounted for these aspects by, for example, estimating the minimum required lake surface area needed for the safe operation of a single boat type (e.g. Jaakson *et al.* 1990; ERM Inc 2004; Rajan *et al.* 2011). The size of this area has mostly been determined based on the social perception of boaters, recorded through interviews and surveys. This specification varies greatly between low and high horsepower boats, as well as for muscle-driven vessels (i.e. kayaks, canoes). An approach to determine the touristic carrying capacity of rivers has been developed, which is based on boat counts and user

doi: 10.2166/wst.2013.418

surveys (Rebellato 2007). However, existing estimates of the touristic carrying capacity of lakes have primarily focused on the human component, thus neglecting the potential ecological impacts of tourism. For a more comprehensive approach, four different components have been recommended to estimate the recreational carrying capacity (O'Reilly 1986; Shelby & Heberlein 1986). These comprise ecological, spatial, economic as well as social aspects. Among these, environmental aspects deserve special attention, as these variables are often regularly monitored but may exhibit considerable variation depending on, for example, season and river discharge.

While the impacts of human use of inland waters are generally well known (Goudie 2006), less is understood about the specific effects arising from recreational activities. Motorised boating activities were found to constitute a major source of noise and unburned fuel emissions into surface waters, although the amount of these inputs varies greatly among the engine types (see e.g. Jüttner *et al.* 1995, Mosisch & Arthington 1998). Thus, it has been suggested to analyse impacts on water quality in order to assess the pollution impacts of recreational watercrafts, which might affect total suspended solids, turbidity, dissolved oxygen, pH and temperature (Rebellato 2007). However, natural processes may also influence these parameters of water quality. In contrast to pollution, the formation of surface waves by all kinds of watercrafts has largely been neglected so far, even though the ecological impacts are strong enough to determine the ecological carrying capacity for boating tourism.

The main pathways of environmental effects that originate from water tourism link boating activities with: (1) key variables of ecosystem health, (2) related political goals, and (3) respective actions of water management (Figure 1). Consequently, the major effects originating from water touristic activities, such as sediment resuspension, the disturbance of wildlife and impact of waves, may affect water quality and water ecology. Water quality variables like turbidity and dissolved oxygen and ecosystem processes like the water filtration activity of mussels are negatively influenced by these impacts. This leads to a decline in the visitor's perception of nature experience and quality. Furthermore, impacts on wildlife (such as waterfowl, invertebrates or mussel filtration activity) might also cause conflicts with regulatory requirements, e.g. the European Water Framework Directive (EC 2000). Hence, a sustainability analysis may be developed to establish integrated water tourism management by using the thresholds identified for various types of use and the related carrying capacity for water tourism.

As the success of water tourism is significantly influenced by the available water quality (Bockstael *et al.* 1989), touristic development should seek to minimise detrimental effects on the self-purification capacity of surface waters, as this would reduce system resilience (Zeng *et al.* 2011). In an earlier study (Lorenz *et al.* submitted) we described the

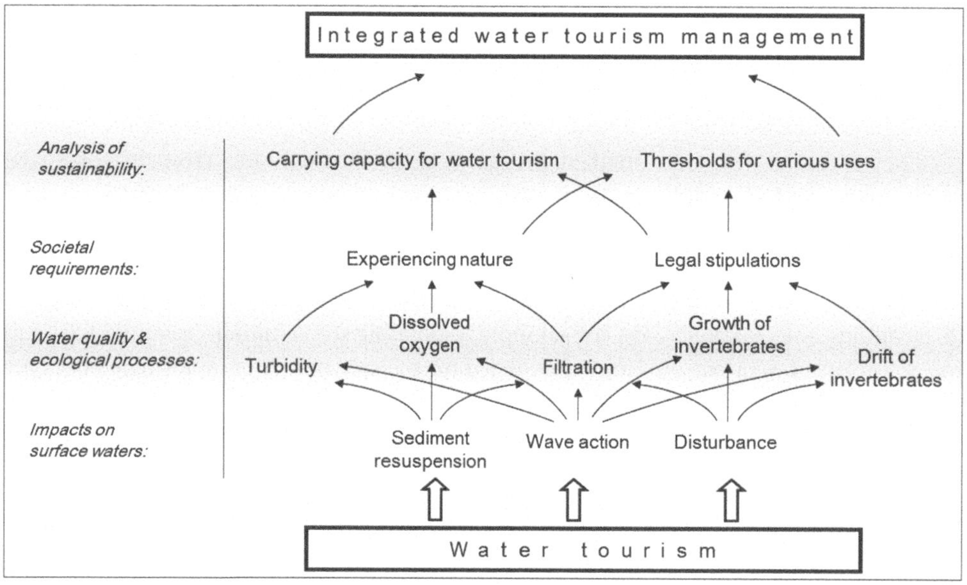

Figure 1 | Conceptual model of the major effects of water tourism on river ecosystems. The effects have impacts on different levels of perception that require consideration in an overall water tourism management strategy. All effects might be additionally impacted if framework conditions, such as water availability, are altered by unswayable events like climate change.

impacts of wave formation from boating on freshwater mussels, which transfer organic matter from the water column to the benthic zone by their filter-feeding activity. Therefore, freshwater mussels directly contribute to the self-purification of surface waters, as most of the organic matter is processed by microbial degradation activity associated with the sediments (Fischer & Pusch 2001). Consequently, any impairment of this self-purification potential will produce negative effects on water quality. Decreasing water quality will in turn affect water tourism; for instance, the attractiveness of boat launching sites is strongly determined by water quality (Bockstael *et al.* 1989).

Hence, we aim to estimate the carrying capacity of a river section used for boating tourism based on an ecological approach that considers self-purification as the key ecosystem process for the preservation of acceptable water quality. By comparing this ecological approach with the spatial and social components that are estimated in parallel the development of integrative concepts to estimate the carrying capacity of surface waters for boating tourism is feasible.

METHODS

Study site

The 21.1 km long river section of the Krumme Spree (Welker & Walz 1998) serves as a connection between two lakes with intensive touristic use, and thus constitutes an ideal area to conduct studies on the impact of boating on water ecosystems. The mean width of this river section is about 25 m, with mean depth ranging from 1.4 to 2.5 m. As calculations of *Usable Lake Areas* suggest the importance of restricted areas (e.g. no-wake areas or shallow water depths) to protect shorelines (Rajan *et al.* 2011), this concept is not as directly applicable to our study river. Therefore, for our calculations we adopted the notion that the shallower portions of lakes (i.e. of less than 1.5 m deep) are the most susceptible to environmental impacts (Wagner 1991). As the Krumme Spree was straightened due to intense building activities between 1906 and 1911, it now has an almost homogeneous trapezoidal cross-section (Pusch & Hoffmann 2000). Hence in this river section, portions of river of less than 1.5 m depth would extend 3 m from the shoreline, reducing the *Total River Area* from 0.528 km^2 to 0.400 km^2 of *Usable River Area*.

Application of the carrying capacity concept to a river section

To apply the current concept of estimating touristic carrying capacity from lakes to rivers, the respective terms and equations required adjustment. The *Optimal Boating Density* describes the recommended space for various boat categories, measured in square metres per boat. We reviewed several carrying capacity studies (Jaakson *et al.* 1990; ERM Inc. 2004; CBSPOR 2011) to obtain values of *Optimal Boating Densities* that were applicable to the watercraft types that are used for boating tourism on the Spree River. Through collating the various published values, we adopted a density of 5,261 m^2/boat for muscle-driven boats, 36,422 m^2/boat for open motorboats (<15 HP), 28,328 m^2/boat for motorised rowing and rafting boats (<10 HP), and 60,703 m^2/boat for yachts (>15 HP). The *Optimal Boating Density* for the observed boating mix of the Krumme Spree was defined by the following equation (after Dearlove 2010):

$$\begin{aligned}&\text{Optimal boating density}\\&\text{of observed boating mix}\\&\quad=\left(\begin{matrix}\text{range of optimal}\\\text{boating densities}\end{matrix}\times\begin{matrix}\text{proportion of}\\\text{motorised boats}\end{matrix}\right)\\&\quad+\begin{matrix}\text{optimal boating density}\\\text{of muscle driven boats}\end{matrix}\end{aligned} \tag{1}$$

The *Range of Optimal Boating Densities* was derived by subtracting the minimum of the *Optimal Boating Densities* for the boat categories from the maximum (60,703–5,261 = 55,442), where the proportions of boating at the Krumme Spree correspond to muscle-driven boating = 0.39, open motorboats = 0.29, motorised rowing and rafting boats = 0.11 and yachts = 0.21. Therefore, the *Optimal Boating Density* for the observed boating mix on the Krumme Spree amounts to an estimated 39,081 m^2/boat. As the areas that are required per boat describe only the water surface area that is needed for the safe operation of this type of boat, we had to recalculate these values for the straight course of rivers instead of lake areas. Assuming that these areas would enclose a boat like a circle, we can take the radius of these circles as the area in running river metres that is required per boat type between consecutive boats. The results of this calculation for the different boat types are: 41 running river metres for muscle-driven boats, 108 m for open motorboats (<15 HP), 95 m for motorised rowing and rafting boats

(<10 HP), 139 m for yachts (>15 HP), and 112 m for the observed boating mix. By using the following equation, this leads to an *Estimated Carrying Capacity* of 188 boats of the observed mix that could be present on the river at any one time:

$$\text{Estimated carrying capacity} = \frac{\text{river length}}{\text{requested area for observed boating mix}} \quad (2)$$

At mean peak use on weekends during July, an average of six boats per hour passes the Krumme Spree (Lorenz *et al.* submitted). To transfer the carrying capacity of 188 boats at the same time on the river into the number of boats per hour, we assume that all boats were moving with a constant velocity of 8 km/h. This leads to a carrying capacity of 257 boats per hour on the whole river section of the Krumme Spree. By using Equation (3), the *Observed Carrying Capacity* of the Krumme Spree is estimated to be 2.3%:

$$\text{Observed carrying capacity}\,[\%] = \frac{\text{total number of boats per hour}}{\text{estimated capacity per hour}} \times 100 \quad (3)$$

RESULTS AND DISCUSSION

Hydrology

Although the water level exceeds the proposed 1.8 m water depth during most of the boating season at the Krumme Spree (Lorenz *et al.* submitted), in the summer months this level may drop by about 0.9 m with receding discharge (Figure 2). An exponential regression model between discharge and water level at the sampling site in the Krumme Spree (Lorenz *et al.* submitted) revealed that with each 1 m^3/s reduction in discharge, the water level falls by about 4.5 cm ($y = 2.061 + 2.027 \times (1\text{–}0.971^x)$, $R^2 = 0.92$, $P < 0.0001$).

Estimation of the ecological carrying capacity

As estimated by Lorenz *et al.* (submitted) by use of coupled hydraulic–ecological modeling, motorboats cause the most adverse effects on the filtration activity of freshwater mussels in the shallow marginal zone by their intense wave production, while effects of motorised rafting and rowing

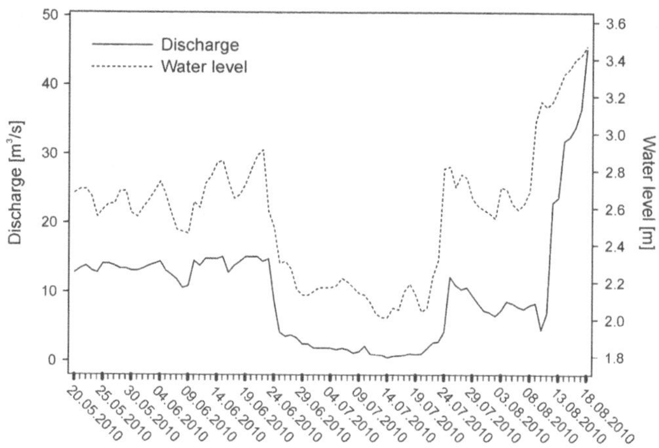

Figure 2 | Dynamics of discharge (records from the nearest gauging station of Leibsch) and related water level fluctuations on the Krumme Spree (Lorenz *et al.* submitted) during the summer months of 2010.

boats or yachts were partially significantly smaller. Therefore, we can state that the number of motorboats on the river constitutes the limiting factor for the estimation of the ecological carrying capacity. This coupled hydraulic–ecological model estimated a number of 93 motorboats passing by within one hour that would be sufficient to prevent filtration activity of all mussel specimens (with boats navigating at assumed 10 km/h). Using Equation (4), this would correspond to 321 boats per hour for a scenario with the observed boating mix:

$$\text{Ecological carrying capacity} = \sum \frac{\text{proportion of boat type} \times \text{maximum no. of motorboats}}{\text{proportion of motorboats}} + \text{maximum no. of motorboats} \quad (4)$$

Under the hydrological conditions of a summer drought as described in Pusch & Hoffmann (2000) with the water depth reduced to 1.5 m, the estimated maximum number of boats would be 62 motorboats per hour until complete mussel filtration would be inhibited, leading to the calculation of 214 boats per hour of the observed boating mix by using Equation (4). By assuming the observed minimal discharge during the summer of 2010 of 0.49 m^3/s, taking into account the above described accompanying drop in water level, the above number would further be reduced to 203 boats. As the number of boats that are able to pass the river within one hour until mussel filtration completely ceases declines with the lowering of the water level (Lorenz *et al.* submitted), we consequently calculated the maximum number of boats from the observed boating mix at each water depth by using Equation (4) (Figure 3). This calculated boating density is further referred to as the *Ecological Carrying Capacity* of the river Krumme Spree.

Estimation of the social carrying capacity

The *Estimated Carrying Capacity* per hour (from now on referred to as the *Social Carrying Capacity*) that is able to pass the river under social and spatial aspects varies by a small number of boats if the water level drops. The almost homogeneous, trapezoidal cross-section of the Krumme Spree results in a reduction of the *Usable River Area* with lowered water levels. With a given 26.6° slope of the riprap, the mean river width would be lowered by about 0.44 m each time the water level drops about 10 cm. Therefore, there is an equal reduction in the *Social Carrying Capacity* at each water level drop. Based on this lowered *Usable River Width*, the *Social Carrying Capacity* was calculated by replacing the *Requested Area for the Observed Boating Mix* from Equation (2) with the *Adjusted Area for the Observed Boating Mix* derived from Equation (5). Within this equation, the loss of river area (in square metres) over one section of the *Requested Area for the Observed Boating Mix* was calculated from the *lowered river width*. Assuming that this area must also be considered for the safe operation of boats when the water level declines, the radius of a circle containing this area must be added to the original *Requested Area for the Observed Boating Mix* for each water level reduction. The dependence of the *Social Carrying Capacity* on varying water levels is also plotted in Figure 3:

$$\text{adjusted area for observed boating mix} = \text{requested area for observed boating mix} + \sqrt{\frac{\text{lowered river width} * \text{requested area for observed} \times \text{boating mix}}{\pi}} \quad (5)$$

Dependence of the carrying capacity on framework conditions

The *Ecological Carrying Capacity* shows a clearly stronger dependency on river water level than the *Social Carrying Capacity*. At water depths greater than 170 cm, the *Ecological Carrying Capacity* exceeds the *Social Carrying Capacity*

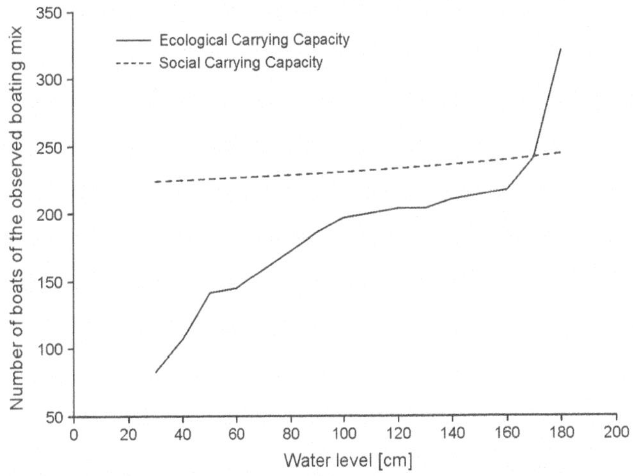

Figure 3 | Dependence of the ecological carrying capacity (black line) and the social carrying capacity (dashed line) of the Krumme Spree on the water level of the river. If the water level drops below 170 cm, the ecological carrying capacity constitutes the limiting component determining the maximum number of boats that is tolerable on the river.

several times. As the effects of open motorboats and motorised rowing and rafting boats vanish at water depths exceeding 190 cm (Lorenz *et al.* submitted), and effects of muscle-driven boats only extend down to water depths of about 100 cm, the *Ecological Carrying Capacity* for water depths between 190 and 275 cm only depends on the number of yachts passing the river. Above this water level, the *Ecological Carrying Capacity* of the Krumme Spree is nearly unlimited with respect to the boating tourism of the mentioned boat types concerning the self-purification activity based on mussel filtration. However, at water levels below 170 cm, the *Ecological Carrying Capacity* drops below the line of the *Social Carrying Capacity*. Thus, the *Ecological Carrying Capacity* becomes the more important component during times of low river discharge, and should then be considered instead of social and spatial aspects.

The ecological status of surface waters is influenced by the natural dynamics of framework conditions, such as climatic variation, and by pressures exerted by humans. Consequently, the carrying capacity for a given human use is influenced by this variability and other uses as well. Restoration efforts in the Krumme Spree section could increase flow velocity during drought periods by decreasing channel depth by 0.5–1 m (Pusch & Hoffmann 2000; Pusch *et al.* 2001), aiming to compensate for the considerable decline in the water levels predicted by the climate change scenarios (Federal Environmental Agency 2005). However, the implementation of these restoration measures would clearly reduce the ecological carrying capacity of this river section for boating activities, which in turn might be compensated for by a potential change in the actual boat mix (ERM Inc. 2004).

CONCLUSIONS

We developed an ecological approach that aims to assure sustainable water tourism in the river section of the Krumme Spree. The approach considers all kinds of watercrafts present in this river section exerting a wave impact on the self-purification activity from freshwater mussels. The ecological carrying capacity showed a marked dependence on the actual water depth of the river, as the self-purification activity of benthic mussels is significantly reduced when shallow river reaches are used for boating. Thus, our approach also allows implications for an integrated water resource management to be derived that might preserve this river section as an area for water tourism. Increased water retention within the catchment area might contribute to maintaining water levels above critical thresholds. Alternatively, a shift in the boating mix on the river during times of low flow, involving a smaller proportion of motorboats, would contribute to the preservation of the river's ecological integrity. Consideration of such results could lead to changes in regulatory policies or to an adaptation of business strategies of the local boating and tourism industry. Our results suggest that specific concepts for water tourism should be developed for each water body, based on detailed information of respective social and ecological requirements. We could demonstrate that the relative importance of social and environmental aspects towards estimating the overall carrying capacity depends on key framework conditions. Therefore, we recommend the consideration of ecological approaches estimating the carrying capacity for boating tourism in sensitive river or lake ecosystems.

ACKNOWLEDGEMENTS

We gratefully acknowledge the financial support provided by the German Federal Ministry of Education and Research (BMBF, FKZ 01LR0803G) through the project INKA-BB.

REFERENCES

Bockstael, N., McConnell, K. & Strand, I. 1989 Measuring the benefits of improvements in water quality: the Chesapeake Bay. *Marine Resource Economics* **6**, 1–18.

CBSPOR *Colorado Board of State Parks and Outdoor Recreation* 2011 Stagecoach State Park Management Plan 2011–2021. Available from: http://parks.state.co.us/SiteCollectionImages/parks/Planning/Strategic%20PLan/2011-2021%20Stagecoach%20Management%20Plan.pdf (accessed 17 August 2011).

Dearlove, P. 2010 Balancing recreational use: lessons learned from Lake Ripley's recreational carrying capacity study. Available from: http://www.wisconsinlakes.org/events/pdf/8%29LakeRipleyRecreationalCarryingCapacityStudy-PaulDearlove.pdf (accessed 17 August 2011).

EC European Parliament and the Council of the European Union 2000 Directive 2000/60/EC Establishing a framework for the community action in the field of water policy. *Official Journal of the European Communities* **L327** (1), 1–72.

ERM Inc. 2004 Deep Creek Lake boating and commercial use carrying capacity study. Available from: http://www.dnr.state.md.us/irc/docs/00015716.pdf (accessed 17 August 2011).

Fischer, H. & Pusch, M. T. 2001 Comparison of bacterial production in sediments, epiphyton and the pelagic zone of a lowland river. *Freshwater Biology* **46**, 1335–1348.

Gabel, F., Garcia, X. -F., Brauns, M., Sukhodolov, A., Leszinski, M. & Pusch, M. T. 2008 Resistance to ship-induced waves of benthic invertebrates in various littoral habitats. *Freshwater Biology* **53** (1), 1567–1578.

Goudie, A. 2006 *The Human Impact on the Natural Environment: Past, Present, and Future*. 6th edition. Wiley-Blackwell, Massachusetts, USA.

Jaakson, R., Buszynski, M. D. & Botting, D. 1990 Carrying capacity and lake recreation planning (Part II). *The Michigan Riparian*, Feb, 7–8.

Jüttner, F., Backhaus, D., Matthias, U., Esser, U., Greiner, R. & Mahr, B. 1995 Emissions of two- and four-stroke outboard engines–I. Quantification of gases and VOC. *Water Research* **29** (8), 1976–1982.

Lorenz, S., Dobra, N. & Pusch, M. T. submitted Modelling the impacts of recreational boating on self-purification activity in a lowland river. *Freshwater Science*.

Mosisch, T. D. & Arthington, A. D. 1998 The impacts of power boating and water skiing on lakes and reservoirs. *Lakes & Reservoirs: Research and Management* **3**, 1–17.

O'Reilly, A. M. 1986 Tourism carrying capacity: concepts and issues. *Tourism Management* **7** (4), 254–258.

Postel, S. & Carpenter, S. 1997 Freshwater ecosystem services. In: *Nature's Services: Societal Dependence on Natural Ecosystems* (G. C. Daily ed.). Island Press, Washington, DC, USA, pp. 195–215.

Pusch, M. T. & Hoffmann, A. 2000 Conservation concept for a river ecosystem (River Spree, Germany) impacted by flow abstraction in a large post-mining area. *Landscape and Urban Planning* **51** (2–4), 165–176.

Pusch, M. T., Köhler, J., Wanner, S., Ockenfeld, K., Hoffmann, A., Brunke, M., Grünert, U. & Kozerski, H. -P. 2001 Ökologisch begründetes Bewirtschaftungskonzept für die Spree unter dem Aspekt der bergbaubedingten Durchflußreduktion (Ecologically justified management approach for the river Spree from the aspect of mining-related flow reduction). Reports of the IGB 11, IGB, Berlin, Germany.

Rajan, B., Varghese, V. M. & Pradeepkumar, A. P. 2011 Recreational boat carrying capacity of Vembanad Lake Ecosystem, Kerala, South India. *Environmental Research, Engineering and Management* **56** (2), 11–19.

Rebellato, B. 2007 Exploring carrying capacity and acceptable change: 2006 River User Satisfaction Surveys. Available from: http://www.whistler2010.com/cms-assets/documents/4757-250985.riverusesurveyreport06.pdf (accessed 17 August 2011).

Shelby, B. & Heberlein, T. A. 1986 *Carrying Capacity in Recreation Settings*. Oregon State University Press, Corvallis, OR, USA.

Federal Environmental Agency 2005 Climate Change in Germany – Vulnerability and Adaption of Climate Sensitive Sectors. Research Report 201 41 253, Federal Environmental Agency (UBA), Dessau, Germany.

Wagar, J. A. 1964 The carrying capacity of wild lands for recreation. *Forest Science Monographs* **7** (1), 1–24.

Wagner, K. J. 1991 Assessing impacts of motorized watercraft on lakes: issue and perceptions. In: Proceedings of a National Conference on Enhancing States' Lake Management Programs, May 1990, Northeastern Illinois Planning Commission, Chicago, IL, USA, pp. 77–93.

Welker, M. & Walz, N. 1998 Can mussels control the plankton in rivers? – a planktological approach applying a Lagrangian sampling strategy. *Limnology and Oceanography* **43** (5), 753–762.

Zeng, C., Liu, Y., Liu, Y., Hu, J., Bai, X. & Yang, X. 2011 An integrated approach for assessing aquatic ecological carrying capacity: a case study of Wujin District in the Tai Lake Basin, China. *International Journal of Environmental Research and Public Health* **8** (1), 264–280.

First received 30 December 2011; accepted in revised form 6 June 2012

Sustainable management of a coupled groundwater–agriculture hydrosystem using multi-criteria simulation based optimisation

Jens Grundmann, Niels Schütze and Franz Lennartz

ABSTRACT

In this paper we present a new simulation-based integrated water management tool for sustainable water resources management in arid coastal environments. This tool delivers optimised groundwater withdrawal scenarios considering saltwater intrusion as a result of agricultural and municipal water abstraction. It also yields a substantially improved water use efficiency of irrigated agriculture. To allow for a robust and fast operation we unified process modelling with artificial intelligence tools and evolutionary optimisation techniques. The aquifer behaviour is represented using an artificial neural network (ANN) which emulates a numerical density-dependent groundwater flow model. The impact of agriculture is represented by stochastic crop water production functions (SCWPF). Simulation-based optimisation techniques together with the SCWPF and ANN deliver optimal groundwater abstraction and cropping patterns. To address contradicting objectives, e.g. profit-oriented agriculture vs. sustainable abstraction scenarios, we performed multi-objective optimisations using a multi-criteria optimisation algorithm.

Key words | artificial intelligence, density-dependent groundwater flow modelling, integrated water resources management, irrigation, multi-criteria optimisation

Jens Grundmann (corresponding author)
Niels Schütze
Franz Lennartz
TU Dresden,
Institute of Hydrology and Meteorology,
01062 Dresden,
Germany
E-mail: *jens.grundmann@tu-dresden.de*

INTRODUCTION

The overexploitation of groundwater resources by agriculture is a growing concern worldwide. Coastal regions in arid climates are particularly vulnerable to excessive groundwater mining because declining groundwater levels lead to saltwater intrusion into the aquifer systems. Using this increasingly saline water for irrigation destroys the agricultural resources, which are the economic basis for the farmers and their communities. The worldwide relevance of this mechanism, e.g. at the coasts of Mexico, Australia, Oman, and along the Mediterranean Sea, was discussed by several authors (Cardona *et al.* 2004; Narayan *et al.* 2007; Kacimov *et al.* 2009).

Up to now, three different types of modelling approach have been used in water resources management: (1) employing simplistic optimisation models (e.g. linear programming), (2) replacing the optimisation procedure with the simulation of several – subjectively selected – 'what if' scenarios and then selecting the best alternatives that satisfy certain objective criteria (Harou *et al.* 2009), and (3) simulation-optimisation. While the second approach results in suboptimal solutions, the first approach suffers from inadequate process descriptions. A general shortcoming of current management models is that they do not reliably portray nonlinear, interdependent phenomena of complex hydrological systems and their interdependencies. In addition, the impact of uncertainties of a changing environment associated with global change on the sustainable management of coupled groundwater–agriculture systems cannot be quantified. To overcome these shortcomings, the third approach – which is referred to as simulation-optimisation – was introduced by Gosavi (2003) who suggested a combination of physically based numerical models and (black-box) optimisation methods. Unfortunately, the optimisation of complex water resources systems often becomes computationally intractable when solving simulation-optimisation problems with large numbers of design variables and/or objectives (Loucks & van Beek 2005).

doi: 10.2166/wst.2013.602

The latest simulation-optimisation modelling approaches extend classical simulation-optimisation strategies by two improvements:

1. customised tuned simulation-optimisation based on surrogate models – these are models which describe the behaviour of process models using only the most relevant features, and
2. the decomposition of complex optimisation problems.

Both improvements are already introduced in management of coastal aquifer systems. Mostly, artificial neural networks (ANN) (Rao *et al.* 2004; Bhattacharjya & Datta 2009; Dhar & Datta 2009) or piecewise linear transformations (Rejani *et al.* 2009) are used as surrogate models to emulate the behaviour of groundwater flow models. The resulting surrogate models help to solve mono-criteria optimisation problems, which are represented by a single objective function (Rao *et al.* 2004), and multi-criteria optimisation problems, if more than one objective function exists (Bhattacharjya & Datta 2009; Dhar & Datta 2009). However, the surrogate models in these studies were designed only for a predefined time period and do not allow for an application of optimisation problems with varying time horizons, which is a major limitation for optimally solving water resources management problems. Another limitation of these studies is that they used detailed model descriptions for the aquifer only, but did not realistically consider the demand side, i.e. the agricultural system and its interdependencies with the aquifer.

The decomposition of optimisation problems is aimed at the separation of a complex optimisation problem characterised by a high number of decision variables into smaller, independent optimisation problems which allow for a faster and more reliable solution of the complex optimisation problem. As an example, Schmitz *et al.* (2007) applied decomposition for optimising furrow irrigation systems by separating the problem of optimal irrigation control from the general problem of optimal irrigation scheduling. More recently, Schütze (2012c) proposed a decomposition solution strategy which allows for the use of numerical process models and Monte Carlo simulations for an optimal water demand management at plot and regional scale.

Within this paper we present key issues of a new simulation-based integrated water management system for sustainable water management in arid zones. We use physically based models to simulate the interacting processes of a fully interacting groundwater–agriculture system, which is affected by water quality due to saltwater intrusion. This model system is linked with multi-criteria optimisation to determine optimal groundwater abstraction schemes and cropping patterns. To get optimal results within reasonable time, surrogate models are introduced which extend the latest approaches by also considering the agricultural water demand, and allow for an application of different time horizons. The combination of a multi-criteria optimisation framework and the derived surrogate models forms a new prototype of a simulation-based integrated water management tool. The functionality of the proposed prototype is demonstrated by a real world example of the Al Batinah coastal region in Oman.

MATERIALS AND METHODS

Basic components of the simulation-based integrated water management system are shown in Figure 1. This combines coupled simulation models, which are describing the considered hydrological systems, one or more objective function(s) and an adequate simulation-optimisation (S-O) method, which are all to be assumed to be fully interoperable.

In the following, an overview about a real application case is given as well as the three elementary steps which are necessary to apply the simulation-optimisation approach in an extended optimisation framework based on surrogate modelling. These steps are as follows:

1. The setup of process models for simulating the behaviour of the agricultural production as well as the aquifer (see subsection "Data and process model setup").
2. The development of appropriate surrogate models. These models describe the behaviour of the process models – and at last the behaviour of the natural system. However, the surrogate models only use the most important state and decision variables which are relevant for an optimal management (see subsection "Methods to generate surrogate model").

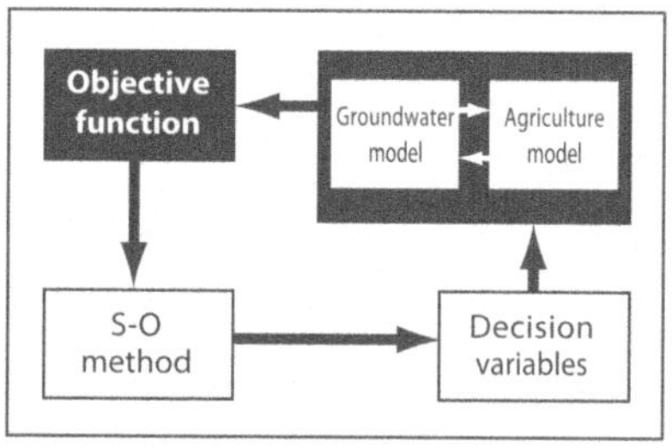

Figure 1 | Simulation-based integrated water management system (for details see Figure 3).

3. The application of the surrogates within a multi-criteria optimisation framework for evaluating Pareto-optimal data sets of the decision variables according to contradicting objectives (see subsection "The optimisation framework for multi-criteria simulation-based optimisation").

The study site

The study site is located in the north of the Sultanate of Oman in the Al Batinah region. Al Batinah is a roughly 30 km broad coastal plain which extends over 250 km along the coast of the Gulf of Oman north-westward of the capital Muscat. The area comprises the coastal zone of the Wadi Ma'awil catchment, as shown in Figure 2.

Oman's agricultural production requires irrigation due to arid climate conditions. Most of the farms are located near the coastline and the farmers abstract the water for irrigation from a coastal aquifer by numerous decentralised and often uncontrolled wells (Figure 2). A progressive intensification of the agriculture since the 1980s along with inefficient irrigation practices led to an increasing water demand and to continuously declining groundwater levels. Ongoing overexploitation of the aquifer enforced the intrusion of marine salt water. As a result, this saltwater intrusion caused the closure of numerous coastline farms due to the increasing irrigation water and soil salinities, and therefore it threatens the economic basis for the farmers and their communities.

The coastal aquifer system is replenished mainly by the lateral inflow caused by the rainfall in the upper mountainous part of the catchment. Long-term measurements show mean average precipitation values up to 250 mm/y in the upper mountains and about 50 mm/y along the coastline. The rainfall variability in time and space is extremely high, which means that relatively wet years can be followed by long dry periods over a couple of years.

Data and process model setup

Agricultural production

Local agriculture is practiced by numerous small-scale farms of less than a hectare of land employing irrigation due to local arid conditions. This permits a variety of crops (field crops to fruit trees) to be grown (Sultanate of Oman 2005). According to the aim of the study, the long-term evaluation of the environmental and economical development of a

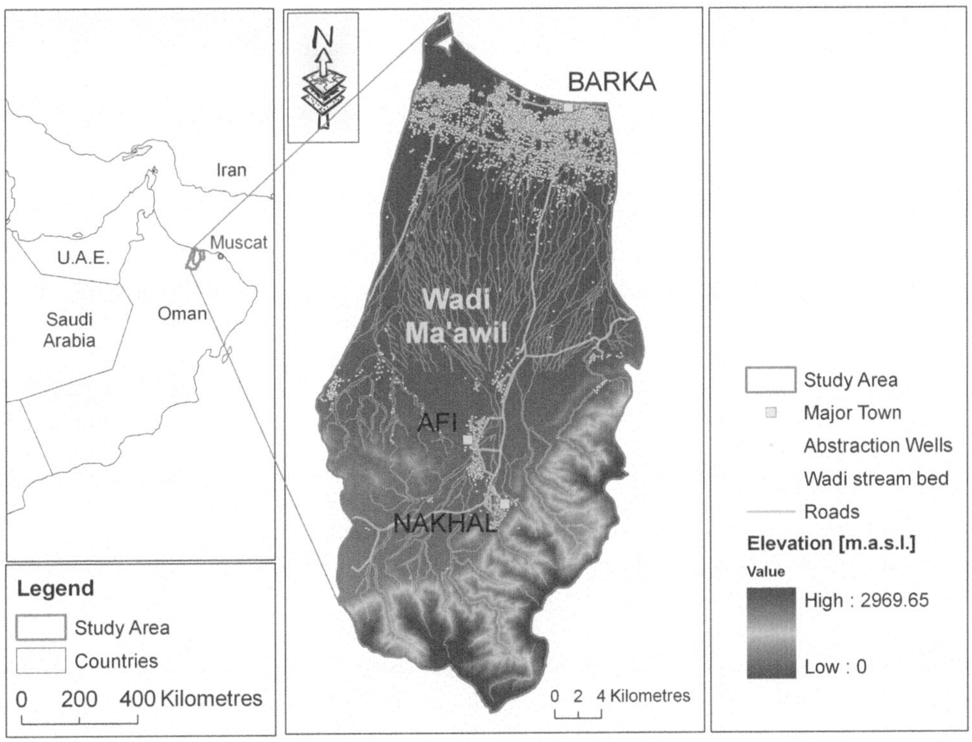

Figure 2 | Study site of the Wadi Ma'awil, Sultanate of Oman.

hypothetical farm – located in the vicinity of the sea – under different management strategies is investigated. For the sake of simplicity the cultivation of two different crops, maize – a salt-sensitive cash crop – and sorghum – a lower priced but more salt-resistant crop, growing in two seasons per year is assumed.

For simulating the agricultural system we used observed daily weather data taken over the last 16 years (1991–2006) from Seeb weather station (International airport, Muscat). A sandy loam soil is typical for the region (50, 36 and 14% sand, silt and clay, respectively) and soil hydraulic properties were taken from Abdelrahman *et al.* (1993). These data are used to parameterise the Agricultural Production Systems Simulator (APSIM by Keating *et al.* (2003)) for simulating soil water transport and crop growth.

Coastal aquifer

The spatial extension of the considered aquifer system covers the coastal area where agriculture is practiced. The hydrogeology of the study site is characterised by a complex heterogeneous structure of fluviatile, aeolian, and marine deposits, consisting of locally cemented gravels, differently sized sand, silt, loam and clay (Walther *et al.* 2012). Walther *et al.* (2012) analysed the available data which were provided by the Ministry of Regional Municipalities and Water Resources and set up a heterogeneous 3D density-dependent groundwater flow model of the alluvial coastal aquifer based on the OpenGeoSys software package (Kolditz *et al.* 2008) for the study site.

To demonstrate the features of the optimisation framework, a 2D vertical slice of the unconfined 3D alluvial coastal aquifer model was constructed to simulate groundwater flow and to account for saline intrusion and transport. Dimensions and parameters of the model are derived and generalised from the upper layer of a heterogeneous 3D groundwater flow model. The model domain is 6 km long with an aquifer thickness of 57 m at the sea boundary and 73 m at the upstream plain side, respectively. It consists of a single unconfined layer with a hydraulic conductivity $k = 1 \times 10^{-3}$ m/s and a porosity of $n = 0.3$. At the upstream boundary a constant flux boundary condition with an estimated inflow q_0, which is equal to the replenishment of the aquifer, is used. The sea side boundary condition consists of a constant pressure (sea water level) together with a constant salt concentration. Source terms within the model domain are applied to represent the extraction behaviour of the farmers using constant extraction rates during one cultivation period.

Methods to generate surrogate models

Estimation of 2D crop water production functions for modelling of impacts of water and salt stress on crop yield

The surrogate model for the agricultural production, called 2D-crop water production function (2D-CWPF), is derived by an extension of the OCCASION methodology (Optimal Climate Change Adaption Strategies in Irrigation by Schütze & Schmitz (2010)) as shown in Figure 3(a).

At first, the physically based SVAT (soil, vegetation and atmosphere) model APSIM is coupled with the GET-OPTIS tailor-made evolutionary optimisation algorithm for optimal irrigation scheduling (Schütze *et al.* 2012a) in order to calculate the maximum yield for a given amount of water and the corresponding optimal irrigation schedule (Loop 1 in Figure 3(a)).

Secondly, the given amounts of water for irrigation are varied (Loop 2 in Figure 3(a)) which results in a single crop water production function (CWPF). The CWPF describes the relationship between the amount of irrigated water and the potential crop yield which can be achieved for each crop and season assuming an optimal water application.

Within a third loop (Monte Carlo Loop in Figure 3(a)) different realisations of local weather characteristics are incorporated in order to derive more robust irrigation schedules. This allows for an agricultural production with a certain reliability. Therefore, 100 synthetic weather realisations were generated by the LARS-WG weather generator (Semenov *et al.* 1998) on the basis of the aforementioned observed daily weather data from Seeb weather station. Each weather realisation results in a single CWPF. A statistical analysis of the whole set of single CWPFs provides an empirical joint probability function which is referred to as the stochastic crop water production function (SCWPF).

In order to include salinity stress, the SCWPF is extended by a second dimension, namely the salinity of the irrigation water using the salinity tolerance index (Steppuhn *et al.* 2005). Combining this index with the SCWPF results in a 2D-SCWPF which additionally considers the consequences of the water quality and allows for an appropriate choice of the crop patterns regarding their salinity tolerance. The SCWPF considers climate variability and provides general applicable irrigation schedules which may achieve a certain amount of yield with a given reliability (e.g. 90%), i.e. a given quantile. Once the quantile is defined it is possible to derive a conditional CWPF which is introduced here as 2D-CWPF$^{0.9}$. The functional relationships of the 2D-SCWPF for defined

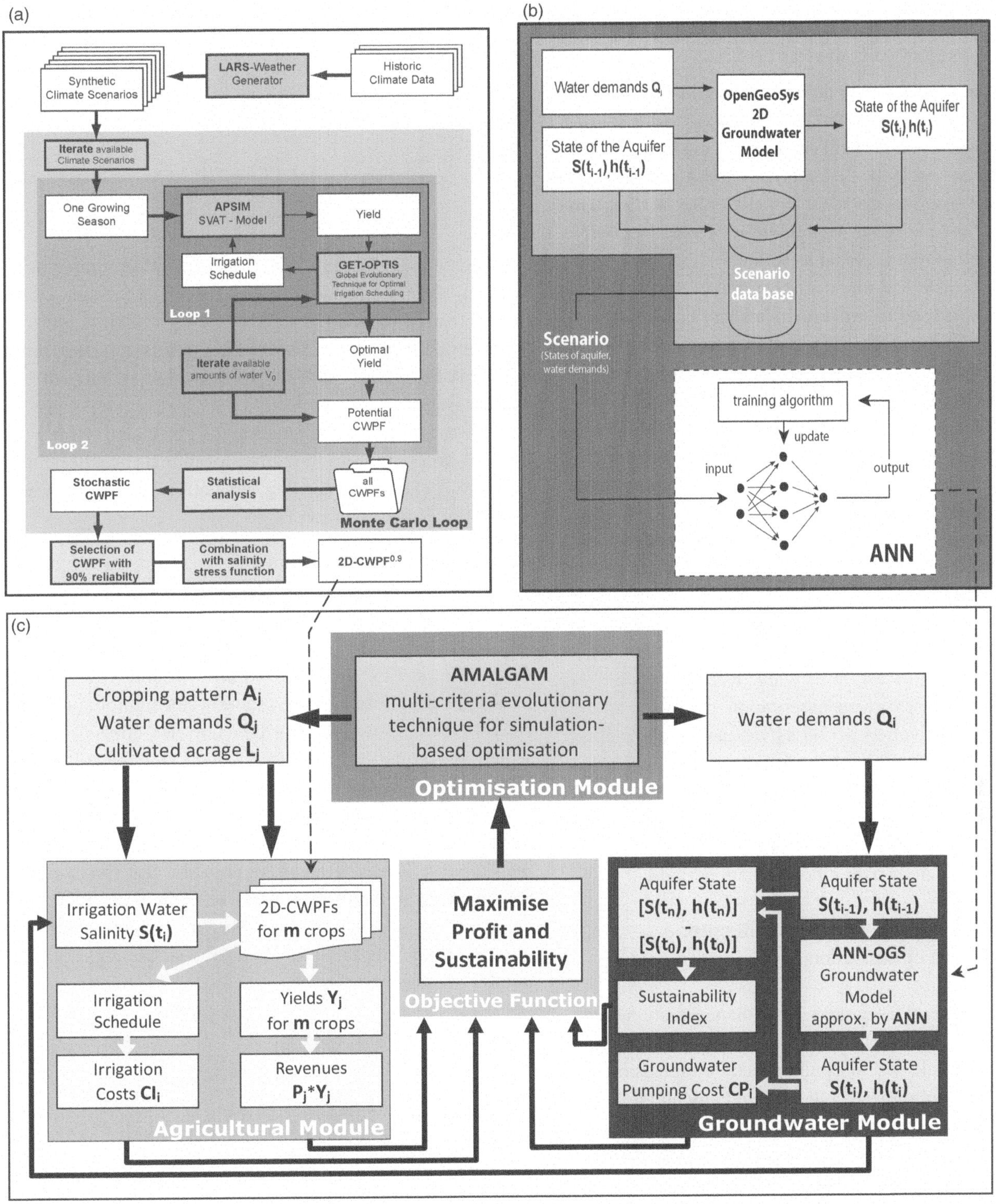

Figure 3 | Simulation-based integrated water management system consisting of: (a) the estimation procedure for the 2D crop water production function, the surrogate model which describes the agricultural production (2D-CWPF), (b) the procedure of the setup of the surrogate model for describing the aquifer behaviour (ANN-OGS) and (c) the optimisation framework, which links the derived surrogate models with the multi-criteria optimisation algorithm for multi-objective optimisation (AMALGAM) for maximising agricultural profit and aquifer sustainability.

soils, crops and irrigation methods as well as the corresponding optimal irrigation schedules are stored in a database.

The surrogate model for modelling the aquifer behaviour

For describing the aquifer behaviour, including the phenomena of seawater intrusion, we use an ANN as a surrogate for the numerical groundwater model OpenGeoSys as shown in Figure 3(b). The final surrogate model is referred to as ANN-OGS. The ANN-OGS is set up as follows.

At first, the above outlined model setup of the 2D OpenGeoSys model is used to generate a database containing the responses of the numerical density-dependent groundwater flow model for all realistically feasible abstraction scenarios. Abstraction rates Q_i for each cultivation period $i = 1 \ldots n$ as well as the state variables, average salinity concentration S and water levels h before (t_{i-1}) and after abstraction (t_i), are stored in a scenario database for selected observation points x_k. All together 2,546 scenarios over 20 years are simulated on a PC-cluster with the 2D OpenGeoSys model.

In a second step, the training of an ANN – more specifically a multilayer perceptron net (Nabney 2002) – is performed with a back propagation learning algorithm (Yan & Minsker 2006). The chosen structure of the ANN-OGS comprises input values of water level and values of average salinity at 13 given observation points x_k at the beginning of a cultivation period as well as the abstraction rate over the period (26 input nodes). As output the ANN-OGS provides the values for the same state variables at the end of the cultivation period at 25 output nodes. This state-space ANN architecture makes it possible to simulate the aquifer's behaviour over different time periods.

The optimisation framework for multi-criteria simulation-based optimisation

The simulation-based integrated water management system, illustrated in Figure 3(c), combines the agricultural and groundwater module within a multi-criteria optimisation framework. It aims to manage both water quality and quantity according to the given objectives. However, the management objectives depend on the interests of the decision maker. The AMALGAM multi-criteria optimisation algorithm which can be seen in Figure 3(c) evaluates Pareto-optimal solutions which allow for identification of best compromises between the given objectives.

Considering a sustainable aquifer management a first objective function (of_1) is formulated as sustainability index (Equation (1)) which describes the stability of the water balance and distribution of the salt concentration in the groundwater system:

$$of_1(t_n) = \sum_{k=1}^{\text{end}} \frac{\text{abs}(S(t_1, x_k) - S(t_n, x_k))}{S_{\max}} + \sum_{k=1}^{\text{end}} \frac{\text{abs}(h(t_1, x_k) - h(t_n, x_k))}{h_{\max}} \quad (1)$$

The index evaluates the change of the aquifer state between the end t_n and the initial state t_1 of the simulation period using the average salinity concentration S and the water levels h on the x_k observation points, where a small sustainability index value indicates a high stability of the aquifer system.

The second objective function (of_2) is related to the agricultural production and assumes a rational, profit-oriented behaviour of the farmers by evaluating all the revenues minus the costs (Equation (2)).

$$of_2 = \sum_{i=1}^{n} \left[\left(\sum_{j=1}^{m} P_j Y_j(t_i) L_j(t_i) - CI_j(t_i) \right) - CP(t_i) \right] \quad (2)$$

where, CI_j and CP are the fixed and variable costs for the irrigated agriculture and the groundwater pumping respectively, P_j the current prices for the cultivated crops $j = 1 \ldots m$ which are produced from the acreage L_j for the cultivation period $i = 1 \ldots n$, and Y_j the crop yield.

However, both objective functions are contradictory. If the farmers aim at earning high profit – to maximise of_2 – they tend to pump as much as possible. But this worsens the aquifer's situation due to a declining water table and an enforced salt water intrusion which has a negative impact on objective function of_1 – the sustainability of the aquifer management.

$$\max(OF^*) = \{(-of_1, of_2)\} \quad (3)$$

$$(A_1(t_1) \ldots A_1(t_n), A_m(t_1) \ldots A_m(t_n), L(t_1) \ldots L(t_n), Q(t_1) \ldots Q(t_n)) = \arg(OF^*) \quad (4)$$

To solve the formulated multi-objective optimisation problem (Equations (3) and (4)) the multi-criteria AMALGAM algorithm (Vrugt *et al.* 2009) is used which can be seen in Figure 3(c). The AMALGAM algorithm evaluates Pareto-optimal solutions which allow for identification of best compromises between the given objectives.

Decision variables within the optimisation problem (Equation (4)) are: A_j the percentage of the cultivated crops – the cropping pattern, the acreage L and the groundwater pumping rates Q for every cultivation period $i = 1 \ldots n$.

The optimisation framework (Figure 3(c)) works as follows. At first, estimates of the decision variables are provided by the optimisation algorithm. Secondly, the consequences, in terms of water level and salinity changes, of the given decision variable, i.e. the pumping rates, are evaluated by the groundwater surrogate model ANN-OGS (Equation (5)). Afterwards, the calculated salinity value of the pumped water for irrigation and the pumping rate are used to calculate the crop yields using the irrigation module (i.e. the 2D-CWPF$^{0.9}$ surrogate model) according to the given values of crop pattern and acreage (Equation (6)).

$$(h(t_i, x), S(t_i, x)) = \text{ANN-OGS}(h(t_{i-1}, x), S(t_{i-1}, x), Q(t_i)) \quad (5)$$

$$Y_j(t_i) = A_j(t_i) * \text{2D-CWPF}^{0.9}(Q(t_i), S(t_i, x_p)) \quad (6)$$

Finally, by incorporating regional specific cost models for the irrigated agriculture and the groundwater abstraction (Malik & Al-Zubeidi 2006), the net profit (Equation (2)) is calculated at the end of the simulation period as well as the sustainability index (Equation (1)). Iterative simulations over improved estimates of the decision variables – given by the optimisation algorithm according to the multi-objective problem (Equations (3) and (4)) – allow for a successive approximation of the true Pareto front.

RESULTS AND DISCUSSION

Generation of surrogate models

The derived surrogate model for the agricultural production (i.e. the database of 2D-CWPF$^{0.9}$, see Figure 3(a)) contains water demand functions for a yield reliability of 90% assuming an irrigation efficiency of 100%. The functions indicate that sorghum achieves a higher yield for a given amount of water and has a higher salt tolerance than maize (9.9 t/ha vs. 7.1 t/ha at 400 mm irrigation volume). However, the price per ton for maize is higher than for sorghum (US$150/t vs. US$100/t), which offers a higher profit, if the water quality is good. To account for different irrigation methods, the water demand has to be divided by the irrigation efficiency of the applied irrigation method (for more details see Schütze et al. 2012b and Grundmann et al. (2012)).

To validate the results of the groundwater surrogate ANN-OGS, 625 of the simulated scenarios are used to calculate the root mean square error (RMSE) between simulation results of the surrogate model and the numerical simulation model. The very low RMSE values of 0.0042 m for the groundwater levels and 44 µS/cm for the salinity values demonstrate the high ability of the derived surrogate model to simulate the aquifer behaviour instead of the physically based 2D OpenGeoSys model. Furthermore, the ANN-OGS operates approximately 1,000 times faster than the physically based OpenGeoSys model, which allows for multi-criteria optimisation within a reasonable time (a couple of hours instead a couple of months).

Application of the multi-criteria optimisation framework

Figure 4 shows results for multi-criteria optimisation run where the 'Sustainability Index (of_1)' is plotted over the 'Profit (of_2)' for all solutions which are evaluated during the optimisation procedure. Therefore, the dark dots represent all solutions found by the optimisation algorithm. The crosses comprise the Pareto-optimal data set, i.e. the Pareto front. In addition, the circles represent specific optimal solutions assuming certain weights assigned to the single objective functions of_1 and of_2. Varied solutions along the Pareto front are of multi-objective type and represent potential compromises between contradicting objectives, depending on the relationship or weighting between both objectives.

Within this application the Pareto-optimal data set consists of 1,743 solutions. The shape of the Pareto front may be divided into two different ranges between specific optimal solutions (circles). Part one comprises the range between the single solution, where only of_1 is considered, and the solution labelled as 'multi-objective scenario'. In this range, only a moderate change in the hydrological system stability occurs if the profit of the agricultural production increases. Within the second part – between the solutions labelled as 'multi-objective scenario' and 'the long-term profit scenario' (of_2), respectively – an increase of the agricultural production yields to a stronger change in the groundwater system's stability, which means a higher risk for damaging the groundwater system.

Figure 5 presents the temporal variability in profit, crop pattern percentages, and the irrigation water salinity for the specific solutions 'multi-objective scenario' and 'the

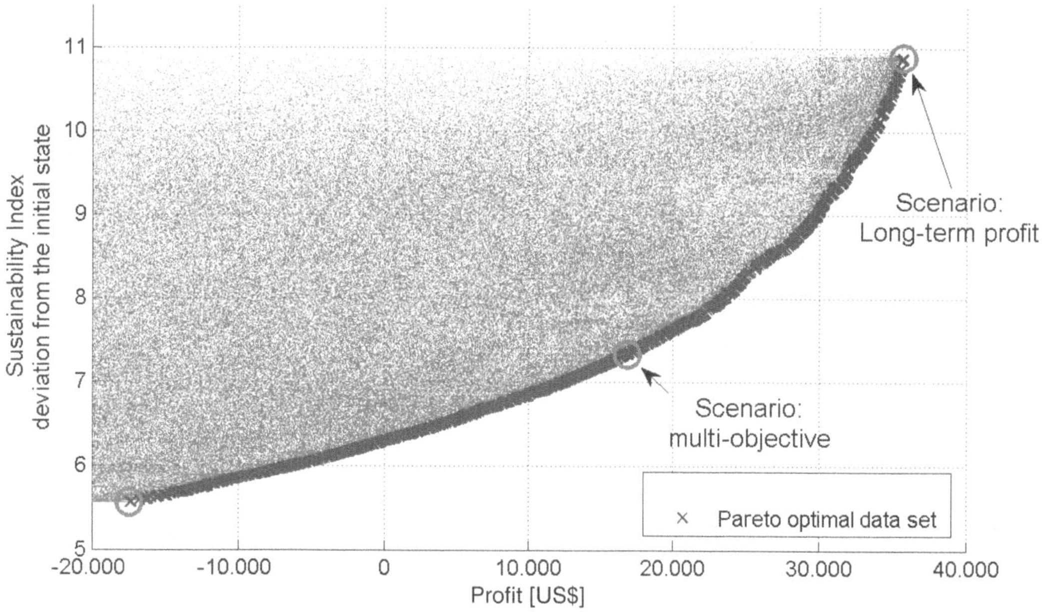

Figure 4 | Pareto front of a multi-criteria optimisation run for the objective functions 'Profit' and 'Sustainability Index'.

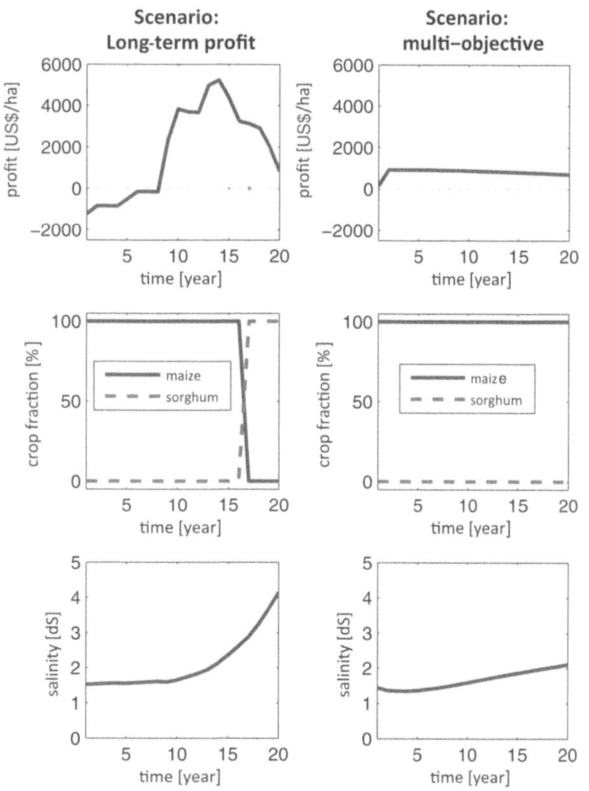

Figure 5 | Results of the multi-criteria optimisation run illustrated by the profit, the percentage of crop pattern and the irrigation water salinity and for the long-term profit scenario (left) and the multi-objective scenario (right).

long-term profit scenario' (circles in Figure 4). The 'long-term profit scenario' shows an increase in profit during the first 15 years and then diminishes with time, necessitating changes in the crop pattern caused by steadily increasing aquifer salinity. In contrast, the 'multi-objective scenario' generates relatively smoother graphs. In this scenario, maize is grown consistently, but on a smaller relative acreage than in the first years of the profit scenario. As a result, the salinity in irrigation water rises only slightly and a modest, sustainable profit is achieved over the entire 20-year time frame. Due to smaller acreage and pumping rates of the 'multi-objective scenario', the fixed initial costs for pumping and irrigation equipment are lower than for the long-term profit scenario. This is shown by different initial values of the profit graphs for the given configuration of the prototype application. Nevertheless, the cumulative profit over the time of 20 years is more than two times higher for the 'long-term profit scenario' as displayed in Figure 4. However, a comparison of the gradients of the profit and salinity graphs at the end of the 20-year period between the selected scenarios indicates that under the 'long-term profit scenario', the farmer will no longer find agricultural production to be profitable due to increased salinity values. In contrast, for the 'multi-objective scenario', the farmer can work sustainably for further years because of a more sustainable aquifer management.

In general, multi-objective optimisation problems can be solved either by means of a multi-criteria optimisation shown in this paper or in a simplified manner by a weighted sum of the objective functions using a mono-criteria optimisation algorithm. Grundmann *et al.* (2012) applied the latter approach and evaluated solutions for the 'long-term profit scenario' and the 'multi-objective scenario'. A comparison of the results achieved by Grundmann *et al.* (2012) with those of this study shows similar results for both scenarios. Especially the objective function values are in agreement with the corresponding points on the Pareto front. This indicates that the multi-objective solution converges to the true Pareto front. However, the multi-criteria optimisation approach provides a larger set of Pareto-optimal solutions, which allows for identification of best compromises between the given objectives. Therefore, the Pareto-optimal data sets can potentially serve as tools for stakeholders and decision makers in order to evaluate management decisions according to their management goals. As shown in this study, these goals may include sustainable environmental and socio-economic development parameters or criteria.

CONCLUSION AND OUTLOOK

The presented prototype of a simulation-based integrated water management model allows for the management of both water quality and quantity of a coupled, dynamic agriculture–groundwater system. Thereby, the modelling of the density-driven groundwater flow is mandatory for calculating crop yield damages due to aquifer overpumping and irrigation with salty water. The application, dealing with a typical management problem in coastal arid regions, shows that farm operation focusing exclusively on profit maximisation leads sooner or later to further progress of the saltwater front. Therefore, the objective of sustainability must explicitly be considered as a second optimisation objective to provide sustainable solutions both in an environmental and social sense. The performed multi-criteria optimisation run helps to understand the interplay and feedback mechanisms between a groundwater management and the agricultural production in terms of water quantity and quality. With the application of a multi-criteria optimisation algorithm we demonstrate the capability of Pareto-optimal solutions to find best compromises between contradicting management objectives. To find the true Pareto front within reasonable time a rigorous application of artificial intelligence methods is necessary in order to speed up the simulations. Along these lines, the presented prototype forms the basis for the development of a large-scale management and planning system focusing on a sustainable arid-zone water resources management. Nevertheless there is still a need for a careful evaluation of the results and preselection of appropriate management scenarios (e.g. due to large number of members of the Pareto-optimal data set). In this context further efforts are necessary to analyse and formulate the optimisation problem and its constraints as well as to compare appropriate optimisation algorithms which are able to solve the multi-criteria optimisation problem.

ACKNOWLEDGEMENT

The manuscript was prepared within the research project IWAS funded by the German Federal Ministry of Education and Research (BMBF) under grant no. 02WM1166. Additionally we wish to thank the Ministry of Regional Municipalities and Water Resources of the Sultanate of Oman for supporting the IWAS-IWRM project.

REFERENCES

Abdelrahman, H., Lepiece, A. & Macalinga, V. 1993 Some physical and chemical characteristics of the batinah soils. *Communications in Soil Science and Plant Analysis* **24** (17–18), 2293–2305.

Bhattacharjya, R. K. & Datta, B. 2009 ANN-GA-based model for multiple objective management of coastal aquifers. *Journal of Water Resources Planning and Management-ASCE* **135** (5), 314–322.

Cardona, A., Carrillo-Rivera, J., Huizar-Alvarez, R. & Graniel-Castro, E. 2004 Salinization in coastal aquifers of arid zones: an example from Santo Domingo, Baja California Sur, Mexico. *Environmental Geology* **45** (3), 350–366.

Dhar, A. & Datta, B. 2009 Saltwater intrusion management of coastal aquifers. I: Linked simulation-optimization. *Journal of Hydrologic Engineering* **14** (12), 1263–1272.

Gosavi, A. 2003 *Simulation-Based Optimization: Parametric Optimization Techniques and Reinforcement Learning*. Kluwer Academics Publishers, Boston.

Grundmann, J., Schütze, N., Schmitz, G.-H. & Al-Shaqsi, S. 2012 Towards an integrated arid zone water management using simulation based optimisation. *Environmental Earth Sciences* **65** (5), 1381–1394.

Harou, J. J., Pulido-Velazquez, M., Rosenberg, D. E., Medellín-Azuara, J., Lund, J. R. & Howitt, R. E. 2009 Hydro-economic models: concepts, design, applications, and future prospects. *Journal of Hydrology* **375** (3–4), 627–643.

Kacimov, A. R., Sherif, M. M., Perret, J. S. & Al-Mushikhi, A. 2009 Control of sea-water intrusion by salt-water pumping: coast of Oman. *Hydrogeology Journal* **17**, 541–558.

Keating, B., Carberry, P., Hammer, G., Probert, M., Robertson, M., Holzworth, D., Huth, N., Hargreaves, J., Meinke, H., Hochman, Z., McLean, G., Verburg, K., Snow, V., Dimes, J., Silburn, M., Wang, E., Brown, S., Bristow, K., Asseng, S., Chapman, S., McCown, R., Freebairn, D. & Smith, C. 2003 An overview of APSIM, a model designed for farming systems simulation. *European Journal of Agronomy* **18** (3–4), 267–288.

Kolditz, O., Delfs, J.-O., Bürger, C.-M., Beinhorn, M. & Park, C.-H. 2008 Numerical analysis of coupled hydrosystems based on an object-oriented compartment approach. *Journal of Hydroinformatics* **10** (3), 227–244.

Loucks, D. P. & van Beek, E. 2005 *Water Resources Systems Planning and Management: An Introduction to Methods, Models and Applications*. UNESCO, Paris.

Malik, A. S. & Al-Zubeidi, S. 2006 Electricity tariffs based on long-run marginal costs for central grid system of Oman. *Energy* **31** (12), 1703–1714.

Nabney, I. 2002 *Netlab. Algorithms for Pattern Recognition*. Springer, London.

Narayan, K. A., Schleeberger, C. & Bristow, K. L. 2007 Modelling seawater intrusion in the Burdekin delta irrigation area, North Queensland, Australia. *Agricultural Water Management* **89** (3), 217–228.

Rao, S., Sreenivasulu, V., Bhallamudi, S., Thandaveswara, B. & Sudheer, K. 2004 Planning groundwater development in coastal aquifers. *Hydrological Sciences Journal - Journal des Sciences Hydrologiques* **49** (1), 155–170.

Rejani, R., Jha, M. K. & Panda, S. N. 2009 Simulation-optimization modelling for sustainable groundwater management in a coastal basin of Orissa, India. *Water Resources Management* **23** (2), 235–263.

Schmitz, G. H., Woehling, T., de Paly, M. & Schütze, N. 2007 GAIN-P: a new strategy to increase furrow irrigation efficiency. *Arabian Journal of Science & Engineering* **32** (1C), 103–114.

Schütze, N., de Paly, M. & Shamir, U. 2012a Novel simulation-based algorithms for optimal open-loop and closed-loop scheduling of deficit irrigation systems. *Journal of Hydroinformatics* **14** (1), 136–151.

Schütze, N., Kloss, S., Lennartz, F., Al-Bakri, A. & Schmitz, G. 2012b Optimal planning and operation of irrigation systems under water resource constraints in Oman considering climatic uncertainty. *Environmental Earth Sciences* **65** (5), 1511–1521.

Schütze, N. 2012c Stochastic optimization of irrigation systems under water resource constraints from plot to regional scale using decomposition. *XIX International Conference on Computational Methods in Water Resources*. Champaign, Illinois, USA.

Schütze, N. & Schmitz, G. H. 2010 OCCASION: a new planning tool for optimal climate change adaption strategies in irrigation. *Journal of Irrigation and Drainage Engineering* **136** (12), 836–846.

Semenov, M., Brooks, R., Barrow, E. & Richardson, C. 1998 Comparison of the WGEN and LARS-WG stochastic weather generators for diverse climates. *Climate Research* **10** (2), 95–107.

Steppuhn, H., van Genuchten, M. & Grieve, C. 2005 Root-zone salinity: II. Indices for tolerance in agricultural crops. *Crop Science* **45** (1), 221–232.

Sultanate of Oman 2005 *Agricultural Census 2005*. Ministry of Agriculture and Fisheries, Director General of Planning & Investment Promotion, Department of Statistics & Information.

Vrugt, J., Robinson, B. & Hyman, J. 2009 Self-adaptive multimethod search for global optimization in real-parameter spaces. *IEEE Transactions on Evolutionary Computation* **13** (2), 243–259.

Yan, S. & Minsker, B. 2006 Optimal groundwater remediation design using an adaptive neural network genetic algorithm. *Water Resources Research* **42** (5), W05407, 14 pp.

Walther, M., Delfs, J.-O., Grundmann, J., Kolditz, O. & Liedl, R. 2012 Saltwater intrusion modeling: verification and application to an agricultural coastal arid region in Oman. *Journal of Computational and Applied Mathematics* **236** (18), 4798–4809.

First received 9 January 2012; accepted in revised form 10 September 2012

Can hydro-economic river basin models simulate water shadow prices under asymmetric access?

A. Kuhn and W. Britz

ABSTRACT

Hydro-economic river basin models (HERBM) based on mathematical programming are conventionally formulated as explicit 'aggregate optimization' problems with a single, aggregate objective function. Often unintended, this format implicitly assumes that decisions on water allocation are made via central planning or functioning markets such as to maximize social welfare. In the absence of perfect water markets, however, individually optimal decisions by water users will differ from the social optimum. Classical aggregate HERBMs cannot simulate that situation and thus might be unable to describe existing institutions governing access to water and might produce biased results for alternative ones. We propose a new solution format for HERBMs, based on the format of the mixed complementarity problem (MCP), where modified shadow price relations express spatial externalities resulting from asymmetric access to water use. This new problem format, as opposed to commonly used linear (LP) or non-linear programming (NLP) approaches, enables the simultaneous simulation of numerous 'independent optimization' decisions by multiple water users while maintaining physical interdependences based on water use and flow in the river basin. We show that the alternative problem format allows the formulation HERBMs that yield more realistic results when comparing different water management institutions.

Key words | externalities, hydro-economic river basin model, mixed complementarity programming, water institutions

A. Kuhn (corresponding author)
W. Britz
Institute for Food and Resource Economics,
University of Bonn,
Nussallee 21,
53115 Bonn,
Germany
E-mail: *arnim.kuhn@ilr.uni-bonn.de*

INTRODUCTION

There is growing awareness that availability of water and its efficient management will become one of the key questions in the 21st century (Chartres & Varma 2010). An often discussed issue in that context is that of an appropriate institutional design (Livingston 1995) to improve resource allocation resulting from unregulated use of water or inefficient assignments of water use rights. The economic assessment of water management institutions is often carried out by running simulations with hydro-economic river basin models (HERBMs) that are based on mathematical programming. Among the crucial outputs of HERBMs are local water 'shadow prices' which are used to estimate appropriate water prices (Conradie & Hoag 2004). This paper proposes a specific formulation of a mixed complementarity problem (MCP), which allows the capture of the multiplicity of individual (i.e. economically independent) but still physically coupled optimization problems which are difficult or simply impossible to implement in a standard non-linear (NLP) or linear programming (LP) format based on optimizing a social welfare criterion. We argue that water shadow prices in standard HERBMs are biased towards uniformity, which can be overcome by using the MCP format.

The rest of the paper is organized as follows. After a brief discussion of the current state-of-the-art in river basin modeling, we describe our problem setting – several firms positioned along a water distribution system competing for water use – and present algebraic formulations and solution algorithms to evaluate a range of typical institutional designs related to water use proposed and analyzed in the literature. Subsequently, we develop a small numerical example and apply the proposed solutions to highlight differences in the results between the two principal optimization formats.

doi: 10.2166/wst.2013.251

STATE-OF-THE-ART AND PITFALLS IN RIVER BASIN MODELING

Current HERBMs that are used to evaluate water management tools such as water pricing, water use rights, water trade, planned water allocation, and physical infrastructure offer hydrological and bio-physical relations in rich detail. Recent examples are integrated river basin models for the Maipo river in Chile (Rosegrant *et al.* 2000), the Jordan Valley (Doppler *et al.* 2002, water pricing), and the Drâa Valley in Southern Morocco (Heidecke & Kuhn 2008), where water pricing options under conjunctive use of water resources are evaluated. These models are typically formulated as optimization problems (LP, NLP), allocating water among users, uses, locations, and points in time such that an *aggregate* social welfare criterion is maximized (we will further call this format '*aggregate optimization*', AO). The welfare measure is typically based on profits or utility of the different water users. This model formulation thus tacitly assumes that agents will allocate the water among each other such as to maximize their *aggregate* welfare.

From an institutional perspective, AO is a very strong and restrictive assumption that does not fit many real-world situations. In river basins with weak or absent water management institutions, each agent optimizes her individual welfare independently while being influenced by decisions of other users who withdraw water from the same common resource. Upstream users, to give an example, will continue to extract water as long as this is feasible, permitted, and increases their individual welfare. But their – perhaps excessive – water use may reduce aggregate welfare, because water availability for economically more efficient downstream users will be reduced. Achieving the social optimum would require that agents either are forced to a more efficient water allocation (central planning), or agree to it, in the latter case asking for compensation of losers out of the gains of winners (water markets). The costs to achieve the socially optimal solution against the background of the simulated institutional setting (such as unregulated water use, assignments of water rights etc.) will be underestimated when resource allocation in the model is driven by an aggregate welfare function. Basically, this problem format implicitly assumes that basic institutional deficiencies often found at the root of water allocation problems are already solved. Indeed, the only institutional settings where aggregate welfare maximization always leads to the same water allocation as with independent optimization (IO) is that of a perfectly functioning water market, i.e. one with zero transaction costs. A HERBM formulated as a social welfare optimization problem is hence not only unsuitable to correctly simulate the reference situation (such as unregulated water use), but also unable to spell out the true reaction of agents to alternative institutional settings and the related welfare gains or losses, both for individual agents and the aggregate.

The allocation of water in a river basin with 'upstream-downstream problems' is the outcome of a multiplicity of individual (i.e. economically independent) but still physically coupled optimization problems which are difficult or simply impossible to implement in a standard NLP format with AO. Modeling the use of water and other resources in such a context of asymmetric access requires a differently structured optimization format. In such an alternative setting, coupled inter-spatial and/or inter-temporal physical balances have to hold, but not the complementary *water shadow price* relations (i.e. economic scarcities) evolving from maximizing aggregate welfare. In mathematical programming, the 'shadow price' of a constraint is a dual variable that represents the amount by which the objective function of the problem changes when the constraint is relaxed or tightened. In this study, local water availabilities wo_i are the constraints facing water users. When water is scarce, it has a marginal economic value per unit to the users which is called the 'shadow price' and which can be interpreted as the user's willingness-to-pay (WTP) for one additional unit of water. AO leads to a water allocation pattern that equalizes water shadow prices across locations in order to maximize aggregate profits. This outcome neglects the lower WTP of upstream users under the absence of joint water management institutions. A model of individual optimization therefore must not impose uniformity of water shadow prices across locations. Unfortunately, the modeler cannot control independently both the physical balances and their complementary shadow price relations simultaneously when using an LP/NLP or other AO format. We will therefore propose and apply a different approach based on the MCP format.

ALGEBRAIC FORMULATIONS CAPTURING INSTITUTIONAL DESIGNS OF WATER USE

Problem setting

Assume n water-using firms positioned along a water distribution system such as a natural river or an irrigation system. Firms are indexed according to their positions, the first one being situated in the uppermost upstream position. Figure 1 illustrates two alternative problem formats which will be applied to the stylized river system.

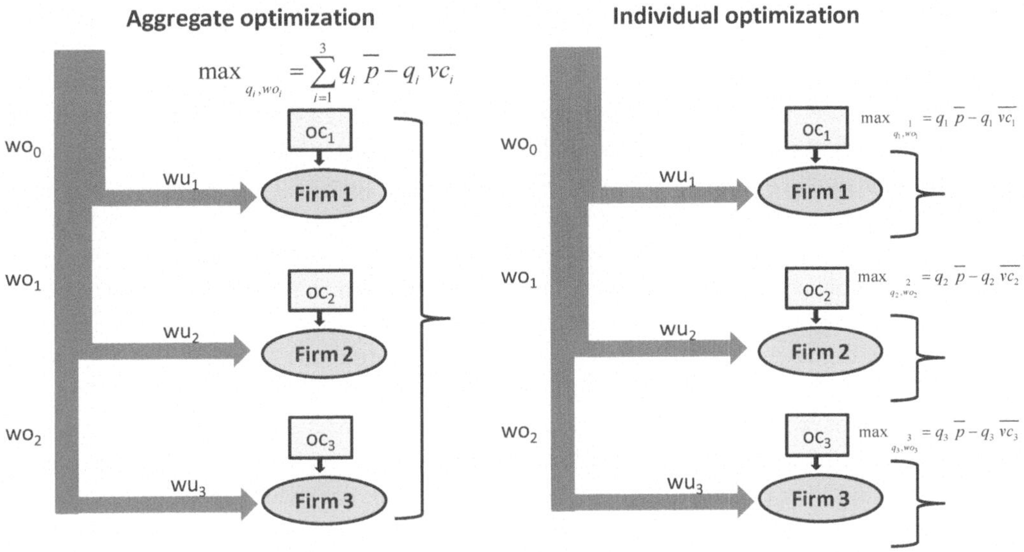

Figure 1 | The difference between aggregate (AO) and independent optimization (IO).

Each firm is characterized by the following attributes (see also Figure 1): (1) its position in the water distribution system expressed by its index i, (2) its operating capacity oc_i, (3) its variable production costs per unit of output vc_i, and finally (4) the water available wo_i (see also graphic). The decision variable of the firm is its output quantity q_i, which depends on its water use wu_i.

Furthermore, a Leontief technology is assumed; for notational simplicity one unit of output requires one unit of water and one unit of operating capacity. The latter is treated as a firm-specific fixed factor and could, for example, capture the firm's endowment with capital, land and production rights or its management capacity.

Water allocation through 'aggregated optimization'

The AO formulation using the above setting allocates water quantities to the individual firms such that the *sum* of all firms' profits, π, reaches its maximum. This is the problem format that existing programming-based HERBMs use. AO would solve the following problem:

$$\max_{q_i, wo_i} \pi = \sum_i q_i \overline{p} - q_i \overline{vc_i}$$

$$s.t. \quad q_i \leq \overline{oc_i} \, [\lambda_i]$$

$$q_i = wu_i$$

$$wo_{i-1} = wu_i + wo_i[\mu_i]$$

$$q_i \geq 0, wo_i \geq 0, wu_i \geq 0 \tag{1}$$

The related Karush–Kuhn–Tucker (KKT, see Kuhn & Tucker 1951) conditions are:

$$\overline{vc_i} + \lambda_i + \mu_i \geq \overline{p} \bot q_i \geq 0$$

$$\overline{oc_i} \geq q_i \bot \lambda_i \geq 0$$

$$wu_i = q_i \bot wu_i \geq 0$$

$$wo_{i-1} = wo_i + wu_i \bot \mu_i$$

$$\mu_i \geq \mu_{i+1} \bot wo_i \geq 0 \tag{2}$$

The first line in set (2) expresses that firms will only operate if their marginal production costs do not exceed marginal revenues per unit of output, the latter being equal to the price p. The marginal production costs consist of the variable costs vc and marginal resource use costs related to operating capacity oc and water use wu. The Lagrange multiplier λ is a shadow price that captures marginal economic returns to operating capacity. The third line is a water production function in its simplest form. The water resource constraint is defined by the last two equations, which links the firms in the river basin by describing the physical flows of water from firm to firm and its use in the distribution system. It expresses that the sum of water use wu and outflow wo at location i of the firm (i.e. the section of the river where the firm withdraws water) must be equal to water inflow; the latter is the water not used by its upstream neighbor. Accordingly, μ is the maximal WTP of the firm for additional water, the *water shadow price*. These physical balances must hold for any institutional setting. Consequently, $wo_{i=0}$ denotes the (fixed) inflow into the system. The last line states that the firm will only use all the water at its disposal if its marginal returns to water use exceed that of its next downstream neighbor. These are the usual *shadow price conditions* for an aggregate social welfare optimum: *marginal returns to a fixed resource must be equilibrated between competing users*. The basic problem in water allocation – finding an institutional design that achieves socially optimal returns by properly constraining the behavior of individual agents – is already part of the solution format itself. Equal economic returns to water in setting (1) imply that upstream users are willing to reduce their water use and profits to make room for efficient users downstream, thus increasing basin-wide profits. But if we stick to the basic assumption of profit maximization by *individual* firms, why should upstream users behave that way if they are not compensated for their profits foregone? Such compensation is not included in setting (1). Consequently, AO should not be used to analyze such institutions if some users have asymmetric access to water. What we would need is an approach which reflects the results of IO by individuals, which we describe next.

Water allocation through 'independent optimization' (IO)

We assume that individual firms *independently* maximize their profits π_i at a given uniform output price p:

$$
\begin{aligned}
&\max_{q_i, wo_i} \ \pi_i = q_i\,\overline{p} - q_i\,\overline{vc_i} \\
&s.t. \quad q_i \le \overline{oc_i}\ [\lambda_i] \\
&q_i = wu_i \\
&wo_{i-1} = wu_i + wo_i[\mu_i] \\
&q_i \ge 0,\ wo_i \ge 0,\ wu_i \ge 0
\end{aligned}
\tag{3}
$$

The difference to (1) above is that the summation sign of the profits is missing. Each firms is 'socially myopic' – it takes the water inflow from its upstream neighbor as given, and, based on this, decides on its own profit-maximizing water use. The KKT conditions related to (3) are:

$$
\begin{aligned}
&\overline{vc_i} + \lambda_i + \mu_i \ge \overline{p} \perp q_i \ge 0 \\
&\overline{oc_i} \ge q_i \perp \lambda_i \ge 0 \\
&wu_i = q_i \perp wu_i \ge 0 \\
&wo_{i-1} = wu_i + wo_i \perp \mu_i \\
&\mu_i \ge 0 \perp wo_i \ge 0
\end{aligned}
\tag{4}
$$

The only difference to the KKT conditions in (2) from the AO problem is the last line. Each firm will continue to use the water at its disposal as long its increases its profit. An outflow to the next neighbor will only appear if any additional profits from water are exhausted; the water shadow price must drop to *zero* then. What we need is to combine the decision logic underlying (4) with the hydrologic relations linking the users in the basin. Obviously, we cannot do that based on explicit optimization – we end up with AO. However, MCP (see Ferris & Munson 2000) offers a format to solve problems formulated in KKT form. We can simply stack equations in (4) together, so that all wo (besides the inflow in the basin) are endogenous variables. The shadow price of each agent is allowed to drop to zero, independent from the shadow price of each neighbor, while the water flow from node to node is correctly accounted for. There exists no equivalent explicit optimization problem to that MCP formulation. The small change in the complementarity condition linking outflow and shadow price of water at each node from (2) to (4) now allows us to analyze different institutions governing the basin without implicitly assuming that all agents voluntarily maximize aggregate welfare. We have constructed the core of a true agent-based model which can be solved simultaneously.

In the following, we aim to at analyze in an elegant way different institutions that manage the available water, i.e. which govern the water use of individual firms. With elegant we mean algebraic formulations and computable solutions which (a) are compact and transparent, (b) require only slight changes in the overall layout of the problem to model different institutions, and (c) avoid switching solution algorithms or sequential solution strategies. Specifically, we compare *two fundamentally different approaches*: (I) our newly proposed approach describing IO of water allocation, i.e. rational, utility-maximizing agents as usually assumed in standard micro-economics, (II) the classical AO format where water allocation is simulated as it would be achieved by letting a fully informed social planner decide about each firm's production program (= socially optimal allocation). These two approaches are both expressed as MCP problems as explained above in (2) and (4).

In order to show the ability of our new approach to handle different institutions and support our claim that the aggregate programming format is not suitable to model these correctly, we apply both approaches to *four water management options* which themselves can be said to be institutional designs, namely (i) unregulated access to water, (ii) water pricing by a water agency with the policy objective of tax revenue generation, (iii) assignment of equitable water rights to improve basin-wide income distribution, and finally (iv) assignment of tradable water rights under consideration of transaction costs, i.e. the establishment of water markets to achieve an economically efficient allocation of water.

Capturing different institutions governing access to water

In the following paragraphs we will outline the water management institutions we want to compare. We start with water pricing and continue with non-tradable and tradable water use rights. The unregulated access case was already introduced above.

Water pricing

It is commonly proposed to use water pricing to reduce inefficient or excessive use of water (Tsur 2005). Assuming a suitable control and enforcement strategy, firms will thus face a common water price. Under water pricing, firms will only use water as long as marginal economic returns per unit of water do not fall below the water price. The water pricing case can hence be described by the following KKT condition in (5), where the only difference to both (2) and (4) is that the exogenously set water price wp is added to the left-hand side of the first line:

$$\overline{vc_i} + \lambda_i + \mu_i + \overline{wp} \geq \overline{p} \perp q_i \geq 0 \tag{5}$$

The change expressed in (5) guarantees that the firms' marginal returns to water use are at least as high as the price charged for water. The marginal returns might still exceed the water price once all available water is used.

Assignment of water rights

The second institution analyzed assigns non-tradable water rights wr to each firm. Each firm's maximal water use is now bounded by either the property rights to use a certain amount of water or by available water inflows. The marginal costs of the firm now additionally comprise the shadow price ν related to the firm's water use right (first line in (6)) which is complementary to water use being constrained by the water right (last line in (6)):

$$\begin{aligned}
&\overline{vc_i} + \lambda_i + \mu_i + \nu_i \geq \overline{p} \perp q_i \geq 0 \\
&\overline{oc_i} \geq q_i \perp \lambda_i \geq 0 \\
&wu_i \geq q_i \perp wu_i \geq 0 \\
&wo_{i-1} = wo_i + wu_i \perp \mu_i \\
&\mu_i \geq 0 \perp wo_i \geq 0 (\text{IO}) \text{ or } \mu_i \geq \mu_{i+1} \perp wo_i \geq 0 (\text{AO}) \\
&\overline{wr_i} \geq wu_i \perp \nu_i \geq 0
\end{aligned} \tag{6}$$

Under decentralized water use (expressed through IO), even an equitable assignment of water rights does not necessarily abolish asymmetric access to the resource. Assume, for instance, that water rights refer to average water availability in the network and are not adjusted dynamically to the actual inflow: if total inflow is lower than the sum of water rights, upstream firms still benefit from their position.

Tradable water rights

We finally expand our framework by allowing the firms to trade their water rights. Now not only must water use be measured and over-use punished, as in the case of water pricing and assignments of water rights, but also a trading system for the water rights has to be introduced:

$$\begin{aligned}
&\overline{vc_i} + \lambda_i + \mu_i + \nu_i \geq \overline{p} \perp q_i \geq 0 \\
&\overline{oc_i} \geq q_i \perp \lambda_i \geq 0 \\
&wu_i \geq q_i \perp wu_i \geq 0 \\
&wo_{i-1} = wo_i + wu_i \perp \mu_i \\
&\mu_i \geq 0 \perp wo_i \geq 0 (\text{IO}) \text{ or } \mu_i \geq \mu_{i+1} \perp wo_i \geq 0 (\text{AO}) \\
&\overline{wr_i} + \sum_j wt_{ji} \geq wu_i + \sum_j wt_{ij} \perp \nu_i \geq 0 \\
&tc \geq \mu_j + \nu_j - \mu_i - \nu_i \perp wt_{ij} \geq 0
\end{aligned} \tag{7}$$

The constraint related to water use and water rights from (6) is now expanded to allow for bilateral water right exchanges wt (see second to last line). The total water rights of the firm shown on the left-hand-side consist of its original assignment plus any rights bought from other firms. They must exceed water use plus selling of water rights to other firms. Accordingly, the last line adds the typical arbitrage conditions for trade: price differences cannot exceed the per unit transaction costs; no trade will occur when price differences are too small. The total WTP for water use rights is composed of the water shadow price μ plus the additional price ν for the water use right.

Allowing the firms to trade their water rights will lead to mutual benefits both for firms with marginal economic returns to water above and below the median. Firms with low returns can sell part of their rights, receiving additional revenues, whereas firms with high returns will buy rights as long the related profit increase per unit of water exceeds the price of the water right. Water trade should theoretically eliminate the difference between unregulated and socially planned water use regarding economic efficiency, at least at zero transaction costs.

TWO ILLUSTRATIVE EXAMPLES

We encoded the basic model described above in GAMS (General Algebraic Modeling System) and used the PATH solver to declare and solve the different MCPs. The declarative approach in GAMS to equations and models and built-in MCP functionalities support a smooth translation of the MCP problems.

A didactic example with three firms

For our first illustrative example, the results, we only simulate three firms. The simulation results are shown in Table 1 below. We choose a total water inflow into the system of 120 units and an output price p of 100 per unit. Each firm has an operating capacity oc of 50 units, which means that total operating capacity in the basin exceeds the amount of water available. Variable costs vc for the three firms are 60, 50, and 40 per unit from upstream to downstream: the more efficient a firm is, the less privileged it is regarding its access to water. This guarantees substantial welfare losses under decentralized water allocation without further management measures. This setting allows the highlighting of differences between the two optimization formats in our didactic example.

We apply four management options, (i) to (iv), to the two fundamental modeling formats, i.e. IO and AO, resulting in eight experiments. The management options were designed as follows:

No management: i.e. unregulated access: firms can use any water available at their node without costs.

Water pricing: a water price wp is charged by a water agency; the price level is the water shadow price of the pure 'social planner' solution (AO applied to i.) minus 1 (i.e. $40 - 1 = 39$).

Non-tradable water rights: a water use right wr of 45 units is granted to each firm. This is more than the total available water, but less than the operating capacity of each firm.

Tradable water rights: the water use rights of 45 for each firm can be traded among the firms at transaction costs tc of 1 for each unit of wr traded. Firms can now buy water rights to fully exploit their operating capacity.

Under the parameterization chosen for these experiments, all results differ more or less strongly between IO and AO. The first two columns ('*no management*') compare the effect of IO versus AO on water allocation without specific management options, i.e. under unregulated access to water. The difference in basin-wide profits is large, as under IO the two most inefficient firms are located upstream and use water up to their operating capacity, leaving only 20 units to the most efficient firm 3 downstream. Under AO, as usually applied in existing HERBMs based on mathematical programming, water is first allocated to the two most efficient firms (3 and 2). The distribution of profits among firms is thus dramatically different, leaving firm 1 with only 40% of the profits under IO, while firm 3 can increase profits by 150%. While the water shadow prices are 40 for all firms under AO, they differ markedly under IO, being zero for firm 1 and 2, and 60 for firm 3.

Under '*water pricing*' at 39 per unit, the water use patterns found under 'no management' will not change: the water price does not exceed the variable costs of the least efficient firm (40 in the case of firm 1). At first glance, this result appears to contradict the common belief that water pricing helps improve the efficiency of water use compared with unregulated access. The latter would only be the case if the water production function is strictly concave, or alternative uses of water within one firm enable substitution effects, as for instance a shift to more water-efficient cropping patterns by irrigators. Water pricing in our restricted representation, if it were carried out by a public agency, only serves the purpose of raising tax revenues, as long as the price is low enough to let all firms use water. The reader should note that the AO problem format again produces a fundamentally different solution in our example.

Alternatively, the public agency may assign fixed (i.e. non-tradable) *water use rights* to improve access of downstream firms, and, depending on the ratio of use rights to average stream flow, to ensure a minimum basin outflow for example, ecological reasons. In our example, water use rights of 45 are assigned to each firm, being below the operating capacity of 50 for each firm, but exceeding in sum the inflow into the river basin, which means that water rights

Table 1 | Results of the simulation experiments

		No management		Water pricing		Non-tradable rights		Tradable rights	
		IO	AO	IO	AO	IO	AO	IO	AO
Water use *wu*	Firm 1	50	20	50	20	45	30	20	20
	Firm 2	50	50	50	50	45	45	50	50
	Firm 3	20	50	20	50	30	45	50	50
Outflows *wo*									
	Firm 1	70	100	70	100	75	90	100	100
	Firm 2	20	50	20	50	30	45	50	50
	Firm 3	–	–	–	–	–	–	–	–
Water rights (when tradable, after trade) *wr*									
	Firm 1	–	–	–	–	45	45	20	35
	Firm 2	–	–	–	–	45	45	50	50
	Firm 3	–	–	–	–	45	45	65	50
Shadow prices μ for water use *wu*	Firm 1	–	40	–	1	–	40	–	40
	Firm 2	–	40	–	1	–	40	–	40
	Firm 3	60	40	21	1	60	40	41	40
Shadow prices ν for water use right *wr*	Firm 1	–	–	–	–	40	–	40	–
	Firm 2	–	–	–	–	50	10	41	1
	Firm 3	–	–	–	–	–	20	–	1
Shadow prices λ for operating capacity *oc*									
	Firm 1	40	–	1	–	–	–	–	–
	Firm 2	50	10	11	10	–	–	9	9
	Firm 3	–	20	–	20	–	–	19	19
Profits of firms	Firm 1	2,000	800	50	20	1,800	1,200	1,800	1,200
	Firm 2	2,500	2,500	550	550	2,250	2,250	2,295	2,295
	Firm 3	1,200	3,000	420	1,050	1,800	2,700	2,180	2,795
Total profits		5,700	6,300	1,020	1,620	5,850	6,150	6,275	6,290
Water pricing revenues		–	–	4680	4680	–	–	–	–
Total profits plus revenues		5,700	6,300	5,700	6,300	5,850	6,150	6,275	6,290

will not be fully exploitable for all firms. The results show that granting water rights exceeding total water availability is not suitable to totally eliminate allocative inefficiencies: firms 1 and 2 can still exploit their locational advantage, leaving only 30 units of water to firm 3, as simulated by IO. But firm 3 is better off than under a situation without water rights as it expands its water use from 20 to 30 units, being able to profit from the reduced water use of upstream firms, which are now constrained by the assigned water use right.

The AO solution again differs, as the inherent social planner ensures that only the two most efficient firms fully exploit their water rights, leaving firm 1 at a water use of 30 units, which is hardly plausible if the real world is characterized by independent firms.

As can be expected, the *tradability of water rights* eliminates allocative inefficiencies. This would not be the case if the transaction costs of water trade (1 per unit traded in the example above) would exceed the smallest difference in variable costs between two firms participating in trade (10 in our case). However, as buying firms have to carry the costs of trade, tradable water rights are not able to fully achieve the maximal profit possible in the basin (as simulated by maximizing social welfare, i.e. AO under 'no management').

Perhaps surprisingly at first glance, the distribution of profits between IO and AO differs markedly despite very small differences in aggregate profits and the same allocation of water to the firms. The reason is that trade in water rights has to be higher under IO in order to prevent firm 1 from still exploiting its locational advantage. Under AO, firm 1 is prevented by the inherent social planner to fully use its remaining water rights (35 units) after having sold sufficient rights to firm 2 and 3 (5 units each, so that the latter can fully use their operating capacity). Under IO, firm 3 must buy an additional 15 units of water use rights from firm 1 to prevent the latter from exercising its remaining water rights. This strongly diminishes the profit of firm 3, while firm 1 enjoys profits from producing 20 units, and receives revenues from selling 25 units of water rights to its downstream neighbors.

SUMMARY AND CONCLUSION

Our paper contributes twofold to the analysis of resource institutions. First, it demonstrates how a specific problem formulation based on the MCP format can be used to simulate different institutions (unregulated access, water rights, water right trading, water pricing) for a theoretically sound representation of a number of individual, decentralized decision makers and their interactions. The main difference to applications of mathematical programming models in the literature is that we are now able to correctly model simultaneously non-coordinated decisions of individual water users while reflecting the position of each water user in a water distribution system, along with further interdependences between users emerging, for example, from input and output markets. The usual aggregate programming format used in river basin modeling which maximizes social welfare fails to properly express the consequences of isolated utility maximization and its resulting externalities, and is hence not well suited to evaluate different institutional settings governing access to water. Using the MCP problem format allows the modeler to take full control of both physical and co-state variables and to model real-world problems in resource use, which often cannot be expressed as a single optimization problem. Moreover, our approach opens new perspectives for the design of agent-based models that are algebraically simple, theoretically sound, and easier to solve, as iterative procedures to reflect spatial and market interactions are no longer required.

As demonstrated in our example, our framework allows analysis of the interaction between asymmetric access and different institutions, such as resource pricing. We conclude that MCP is indeed a concise, elegant and highly flexible format to model these institutions. As physical water flow balances are integrated in the system, the MCP framework also can be employed for much more complex hydro- and bio-economic simulation models. Our presentation of a system of independent, utility-maximizing decision units linked via resource and other interdependences in MCP can thus be used as a nucleus for a new class of agent-based models, and opens new opportunities, for instance, to simulate the dynamic emergence of institutions between independent spatial agents in a concise and theory-consistent manner.

ACKNOWLEDGEMENTS

This study was supported by the German Research Foundation (DFG), Research Unit 1501 'Resilience, Collapse and Reorganization in Social-Ecological Systems of African Savannas'. We would like to thank Prof. Thomas Heckelei, Bonn, for a careful review of our paper.

REFERENCES

Chartres, C. & Varma, S. 2010 *Out of Water: From Abundance to Scarcity and How to Solve the World's Water Problems* Financial Times Press, Upper Saddle River, New Jersey.

Conradie, B. & Hoag, D. 2004 A review of mathematical programming models of irrigation water values. *Water SA* **30** (3), 287–292.

Doppler, W., Salman, A. Z., Al-Karablieh, E. K. & Wolff, H.-P. 2002 The impact of water price strategies on the allocation of irrigation water: the case of the Jordan valley. *Agricultural Water Management* **55** (3), 171–182.

Ferris, M. C. & Munson, T. S. 2000 Complementarity problems in GAMS and the PATH solver. *Journal of Economic Dynamics and Control* **24** (2), 165–188.

Heidecke, C. & Kuhn, A. 2008 Water pricing options for the Middle Drâa River Basin in Morocco. *African Journal of Agricultural and Resource Economics* **2** (2), 170–187.

Kuhn, H. W. & Tucker, W. 1951 Nonlinear Programming. In: *Proceedings of the Second Berkeley Symposium on Mathematical Statistics and Probability*. University of California Press, Berkeley, CA, pp. 481–492.

Livingston, M. L. 1995 Designing Water Institutions: Market Failures and Institutional Response. *Water Resources Management* **9** (3), 203–220.

Rosegrant, M. W., Ringler, C., McKinney, D. C., Cai, X., Keller, A. & Donoso, G. 2000 Integrated economic-hydrologic water modeling at the basin scale: The Maipo River basin. *Agricultural Economics* **24** (1), 33–46.

Tsur, Y. 2005 Economic aspects of irrigation water pricing. *Canadian Water Resources Journal* **30** (1), 31–46.

First received 4 January 2012; accepted in revised form 3 April 2012

Theme VI

Water governance: actors and institutions

The water governance challenge: the discrepancy between what is and what should be

H. M. Ravnborg and K. M. Jensen

ABSTRACT

In 2010, the UN General Assembly declared the right to safe and clean drinking water and sanitation as a human right that is essential for the full enjoyment of life and all human rights. Yet, findings from the Competing for Water research programme suggest that all too often, people in need of water for domestic purposes lose out to people and companies who claim access to water for productive purposes. Likewise, in many countries, specific water authorities at national as well as basin and watershed level have been formed and assigned the responsibility to allocate water according to the water policy and the associated legal framework. Yet, findings from the Competing for Water research programme show that real-world water allocation takes place through a wide array of institutions, ranging from the rural community, over agreements mediated by local lawyers, district officials and non-governmental organisations, to decisions made in the president's office. The Competing for Water programme entails empirical research conducted in Bolivia, Mali, Nicaragua, Vietnam and Zambia. Based on findings from this research, this paper identifies the discrepancies between statutory and actual water governance, analyses the underlying causes and explores the implications for ongoing water governance reform.

Key words | competition, conflict, governance, inequality, integrated water resources management, poverty

H. M. Ravnborg (corresponding author)
K. M. Jensen
Danish Institute for International Studies (DIIS),
Strandgade 65,
DK-1401 Copenhagen K,
Denmark
E-mail: *hmr@diis.dk*

INTRODUCTION

Many countries, including many developing countries, are currently in a process of water reform. Typically, this process entails both a policy reform and a reform of the legal and administrative framework within which development, allocation and use of water is governed.

A common feature in most countries' water policy is the principle that priority is assigned to domestic uses over productive uses in the allocation of water at any given scale. In this way, the water policies of many countries have anticipated the 2010 UN General Assembly declaration of the right to safe and clean drinking water and sanitation as a human right, essential for the full enjoyment of life and all human rights. (United Nations General Assembly, Res. A/64/L.63/Rev.1, July 26, 2010)

Although progress has been made towards improving sustainable access to safe drinking water and sanitation during the past decades, significant challenges remain towards accomplishing the intentions of assigning priority to human drinking water over other uses of water and ultimately guaranteeing safe and clean water as a human right.

Likewise, in compliance with Agenda 21 (Article 18.8 (here quoted from Molle *et al.* 2007)) and the subsequent recommendation made at the World Summit in Johannesburg in 2002, many countries are adopting integrated water resources management (IWRM) as a basic principle for their water governance. In many countries, this has been interpreted as requiring that water is managed on the basis of river basins (Molle *et al.* 2007) and the principal focus of attention in this respect has been to put in place 'the ideal organizational model for river basin management' (*ibid.*: 609). Thus, various types of organizations have been created at river basin and sub-basin level (watershed, catchment,

doi: 10.2166/ws.2013.056

etc.) to develop, allocate and manage the use of water. Yet, even in places where such organizations have been created, decisions with respect to the development, allocation and use of water are taken in a wide array of institutional settings, ranging all the way from individuals and small groups of individuals, sometimes involving agreements mediated by local lawyers, district officials, etc., over communities, local governments and non-governmental organisations (NGOs), to decisions made in the president's office.

The identification of such discrepancies between 'what is' and 'what should be' is one of the findings from research conducted as part of the Competing for Water programme in Bolivia, Mali, Nicaragua, Vietnam and Zambia. (For information about the programme, its partners and publications, see www.diis.dk/water). Following a brief description of the methods employed and the data upon which this paper is based, the subsequent two sections present and discuss empirical evidence of 'what is' with respect to competition for water for domestic and productive uses and with respect to the institutions called upon and involved in water governance in five rural districts – one district in each of the countries mentioned above. The fifth and final section summarizes the conclusions and suggests possible implications for efforts to improve pro-poor water governance in developing countries.

METHODS AND DATA

The paper draws on three sets of empirical data produced with reference to five rural districts. Apart from being located in three different continents, the five districts – Tiraque district in Bolivia, Douentza district in Mali, Condega district in Nicaragua, Con Cuong district in Vietnam and Namwala district in Zambia – were selected through a maximum variation sampling strategy (Patton 2002) to represent a high level of mutual heterogeneity in terms of rainfall pattern, population density, and the importance of irrigation, livestock keeping and fishing, hydro-power and other water infrastructure development, and formal water allocation mechanisms.

The first data set is an inventory of water-related conflictive and cooperative events (Ravnborg *et al.* 2012). (For more details on the methodology for developing the inventories of water-related events, see Ravnborg *et al.* (2012).) A water-related event is defined as 'an action (or a set of actions) that seeks to secure one or more parties' access to or use of water by (i) challenging other parties' access or use; (ii) confirming own or other parties' access or use; or (iii) collaborating with other parties to secure access or use'. Several events may relate to a single water-related situation, defined as 'a social situation where two or more parties have actual or potentially competing interests in the same water resource'. Water-related events may be either conflictive or cooperative.

Based on interviews with community members as well as a wide range of community-external actors undertaken in or with reference to a geographically stratified sample of 10 rural communities in each district, a total of 714 public, water-related events were identified to have occurred between 1997 and 2007. Using the district-specific proportions that the population of the sample communities constitutes of the total rural population of five districts as the extrapolation factor, a total of 6,089 water-related events is estimated to have occurred in the five districts between 1997 and 2007. These events are associated with a total of 1,954 water-related situations (Table 1).

Inspired by the event intensity scale developed by Wolf and colleagues to characterize the intensity of water events

Table 1 | Number of water-related events identified in sample communities and estimated number of water-related events and situations having occurred 1997–2007, by district

	Tiraque district	Douentza district	Condega district	Con Cuong district	Namwala district	All districts
Number of events identified in sample communities	132	143	198	122	119	714
Estimated number of events having occurred in district	1,605	1,017	860	1,283	1,324	6,089
Estimated number of situations having occurred in district	456	195	351	521	431	1,954

in transboundary water basins (Yoffe *et al.* 2003; Wolf *et al.* 2003) and by Thomasson (2005), a scale was developed according to which to assess the intensity of local-level water events. The intensity of conflictive events ranges from the least conflictive (−1) involving informal expressions of discontent or verbal disputes among conflicting parties to the most conflictive (−7) which entail organised collective violence or warfare. For the cooperative events, the water event intensity scale ranges from the least cooperative (+1) where parties express casual verbal recognition of each others' access rights to the most cooperative (+7) where parties decide to merge formally individual access rights. (For more details on the water event intensity scale, see Ravnborg *et al.* (2012).)

The second data set is provided through a questionnaire-based survey on household poverty, access to water for domestic and productive purposes and access to water governance institutions. The survey was conducted in 2008 and was administered to a random sample of 400 rural households in each of the five districts (Ravnborg 2010). Inspired by the reservations expressed by Sen (1981, 1985) towards understanding and measuring poverty and well-being solely on the basis of income or expenditure data, the poverty profile was developed on the basis of people's own perceptions of poverty, identified through well-being rankings. The rankings were conducted in three communities, selected through a maximum variation sampling strategy with respect to factors which could potentially lead to the existence of different perceptions of well-being. The descriptions of different poverty levels resulting from the rankings were 'translated' into indicators. These indicators formed the basis for constructing a household poverty index. A set of three poverty categories was defined, namely the 'poorest', the 'less poor' and the 'non-poor' households. (For more details on the methodology, see Ravnborg *et al.* (1999) and Ravnborg (2010).) Table 2 shows the distribution of households among poverty categories for the five districts.

The third and last set of empirical data is based on qualitative inquiries conducted in a total of 13 communities in the five districts during 2009. The qualitative inquiries were conducted in order to provide more detailed understanding of the processes and relationships through which access to water is obtained, secured and lost in rural communities. A range of techniques were employed as part of these inquiries, including mapping of social, economic and political relationships, livelihood mapping, and focus group interviews with different types of community members such as community leaders, de-facto single female household heads, new-comers, etc. (The case studies are documented in reports available at http://www.diis.dk/sw33068.asp.)

COMPETITION FOR WATER IN FIVE RURAL DISTRICTS

Competition for water – i.e. the situation that different actors seek access to the same water resource for similar or different purposes which potentially or actually, fully or partly exclude each other – primarily spells out into action during the dry season. Of the 6,089 water-related events that are estimated to have occurred between 1997 and 2007 in the five districts, 80 percent began during the dry season. Shortage of water both accentuates the need for people to develop mechanisms and agreements on how to share and manage the use of the water available, and give rise to contestations and confrontations over access to and use of water among competing parties.

Table 2 | Distribution of households according to household poverty category by district, percent households per poverty category by district

Poverty category	Tiraque district ($N = 400$)	Douentza district ($N = 400$)	Condega district ($N = 395$)	Con Cuong district ($N = 400$)	Namwala district ($N = 400$)
Non-poor	31	14	23	36	27
Less poor	50	50	32	43	31
Poorest	19	37	45	21	43
All poverty categories	100	100	100	100	100

Both from a concern with conflict prevention in a broad sense and from a more specific concern with securing access to drinking water for all even in critical periods of the years, events where different actors compete for access to water for different purposes, i.e. multiple-use events, merit further scrutiny.

Conflict and cooperation related to competing claims from actors who (wish to) use water for different purposes are most common in Douentza and Namwala districts. Here multiple-use situation accounted for 64 and 78 percent, respectively, of the water-related situations within which events had taken place between 1997 and 2007. At the other end of the spectrum, in Tiraque and Con Cuong districts, multiple-use situations constituted 6 and 15 percent, respectively. In between, in Condega district, multiple-use situations constituted 44 percent of the water-related situations. In three of the five districts – Condega, Con Cuong and Namwala – multiple-use events are more likely to be conflictive than single-use events. In Condega district, 68 percent of the multiple-use events are conflictive as compared to 32 percent of the single-use events; in Con Cuong district, 71 percent of the multiple-use events are conflictive as compared to 35 percent of the single-use events and in Namwala district, 74 percent of the multiple-use events are conflictive as compared to 58 percent of the single-use events (significance of Pearson's chi-square < 0.001). In Douentza district, the opposite is the case. Here, 32 percent of the multiple-use events are conflictive compared to 48 percent of the single-use events (significance of Pearson's chi-square <0.001). Moreover, the conflict intensity tends to be higher for the conflictive multiple-use events than for the conflictive single-use events. Based on the water event intensity scale, the average intensity of the conflictive multiple-use events ($n = 1{,}255$) is -2.41 compared to an average intensity of -1.89 for the conflictive single-use events ($n = 1{,}800$) (significance of $F < 0.001$; analysis of variance). Likewise, the intensity of the cooperative single-use events tends to be higher than that of the cooperative multiple-use events. The average event intensity for the cooperative single-use events ($n = 1{,}971$) is $+2.86$ compared to an average intensity of $+2.56$ for the cooperative multiple-use events ($n = 984$) (significance of $F < 0.001$; analysis of variance).

Overall, the majority of these multiple-use events take place between people who wish to use water for domestic purposes and people who wish to use water for productive purposes such as watering livestock and crops and, particularly in Con Cuong district, also for small-scale mining activities (Ravnborg *et al.* 2012). Three video reports (available at www.diis.dk/water/videoreports) produced as part of the Competing for Water programme provide examples of such multiple-use events from Namwala, Condega and Con Cuong districts.

Over the past decades, significant progress has been made towards establishing infrastructure to provide rural people with access to safe drinking water. An average of 72 percent of the households in the five districts indicated to have a public water tap, a deep well or a drilled borehole as their most important source of drinking water during the dry season. In Tiraque, Condega and Con Cuong districts, two-thirds or more of all rural households walk less than 5 minutes to fetch water from their most important drinking water source, while water infrastructure coverage is lagging behind Douentza and Namwala districts. Here, only 36 and 46 percent, respectively, have less than 5 minutes walk to their most important drinking water source and more than a quarter of the households in the two districts have to walk more than 20 minutes to fetch drinking water.

Despite progress made, our data reveal a worrying tendency of a social skewedness with respect to access to safe drinking water made available through public investments during past decades (Figures 1(a) and 2(a)). This is particularly pronounced in Tiraque and Con Cuong districts which in general enjoy the best water infrastructure coverage for domestic as well as for productive (irrigation) purposes. In Tiraque district, most people depend on public, gravity-fed schemes for drinking water. However, while 86 percent of the non-poor households indicated public water supply scheme as their most important dry season drinking water source, this was only the case for 64 percent of the poorest households. In the absence of access to public drinking water supply, more than a third of the poorest households in Tiraque use less safe water sources, e.g. shallow wells (21 percent of the poorest households) and surface water from streams and irrigation canals (14 percent of the poorest households) as drinking water during the dry season. In Con Cuong district, deep wells, often publicly funded, constitute the most common source of drinking water during the dry season, but again access is socially skewed. Three quarters (78 percent)

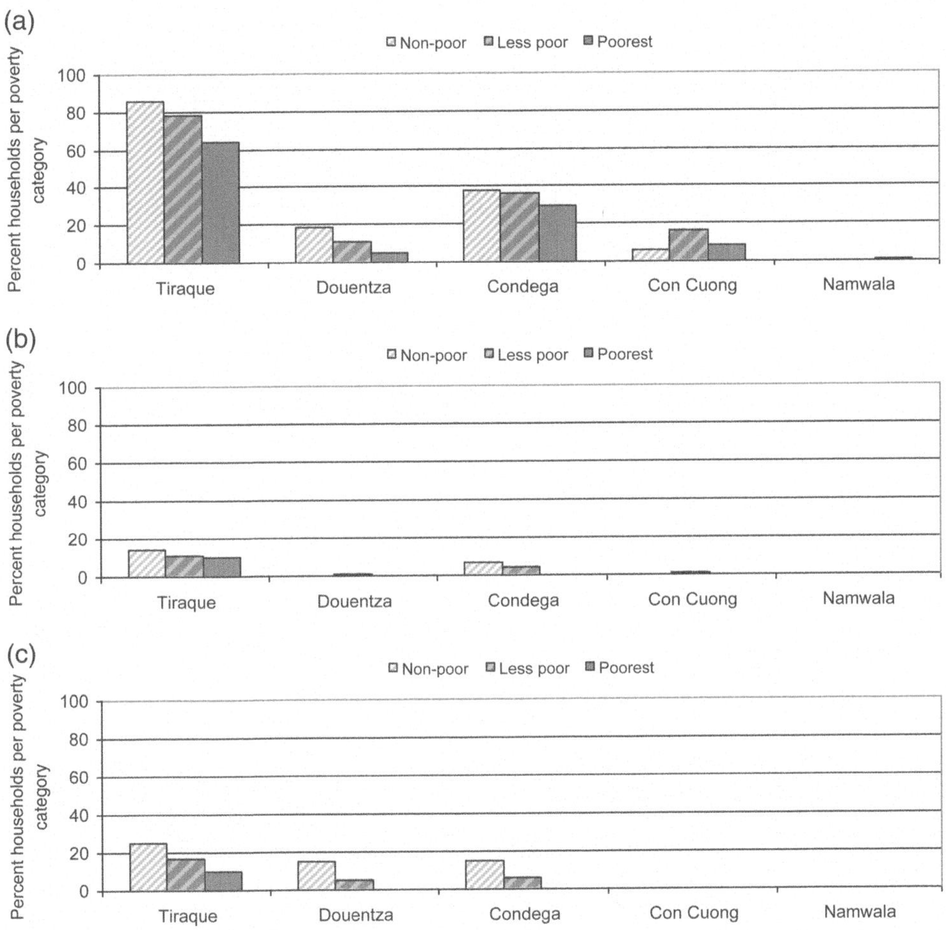

Figure 1 | Public water supply access by household poverty level, by research location (a: for dry season drinking water; b: for wet season animal watering; c: for dry season animal watering), *Percent households by poverty category (see Table 2 for n values).*

of the non-poor households have deep wells as their most important drinking water source during the dry season compared with only a third (34 percent) of the poorest households. Instead the poorest households use surface water from the streams and canals for drinking during the dry season (being the case for 48 percent of the poorest households compared to 8 percent of the non-poor households). Similar patterns exist for public water supply schemes in Condega and Douentza districts (Figures 1(a) and 2(a)).

Moreover, there is a social skewedness with respect to who loses and who gains when competition for access to these water sources intensifies during the dry season. In most places, publicly funded deep wells, boreholes and gravity-fed water taps are primarily intended to supply rural people with safe drinking water while water for other purposes, such as watering of animals and crops, is intended to be drawn from other sources. Yet, as other nearby sources are drying out as the dry season progresses, the temptation grows to use water from the drinking water taps and deep wells for livestock and crops. Because livestock and particularly cattle ownership is socially skewed in favour of the non-poor, this adds to the disadvantaged position that the poorest households face in obtaining secure access to drinking water. However, in Douentza district where cattle ownership is most widespread with 73 percent of the households owning cattle, it is primarily the poorest livestock-owning households that are tempted to use deep wells, boreholes and public water supply schemes to water their livestock, whereas the non-poor households to a larger extent have access to privately owned shallow wells for watering their livestock. As illustrated in Figures 1(b), 1(c), 2(b) and 2(c), the proportion of households using public

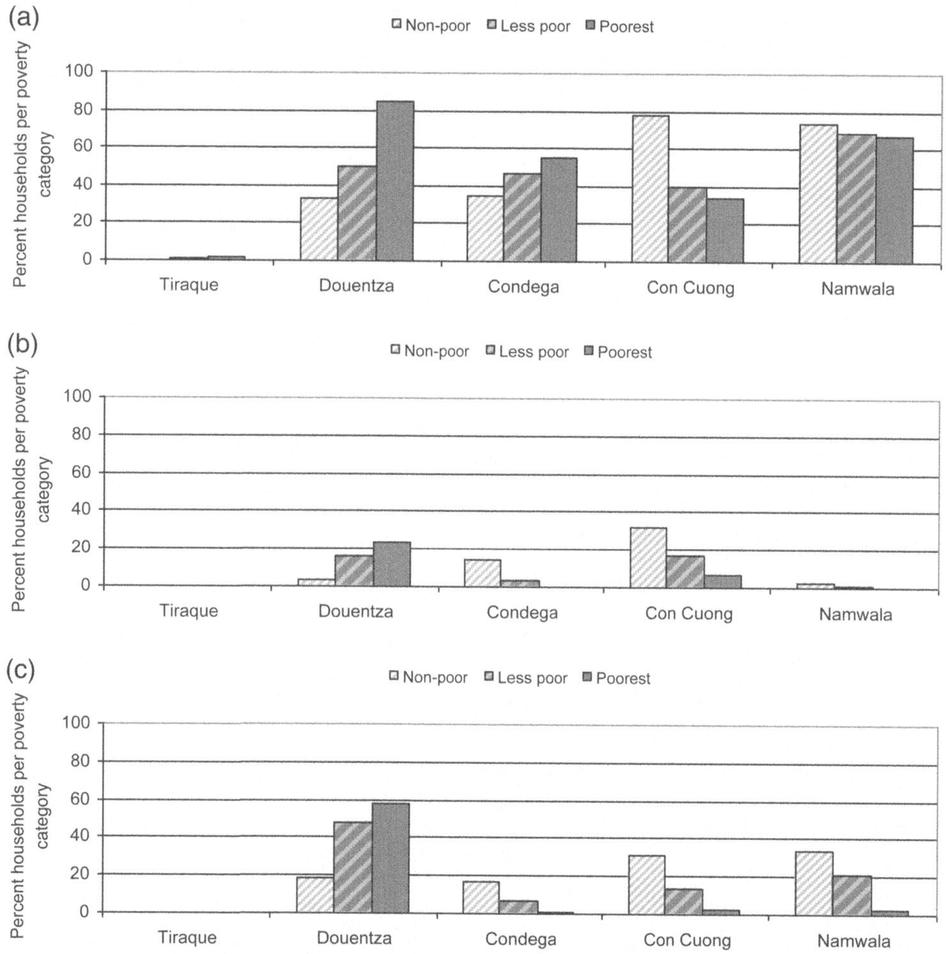

Figure 2 | Ground water (deep wells and boreholes) access by household poverty level, by research location (a: for dry season drinking water; b: for wet season animal watering; c: for dry season animal watering), *Percent households by poverty category (see Table 2 for n values).*

water taps, deep wells and boreholes to supply water for their livestock increases in the dry season as compared to the rainy season in all districts with the exception of Con Cuong. Among the five districts, Con Cuong is where cattle ownership is least common with only 29 percent of the households owning cattle.

Our inventory data indicate that it is often people who use water for drinking and other domestic purposes who have to give in when faced with competing claims for the same water, e.g. from livestock keepers or farmers who use water for irrigation (Figure 3). Only in 11 percent (83 events) of the 728 multiple-use, conflictive events involving rural domestic consumers as one of the parties, rural domestic consumers were assessed to 'win' while they were assessed to lose in 39 percent (282 events) of these multiple-use, conflictive events. Of the 346 multiple-use, conflictive events involving competing claims between users of water for livestock and domestic water users, livestock keepers were assessed to 'win' in 23 percent (81 events) of the events while domestic water users were assessed to 'win' only in 14 percent (49 events) of the events. Finally, of the 247 multiple-use, conflictive events involving competing claims between users of water for irrigation and domestic water users, irrigation farmers were assessed to 'win' in 40 percent (100 events) of the events while domestic water users were assessed to 'win' only in 7 percent (17 events) of the events.

The finding that multiple-use events are most common where water infrastructure coverage is poorest (Douentza, Namwala and to a lesser extent Condega district) is hardly

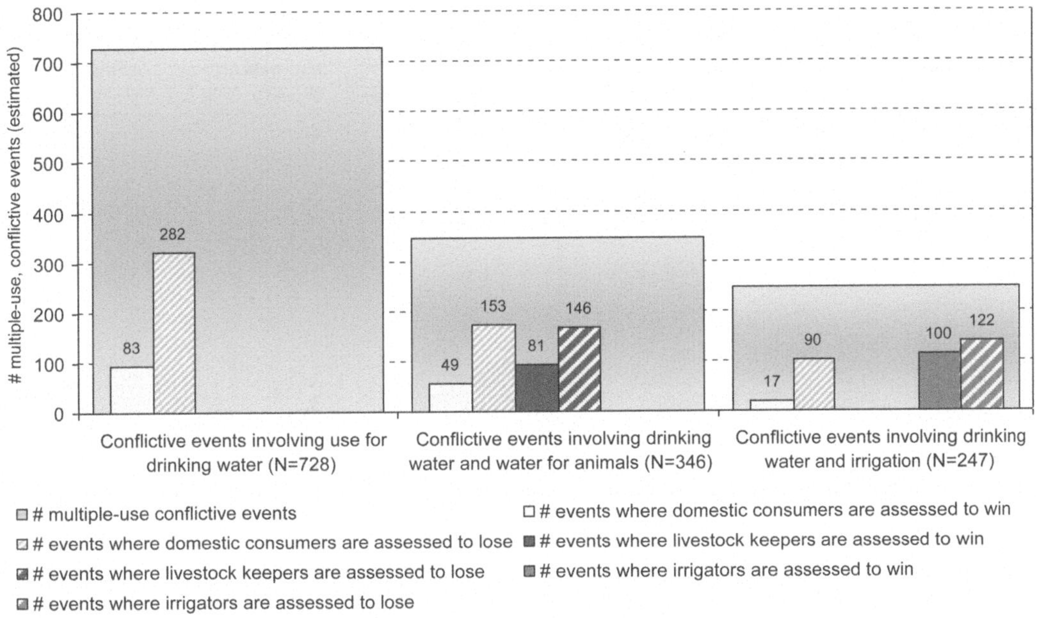

Figure 3 | Winners and losers in multiple-use, conflictive events, all districts, *Estimated number of events*.
Modified on the basis of Figure 5, *Water Policy*, Vol. 14, issue 2, pp 336–357 (2012) with permission from the copyright holders, IWA Publishing.

surprising. However, combined with the insight that poor domestic water consumers are more likely to lose out in such conflictive water events than non-poor households who often also use the water for productive purposes, the results become more alarming, particularly when water has been made available through publicly funded water infrastructure. They suggest that when rural drinking water infrastructure is spread thinly both population-wise and geographically without infrastructural and institutional provisions for meeting demands for water for productive purposes, public water infrastructure investments are likely to exacerbate economic inequality through providing subsidized access to productive water for the non-poor, without providing sustainable access to safe drinking water for the poorest part of the population.

WATER GOVERNANCE IN AN INSTITUTIONAL PERSPECTIVE IN FIVE RURAL DISTRICTS (THIS SECTION DRAWS EXTENSIVELY UPON RAVNBORG *ET AL*. (2012))

Conflict and cooperation about water largely takes place without coming to the knowledge of authorities outside the location of the water-related events, including authorities which hold statutory water governance mandates. Just over a quarter of the 1,954 water-related situations which had given rise to water-related events in the five districts between 1997 and 2007 were recorded by outside authorities (not including community-based authorities such as community leaders, village government, water committee, etc.), the media, NGOs, etc., either in writing (administrative files, internal memos, local newspaper articles, etc.) or in the memory of actual or previous staff members.

Yet, in 42 percent of the water-related situations, one or more of the direct parties to the situation had called upon a third party to mediate in the conflict, to solicit support for improving water-related infrastructure, etc. In six out of 10 of these situations where a third party had been called upon, authorities and other social actors from *outside* the location of the situation had been called upon as the third party, whereas in the remaining four out of 10, only community-based authorities had been called upon as the third party. However, significant differences exist between the five districts in this respect, reflecting their mutually diverging institutional contexts. As shown in Table 3, third parties, particularly community-external third parties are

Table 3 | Third party called upon[a] in water-related situations by research location, percent situations per research location

	Tiraque district ($N = 456$)	Douentza district ($N = 195$)	Condega district ($N = 351$)	Con Cuong district ($N = 521$)	Namwala district ($N = 431$)	All districts ($N = 1,954$)
Third party called upon (any)	21.1	60.0	65.0	30.3	50.8	41.9
Location of third party called upon:						
–community-level authority, only	6.1	11.8	2.8	29.6	28.1	17.2
–community external authority called upon (also)	14.9	48.2	62.4	0.6	22.7	24.7
Type of third party called upon[b]:						
–community-level authority	9.2	28.2	16.0	29.8	28.5	22.1
–district authority	7.9	22.1	31.1	18.8	15.8	18.1
–ministry	3.9	2.1	20.5	–	11.6	7.4
–local lawyer	12.9	13.8	14.8	–	–	7.1
–police	3.1	–	14.0	–	–	3.2
–NGO	–	20.0	13.4	–	5.6	5.6
–water authority	–	4.1	20.5	–	6.0	5.4
–legal authority	6.4	13.8	3.4	–	–	3.5
–provincial or departmental authority	3.3	9.7	1.4	–	5.6	3.2
–international organisation	–	–	5.4	–	–	1.0
–media	–	–	1.7	–	0.5	0.4
–civil society organisation	–	2.1	–	–	–	0.2
–church	–	–	–	–	–	–
–other	–	–	1.4	–	–	0.3

[a]Correlation between type of third party called upon (community-level authority only, community-external authority lso, ano third party) and research location significant at 0.001 level (Pearson's chi-square test).

[b]In some water-related situations, more than one third party was called upon. Therefore, the institution-specific percentages do not add up to the total percentage of situations in which third parties have been called upon.

most likely to be called upon in Condega district. Apart from the proximity of the district headquarters to most rural communities, Condega's district administration has had an explicit policy of public attention and has been favoured by a number of donor-funded interventions related to water supply and watershed management implemented through ministries, water supply agencies, NGOs, sometimes in coordination with the district authorities. Moreover, the district environmental commission in Condega has assumed an active role in water governance issues, particularly since 2005 (Gómez & Ravnborg 2012). At the other extreme, third parties are least likely to be called upon in Tiraque and Con Cuong districts. Both districts have a long history of irrigation farming and are widely served by irrigation canals. Over time developed mechanisms, including community-based committees, for allocating irrigation water among the plots served by the canals. In addition, both districts have separate infrastructure for domestic water. Thus in Tiraque, community-based irrigation committees were themselves a direct party to close to one third of the water-related situations that had taken place in the district. In Con Cuong district which like the rest of Vietnam has a hierarchical political and administrative system, attempts to call upon community-external authorities as third parties have to take place through the community organisation. Thus, almost exclusively community-based authorities were called upon as third parties in Con Cuong.

Despite these differences, the institutions called upon as third parties tend to have mandates that reach beyond water-related issues, whereas institutions which have a specific water-related mandate are much less called upon (Table 3). Only in Condega district, the water supply agency (ENACAL) had been called upon to any notable degree due to executing a donor-funded rural drinking water project during the study period.

The data presented suggest that despite efforts to create institutions with designated water mandates, transcending administrative divides between communities, districts, etc. to enable integrated water governance, people tend to call upon institutions they know and that are within reach when in need of a third party in a water-related situation. This is partly out of convenience. When asked who they would call upon first, second and later in cases of water sources running dry due to excessive withdrawals by others, respondents in all five districts ($N = 1{,}995$ households) indicated that they would first go to community institutions and neighbours and only later to institutions outside the community such as district authorities. That catchment and river-basin organisations are not called upon may partly be out of ignorance of their existence, partly due to their sporadic existence. As IWRM primarily has been promoted by international donor agencies, the functioning of catchment organisations has tended to fluctuate with the availability of external funding.

Moreover, judging from the insights from the five districts, little practical consideration has been given to what IWRM implies at the local level. The role that district authorities play in water governance has been left largely unfunded, if at all recognized. Community-based water committees have received more attention. However, particularly when associated with donor-funded drinking water infrastructure, as in Douentza, Condega and Namwala districts, the focus has been on infrastructure maintenance rather than on allocation issues.

Several of our case studies (Gómez & Ravnborg 2011; Funder *et al.* 2012) found that it is the community-level power imbalances which make the poor refrain from claiming their right to access to drinking water. As a woman explained in a focus group meeting with poor women in a Nicaraguan community:

> 'If they [the non-poor cattle owners] are fetching water, filling their tanks [for watering cattle], I prefer to wait and fetch the little water which is left' (Woman A/B, personal communication held in San Isidro, 2009).

Referring to the fact the non-poor households hold significant statutory as well as economic and physical power, another woman expanded:

> 'Those who are in the water committee are the same people who have constructed tanks; it is better not to say anything in order not to get into trouble' (Woman A/B, personal communication held in San Isidro, 2009).

Although widely recognized, such power imbalances and their implications for poor people's access to safe drinking water have only rarely been addressed in efforts to improve access to drinking water. This neglect contributes to explain the discrepancy between water governance as envisaged in national water policies and regulatory frameworks and water governance as it is practiced at the local level.

CONCLUSIONS – THE IMPLICATIONS OF DISCREPANCIES FOR WATER GOVERNANCE

Despite decades of policy commitment from national governments and the international community to secure sustainable access to safe drinking water for all and to establish the organizational infrastructure for planning and managing water use and development (Molle *et al.* 2007; Lenton & Muller 2009; Peña 2011), our results suggest that the reality in rural areas continues to fall short of these expectations in many developing countries.

When competition for water intensifies in the dry season, poor households are likely to lose their access to safe drinking water to non-poor households who often also use the water for productive purposes. Several factors collude to cause this unfortunate but well-known situation. Water supply infrastructure is often scaled and implemented to satisfy the needs for drinking water only. Rules are established which assign priority to drinking water and often prohibit the use of water for productive purposes. While this approach obviously makes it possible to obtain larger geographical coverage of drinking water supply, empirical evidence from the five districts presented above indicates that larger geographical coverage comes at the cost of smaller socio-economic coverage. The economic gains associated with circumventing the rules that assign priority to domestic over productive use in the allocation of water simply exceed the capacity – and will – of local enforcement of existing rules. When the water supply infrastructure put in place is insufficient to satisfy both domestic and productive demands for water, and when community-level power imbalances are not specifically addressed in efforts to improve water governance, risks are high that water supply infrastructure is appropriated by the local elite and that improvements in poor people's access to safe drinking water fall short of expectations and come at the price of increasing social and economic inequality.

Two options exist to remedy this situation. First, more attention should be placed at developing water infrastructure that provides sufficient water to cover both domestic and productive demands for water. In order not to exacerbate inequality by providing a subsidy in favour of the non-poor who are in a better position to exploit water for productive purposes, ways should be identified to ensure that those who benefit from such public water infrastructure investments for productive purposes beyond a certain minimum level, contribute – proportionally more – to the financing and running of the infrastructure, e.g. as part of general taxation schemes, water use permit systems, etc. The second option is to maintain the exclusive focus on domestic water supply in water infrastructure development, but combined with increased investments in the capacity to provide external support for enforcement of water allocation rules and for conflict mediation and resolution. Both of these options call for skills beyond the scope of hydrology and for political deliberations and democratic representation that lie outside that of designated and often technically oriented basin- and watershed management organizations (Barham 2001; Ravnborg 2004; Molle 2009).

This call is corroborated by the efforts that rural people make to deal with competition for water. The institutions called upon to help provide solutions to water needs or to mediate in water-related conflicts tend to be institutions that are within immediate reach, are known to people due to their permanency over time, or that enjoy a certain democratic legitimacy. Institutions with designated water mandates such as water authorities, basin or watershed committees that have a more sporadic presence depending on donor funding, etc., were much less called upon as third parties to situations of water-related conflict and cooperation.

During the past decade, resentments towards IWRM as a feasible approach to water governance have emerged (e.g. Biswas 2004, 2008; Molle *et al.* 2007; Molle 2008; Lenton & Muller 2009; Butterworth *et al.* 2010). Some authors call for the need to draw upon adaptive management approaches to make IWRM more flexible and sensitive to context by learning from progress and drawbacks as they unfold (Pahl-Wostl & Sendzimir 2005). In places where IWRM approaches are well underway as an organizational approach for water governance, the way forward may be to sustain these efforts in order not to lose gained momentum. However, in places where limited progress has been made towards the practical implementation of IWRM, the findings presented in this paper suggest that recognizing and supporting the role of a wide array of institutions which actually – as we have seen – take part in water governance, may be the way forward. This is in line with current thinking on polycentric governance. According to Ostrom:

> 'Polycentric' connotes many centers of decision making that are formally independent of each other…. To the extent that they take each other into account in competitive relationships, enter into various contractual and cooperative undertakings or have recourse to central mechanisms to resolve conflicts, the various political jurisdictions […] may function in a coherent manner with consistent and predictable patterns in interacting behaviour. To the extent that this is so, they may be said to function as a 'system' (Ostrom *et al.* 1961; here quoted from Ostrom 2010: 411).

Seen from this perspective, the water governance challenge is to close the gap between 'what is' and 'what should be' by recognizing the diverse array of statutory, customary and private institutions that take part in water governance, and make sure that 'central mechanisms' to resolve actual as well as potential conflicts such as the political, legal and administrative framework are in place so that water governance institutions as well as citizens have common points of reference.

ACKNOWLEDGEMENTS

The authors wish to acknowledge the collaboration of Rocio Bustamante, Abdoulaye Cissé, Signe M. Cold-Ravnkilde, Vladimir Cossio, Moussa Djiré, Mikkel Funder, Ligia I. Gómez, Phuong Le, Carol Mweemba, Imasiku Nyambe, Tania Paz, Huong Pham, Roberto Rivas, Thomas Skielboe, and Nguyen T.B. Yen. The Competing for Water programme was undertaken with the financial support from the Danish Council for Development Research.

REFERENCES

Barham, E. 2001 Ecological boundaries as community boundaries: the politics of watersheds. *Society and Natural Resources* **14**, 181–91.

Biswas, A. 2004 Integrated water resources management: a reassessment. *Water International* **29** (2), 248–256.

Biswas, A. 2008 Integrated water resources management: is it working? *International Journal of Water Resources Development* **24** (1), 5–22.

Butterworth, J., Warner, J., Moriarty, P., Smits, S. & Batchelor, C. 2010 Finding practical approaches to integrated water resources management. *Water Alternatives* **3** (1), 68–81.

Funder, M., Bustamante, R., Cossio, V., Huong, P. T. M., van Koppen, B., Mweemba, C., Nyambe, I., Phuong, L. T. T. & Skielboe, T. 2012 Strategies of the poorest in local water conflict and cooperation – Evidence from Vietnam, Bolivia and Zambia. *Water Alternatives* **5** (1), 20–36.

Gómez, L. I. & Ravnborg, H. M. 2011 *Power, Inequality and Water Governance – The Role of Third Party Involvement in Water-Related Conflict and Cooperation*. CAPRi Working Paper 101. International Food Policy Research Institute, Washington, DC.

Lenton, R. & Muller, M. 2009 *Integrated Water Resources Management in Practice: Better Water Management for Development*. Earthscan, London.

Molle, F., Wester, P. & Hirsch, P. 2007 River basin development and management. In: *Water for food – Water for life. Comprehensive Assessment of Water Management in Agriculture* (D. Molden, ed.). EarthScan, London, pp. 585–624.

Molle, F. 2008 Nirvana concepts, narratives and policy models: Insights from the water sector. *Water Alternatives* **1** (1), 131–156.

Molle, F. 2009 Water, politics and river basin governance: repoliticizing approaches to river basin management. *Water International* **34** (1), 62–70.

Ostrom, E. 2010 *Beyond Markets and States: Polycentric Governance of Complex Economic Systems*. Nobel Lecture, December 8, 2009. Les Prix Nobel, pp. 408–444.

Ostrom, V., Tiebout, C. M. & Warren, R. 1961 The organization of government in metropolitan areas: a theoretical inquiry. *American Political Science Review* **55**, 831–842.

Pahl-Wostl, C. & Sendzimir, J. 2005 The relationship between IWRM and Adaptive Water Management. NeWater Working Paper 3.

Patton, M. Q. 2002 *Qualitative Research and Evaluation Methods*. 3rd rev. edition, Sage Publications, Thousand Oaks, CA.

Peña, H. 2011 *Social Equity and Integrated Water Resources Management*. TEC Background Papers 15. Global Water Partnership Technical Committee (TEC), Stockholm.

Ravnborg, H. M. 2004 Conflict prevention and mitigation in water governance: lessons learned and options available. In *Water and Conflict – Conflict Prevention and Mitigation in Water Resources Management* (H. M. Ravnborg, ed.). DIIS Report 2004:2. Danish Institute for International Studies, Copenhagen.

Ravnborg, H. M. with the collaboration of Escolán, Guerrero, M. P., Méndez, M. A., Mendoza, F., de Paez, E. M. & Motta, F. 1999 *Developing regional poverty profiles based on local perceptions* (also available in Spanish). CIAT publication 291. Centro Internacional de Agricultura Tropical, Cali.

Ravnborg, H. M. 2010 Methodology for developing household poverty profiles for the Competing for Water research locations. Mimeo. Available from: http://diis.dk/graphics/Subweb/Water/Methodology%20-%20Competing%20for%20water.pdf.

Ravnborg, H. M., Bustamante, R., Cissé, A., Cold-Ravnkilde, S. M., Cossio, V., Djiré, M., Funder, M., Gómez, L. I., Le, P., Mweemba, C., Nyambe, I., Paz, T., Pham, H., Rivas, R., Skielboe, T. & Yen, N. T. B. 2012 The challenges of local water governance – the extent, nature and intensity of water-related conflict and cooperation. *Water Policy* **14** (2), 336–357.

Sen, A. 1981 *Poverty and Famines: An Essay on Entitlement and Deprivation*. Clarendon Press, Oxford.

Sen, A. 1985 *Commodities and Capabilities*. North Holland, Amsterdam.

Thomasson, F. 2005 *Local Conflict and Water: Addressing Conflicts in Water Projects*. Swedish Water House, Stockholm.

Yoffe, S., Wolf, A.T. & Giordano, M. 2003 Conflict and cooperation over international freshwater resources: indicators of basins at risk. *Journal of the American Water Resources Association* **39** (5), 1109–1126.

Wolf, A. T., Yoffe, S. B. & Giordano, M. 2003 International waters: identifying basis at risk. *Water Policy* **5**, 29–60.

First received 24 February 2012; accepted in revised form 16 May 2012

Towards adaptive and integrated management paradigms to meet the challenges of water governance

J. Halbe, C. Pahl-Wostl, J. Sendzimir and J. Adamowski

ABSTRACT

Integrated Water Resource Management (IWRM) aims at finding practical and sustainable solutions to water resource issues. Research and practice have shown that innovative methods and tools are not sufficient to implement IWRM – the concept needs to also be integrated in prevailing management paradigms and institutions. Water governance science addresses this human dimension by focusing on the analysis of regulatory processes that influence the behavior of actors in water management systems. This paper proposes a new methodology for the integrated analysis of water resources management and governance systems in order to elicit and analyze case-specific management paradigms. It builds on the Management and Transition Framework (MTF) that allows for the examination of structures and processes underlying water management and governance. The new methodology presented in this paper combines participatory modeling and analysis of the governance system by using the MTF to investigate case-specific management paradigms. The linking of participatory modeling and research on complex management and governance systems allows for the transfer of knowledge between scientific, policy, engineering and local communities. In this way, the proposed methodology facilitates assessment and implementation of transformation processes towards IWRM that require also the adoption of adaptive management principles. A case study on flood management in the Tisza River Basin in Hungary is provided to illustrate the application of the proposed methodology.

Key words | IWRM, management and transition framework (MTF), management paradigms, participatory modeling, water governance, water management

J. Halbe (corresponding author)
C. Pahl-Wostl
Institute of Environmental Systems Research,
University of Osnabrück,
Barbarastr. 12,
49076 Osnabrück,
Germany
E-mail: *jhalbe@uos.de*

J. Sendzimir
International Institute for Applied Systems Analysis (IIASA),
Schlossplatz 1,
A-2361 Laxenburg,
Austria

J. Adamowski
Department of Bioresource Engineering,
McGill University,
21 111 Lakeshore,
H9X 3V9,
Ste Anne de Bellevue,
Quebec,
Canada

INTRODUCTION

The persistence of water issues in many parts of the world has given rise to innovative concepts that advocate an integrated approach to address complexity and uncertainty. Integrated Water Resource Management (IWRM) is the most prominent of such concepts that stresses the importance of integrated and participatory management processes and reform of water governance systems (Medema *et al.* 2008). The term 'management' refers to operational activities including the operation, monitoring, strategic planning and implementation of measures, whereas the term 'governance' comprises the rules under which a management system operates and different actors and networks help develop and implement water policies (cf., Pahl-Wostl 2009). However, theories and methods for sustainable water resource management and governance are still in the developmental phase and continuous experiments in application are required to determine effective approaches for research and practice (cf., Galaz 2007; Medema *et al.* 2008). Methodologies are needed that deal with real-world complexity in order to find effective solution strategies, and facilitate knowledge transfer between science, policy, engineering and local communities. Even though IWRM is a concept widely aspired to, it is often still rooted in a traditional 'predict and control paradigm' despite its linkage to the idea of adaptive management (cf., Jeffrey & Gearey 2006). The need for the integration of adaptive approaches like policy experimentation and learning into IWRM is increasingly acknowledged in order to effectively realize the concept of IWRM (Galaz 2007).

Management paradigms are appropriate concepts to systematically and comprehensively analyze the interlinkages between a resource system (e.g., groundwater resources),

doi: 10.2166/wst.2013.146

water management system (e.g., infrastructure) and water governance system (i.e., regulatory structures and processes). A specific analysis of the multiplicity of elements of water resource management paradigms has been discussed by Pahl-Wostl *et al.* (2011). They define a management paradigm as 'a set of basic assumptions about the nature of the system to be managed, the goals of managing the system and the ways in which these goals can be achieved' (Pahl-Wostl *et al.* 2011). By being explicit about underlying paradigms, inconsistencies in water management and governance systems become apparent (as discussed by Jeffrey & Gearey (2006)). For instance, public participation can be applied in a 'predict and control water management paradigm' (i.e., stakeholders are only informed and consulted), as well as in a 'community involvement paradigm' (i.e., co-management of stakeholders). The ignorance of underlying paradigms can lead to miscommunication and subsequent management problems (e.g., stakeholders expect active involvement but can become frustrated due to missing opportunities for engagement).

This paper presents a methodology for the participatory analysis of management and governance systems that supports the design and implementation of transformation processes towards sustainable water management. In this methodology, participatory model building is applied to elicit case specific water management paradigms held by stakeholders. This information is then used to comprehensively analyze the management and governance system through the application of the Management and Transition Framework (MTF) developed by Pahl-Wostl *et al.* (2007, 2010). Based on such an analysis of the status quo, the methodology allows for the participatory envisioning and design of pathways towards sustainable water management and governance. In this way, the proposed methodology facilitates the development, assessment and implementation of strategies towards sustainable water resources governance and management.

The methodology builds upon the management paradigm concept developed by Pahl-Wostl *et al.* (2011) and the use of group model building for the analysis of paradigms (cf., Sendzimir *et al.* 2007). Innovative elements of the methodology proposed in this paper are the conceptualization of sub-system and overall-system paradigms, and the delineation of a structured action research process including elicitation, analysis and assessment of paradigms. Another innovative element of the proposed methodology is the application of the MTF for the design of pathways towards implementation of integrated and adaptive water management. The approach presented in this paper goes beyond theoretical explanations of institutional and policy change (cf., Cashman 2009) by allowing for participatory analysis and active governance of transformation processes.

This paper is structured in two parts. First, the underlying concepts and methods upon which the proposed methodology is built are presented. Following this, a case study on flood management in the Tisza River Basin in Hungary is provided to illustrate the application of the proposed methodology.

METHODS

The proposed methodology combines an analysis of the overall water management and governance system (using the MTF) with an investigation of embedded management paradigms applying the systems thinking method in a participatory modeling process. Participatory modeling supports the analysis of case-specific elements of resource management and governance systems, while the MTF analysis allows for a broader perspective by integrating the detected elements found via participatory modeling into an overall system perspective.

Participatory model building

Different participatory modeling approaches exist that follow different objectives and apply a range of methods (cf., Jonsson *et al.* 2007; Renger *et al.* 2008; Voinov & Bousquet 2010). In this paper, the method of participatory model building using systems thinking and system dynamics methods is proposed for analyzing the perceptions of stakeholders on the management and governance system. Systems thinking is a method for the qualitative analysis of systems and their dynamic behavior through time. System dynamics modeling is based on this qualitative analysis and comprises the quantitative simulation of systems to discover their inherent dynamics as well as allowing for the testing of strategies.

Causal Loop Diagrams (CLDs) are powerful tools of the systems thinking method for the qualitative analysis of systems. They help to depict the system's structure, and mark time delays and feedback processes that are often responsible for difficulties in controlling the inherent dynamics of the management system. In these diagrams, elements of the system are connected by arrows having positive and negative polarities. A positive link indicates the parallel behavior of variables: in the case of an increase in the 'cause' variable, the variable that is affected also increases, while a decrease in the 'cause' variable implies a decrease in the affected one. A negative link indicates an inverse linkage between variables (see Figure 1 as an example of a CLD).

Despite the fact that expert models often offer comprehensive and scientifically validated results, missing ownership and understanding of the model by decision-makers and other stakeholders often impedes the implementation of model-based recommendations. This has led to the development of system dynamics applications that involve stakeholders in quantitative model building. Group model building processes (also called mediated modeling) for collaborative management of complex human-environment problem situations have begun to be applied more frequently over the course of the last decade (e.g., Costanza & Ruth 1998; van den Belt 2004; Tidwell *et al.* 2004; Metcalf *et al.* 2010; Halbe *et al.* accepted).

Management and transition framework (MTF)

Understanding processes of change towards sustainable resource management and governance requires an analytical approach that allows for the analysis of the interdependence between structural context and process characteristics. The MTF developed by Pahl-Wostl *et al.* (2010) supports such analyses of water governance regimes and transition processes towards more adaptive and sustainable systems. The MTF builds upon the three conceptual pillars of adaptive management (cf., Holling 1978), social learning and transformation processes (cf., Pahl-Wostl *et al.* 2007), as well as the Institutional Analysis and Development Framework (which is aimed at the analysis of the role of institutions in collective choice processes, cf., Ostrom (2005)). Specific emphasis is given to the analysis of adaptive capacity and multi-level learning processes. However, the MTF is not constrained to one specific theory; instead it provides a flexible language that can be tailored to specific research questions (for examples of applications of the MTF, see Schlüter *et al.* (2010); Sendzimir *et al.* (2010)).

The MTF helps to formalize structural elements of a water system (which are denoted as 'classes') as well as policy and learning processes (cf., Pahl-Wostl *et al.* 2010). Central classes in the MTF are as follows. An 'Action Situation' refers to formal or informal social

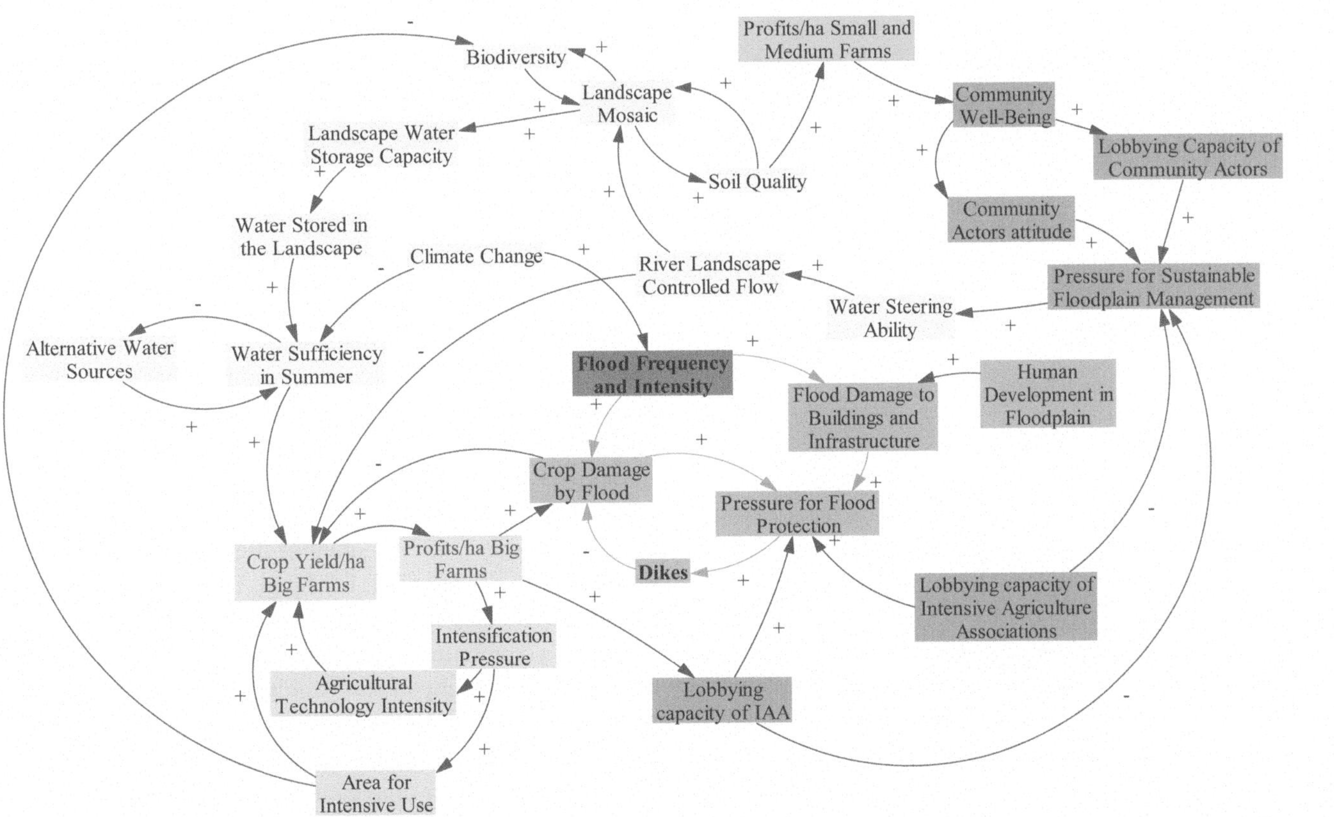

Figure 1 | Causal Loop Diagram regarding the flooding problem in the Tisza River Basin (extended from Sendzimir *et al.* (2007)). The colors reflect the sub-system paradigms: 'Adapt to Floods' paradigm (blue); 'Control Floods' paradigm (orange); 'Economies of Scale' paradigm (green); 'Tradition' paradigm (grey); 'Community Involvement' paradigm (pink). The problem variable is highlighted in red. Please refer to the online version of this paper to see this figure in color: http://www.iwaponline.com/wst/toc.htm.

processes that lead to relevant outcomes for water management. Results can be 'Institutions' (e.g., a new water legislation), 'Knowledge' (e.g., increased understanding of stakeholder problem perspectives) or 'Operational Outcomes' (i.e., direct physical interventions in the system such as the implementation of infrastructure or distribution of water to different uses). The 'Action Arena' class sets the context for the management of a specific water-related problem such as flood management, and is characterized by 'Strategic Management Goals', 'Actors' and a number of 'Action Situations'. In this way, the MTF provides a common language to analyze and discuss complex management and governance systems in research and practice. Relational databases are used to support formalization and standardization of data collection and analysis protocols (cf., Knieper *et al.* 2010). A graphical interface allows for the straightforward presentation and discussion of analyses (http://www.yworks.com/en/products_yed_about.html).

A management process can be depicted as a temporal sequence of action situations that are linked by institutions, knowledge or operational outcomes (see example in Figure 3) and represent different phases in an overall policy cycle (e.g., policy formulation or implementation). Another approach to analyze a management and governance system is the interpretation of action situations as governance and management functions (e.g., water purification or allocation or conflict resolution). While temporal analysis is more suitable to examine the evolution of management issues over time, functional analysis allows for the comprehensive analysis of the status quo of management and governance systems at specific points in time.

RESULTS AND DISCUSSION

In this section, the proposed methodology is described in more detail and its application is illustrated with an example of flood management in the Tisza Basin, Hungary. The Tisza is a transboundary river and extends from the Ukrainian Carpathian mountains along the Romanian border, flows across the great Hungarian plain and enters the Danube in the Serbian Republic. It is the largest tributary of the Danube with a total catchment area of 157,200 km^2. In the Hungarian reach of the Tisza Basin, a centralized water management regime has existed since the 19th century, with a focus on engineered flood protection through the large-scale construction of dikes in order to allow for intensive agriculture and to protect residential and industrial areas. Rising flood intensities and frequencies have resulted in significant challenges for the existing water management paradigm over the past decade. A bottom-up learning process was formed by activists and academics that brought innovative ideas into the flood policy debate. However, a transition towards alternative paradigms has stalled due to weak linkages between the informal learning process and the formal institutions (Sendzimir *et al.* 2010). To address issues such as this, the methodology proposed in this paper supports a structured action research process of water management and governance systems, comprising three steps: (1) elicitation of management paradigms at the sub-system level; (2) analysis of the status quo in management and governance regimes; (3) design of pathways to overcome detected barriers towards sustainable water management.

Methodology

A management paradigm is defined by a specific 'system perspective' regarding the management problem, chosen 'solution strategies', as well as 'risk and uncertainty management strategies' (cf., Pahl-Wostl *et al.* 2011). The proposed methodology differentiates between paradigms linked to the sub-system level (e.g., the social, environmental, or technical system) and the overall system level (i.e., comprising the complete management and governance system). The former are called 'sub-system paradigms' and the latter 'overall-system paradigms'.

Management paradigms can co-exist at the sub-system level, either by being complementary (i.e., reinforcing each other) or competing (i.e., balancing each other). The proposed methodology builds on the notion that a concerted set of paradigms is usually needed, each tailored to the specific sub-system, to find effective and sustainable solutions. For instance, technical sub-systems (e.g., infrastructure) can be managed by a 'control paradigm' that aims at controlling the behavior of the sub-system (different paradigms are presented in detail below, cf., Table 1). Selected social issues (e.g., an allocation system) can be governed by a 'community paradigm' that builds upon the self-organization capacity of stakeholders (e.g., installing irrigation associations). Several paradigms can also belong to the same sub-system by complementing each other. For instance, water pricing can be implemented by applying a 'market paradigm' (i.e., prices are set by demand and supply) as well as a 'control paradigm' (i.e., the range of prices is predetermined). Alternative sub-system paradigms can also

Table 1 | Management paradigms elicited in Figure 1

Dimension	Name: 'Economies of Scale' Paradigm	'Control Floods' Paradigm	'Adapt to Floods' Paradigm	'Community Involvement' Paradigm	'Tradition' Paradigm
System Perspective	Big farms	River and protected values	Floodplain landscape	Flood prone communities	Small farms
Solution Strategies	Economies of scale; rationalization	Build dikes	River-landscape controlled flows	Community involvement	Traditional farming methods
Risk and Uncertainty Management	Reduce flooding risk and uncertainties	Reduction of uncertainty	Accept flood risk; adaptive management (through experimentation)	Uncertainty dialogue	Build on experience from the past

co-exist by being linked to different locations. For example, an 'adapt to floods paradigm' can be applied in rural areas where retention areas are available, while a 'control floods paradigm' might be more likely to be implemented in urban areas due to fewer adaptation options.

An encompassing 'overall-system paradigm' is linked to the overall resource system and can emerge from the sub-system level (e.g., through the supersession of other sub-system paradigms), or can be purposefully implemented by a higher-level institution (e.g., a ministry for water). In the case of a 'control paradigm' at the overall system level, heterogeneity of sub-system paradigms is constrained as only a limited number of paradigms are compatible with this overall-system paradigm. However, an 'integrated and adaptive overall-system paradigm' (cf., Pahl-Wostl *et al.* 2011) allows for the coordination of various sub-system paradigms and increases the adaptive capacity of the overall management system.

The proposed methodology that allows for the elicitation of management paradigms at the sub-system level (Step 1) and overall system level (Step 2), as well as the visioning of pathways towards sustainable water management (Step 3) is presented in the following sections.

Step 1: 'Elicitation of sub-system specific management paradigms'

Participatory model building using systems thinking can support the elicitation of sub-system specific management paradigms from individual participants or groups. The interviewee/group is asked to include the causes and consequences of the particular problem (i.e., the 'system perspective'), as well as preferred 'solution strategies' (e.g., technical approaches like building dams or socio-economic aspects like stakeholder involvement). The resulting CLDs will comprise elements of the resource system (e.g., variables like 'precipitation' or 'vegetation type'), the management system (e.g., 'dams' or 'retention areas'), and the governance system (e.g., 'public participation' or 'water legislation'). Further information about the 'risk or uncertainty management' strategies is needed to derive management paradigms from CLDs. Uncertainties are commonly typified in ontological and epistemological uncertainties. While the former denotes complex phenomena whose behavior cannot be predicted, the latter refers to incomplete knowledge or information about a system that can be attained through scientific research (Walker *et al.* 2003). Relational uncertainties are a third type of uncertainty and acknowledge subjective perceptions of actors. People perceive objects differently depending on personal values, roles, and interests (Brugnach *et al.* 2008). Based on this categorization, possible strategies for the handling of uncertainties and risks comprise the following (cf., Brugnach *et al.* 2008):

- Acceptance of uncertainty since *ontological uncertainties* imply that predictions cannot be made.
- Reduction of uncertainty since *epistemological uncertainties* can be minimized through purposeful research.
- Uncertainty dialogue since relational uncertainties requires a dialogue between stakeholders. For instance, the method of participatory model building facilitates the learning of groups and the revision of mental models and frames of participants. The goal of this reframing process is not to determine a 'true' frame – rather, the process aims at widening individual frames to one that considers multiple values and interests of stakeholders.

Figure 1 shows a CLD that was developed via a group model building processes in the Tisza River Basin (see Sendzimir *et al.* 2007). The CLD contains case-specific elements of the resource (e.g., soil quality), management (e.g., dikes), and governance system (e.g., lobbying capacity of community actors).

The group model in Figure 1 integrates different aspects of the flooding problem in the Tisza Basin that were mentioned by stakeholders including government representatives, local activists, and scientists. The CLD shows two different paradigms that are directly related to the flooding problem (represented by the problem variable 'Flood Frequency and Intensity'), namely the 'Adapt to Floods' paradigm (related elements are marked in blue in Figure 1) and the 'Control Floods' paradigm (marked in orange). For the specific attributes of paradigms, see Table 1. Further paradigms are indirectly related to flooding, including the 'Economies of Scale' paradigm (marked in green) of the intensive agriculture sector, and the 'Tradition' paradigm (marked in grey) that is more applicable to small farms. In addition, there are societal processes that demand more of a 'Community Involvement' paradigm (marked in pink) that relies on dialogue between actors.

The group model presents the interplay of different paradigms held by stakeholders related to the flooding problem in the Hungarian reaches of the Tisza Basin. The model clarifies the specific system elements that are related to paradigms. Thus, the CLD can support a purposeful discussion and handling of management paradigms.

Step 2: 'Analysis of management paradigms embedded in the overall management and governance system'

Management paradigms cannot only belong to the sub-system level but also belong to the overall system. Management paradigms of the overall system can be considered as a general 'mindset' that dominates in the management and governance system. Such overall-system paradigms are not only represented in the way of governing water resources, but are also manifested in infrastructure, information management, and finance, amongst others (cf., Pahl-Wostl *et al.* 2011).

Overall-system paradigms can influence paradigms at the sub-system level. For instance, a 'Control' overall-system paradigm might only be compatible with the 'Economies of Scale' sub-system paradigm due to a similar uncertainty and risk strategy (cf., Figure 1 and Table 1) but usually hampers the functioning of other paradigms at the sub-system level. However, an overall 'Adaptive and Integrated' paradigm, as conceived by Pahl-Wostl *et al.* (2011), allows for the functioning of all paradigms at the sub-system level due to its integrative nature. This permits diversity in management approaches, which increases the resilience of the social-ecological system (cf., Folke 2006).

Management paradigms pertain to both structural as well as process characteristics of management and governance systems. In the MTF, sub-system paradigms are considered as a type of cultural-cognitive institution and can therefore be an input or output of action situations (i.e., sub-system paradigms are social rules that can influence water management processes). Each action situation is characterized by at least one paradigm. The designation of paradigms to an action situation depends, first, on participating stakeholders who include their perspectives, and second, on input factors (e.g., knowledge, institutions, operational outcomes).

Overall-system management paradigms have a more general influence on the water system and are therefore linked to the Action Arena. As mentioned above, an 'Integrated and Adaptive' paradigm can function as such an overall-system paradigm. However, a 'Predict and Control' paradigm might be the most prevalent current overall-system paradigm, which hampers the application of most of the paradigms at the sub-system level (and thereby limits diversity). Figure 2 shows a graphical representation of the management and governance system in the Tisza Basin comprising action situations as management functions (boxes), institutions (circles), knowledge (trapezes), and operational outcomes (octagons).

On the left hand side of Figure 2, the 'status quo' of the management and governance regime in the Hungarian Tisza Basin is depicted. It is characterized by the dominance of a 'predict and control' paradigm at the overall-system and sub-system level that hampers solution strategies other than 'flood protection'. Intensive agriculture is tightly linked to this kind of management and governance system (Sendzimir *et al.* 2007). On the right hand side, a future vision of an alternative system is presented, which is based on the combination of paradigms included in the group model (see Figure 1). In this future system, an 'integrated and adaptive' flood management paradigm has emerged at the overall-system level, which allows for a diversity of paradigms at the sub-system level (cf., Grabs *et al.* 2006). A 'stakeholder platform' coordinates the 'Flood Control' and 'Flood Adaptation' management functions. Small-scale and intensive agriculture are exogenous elements that are linked to this Action Arena.

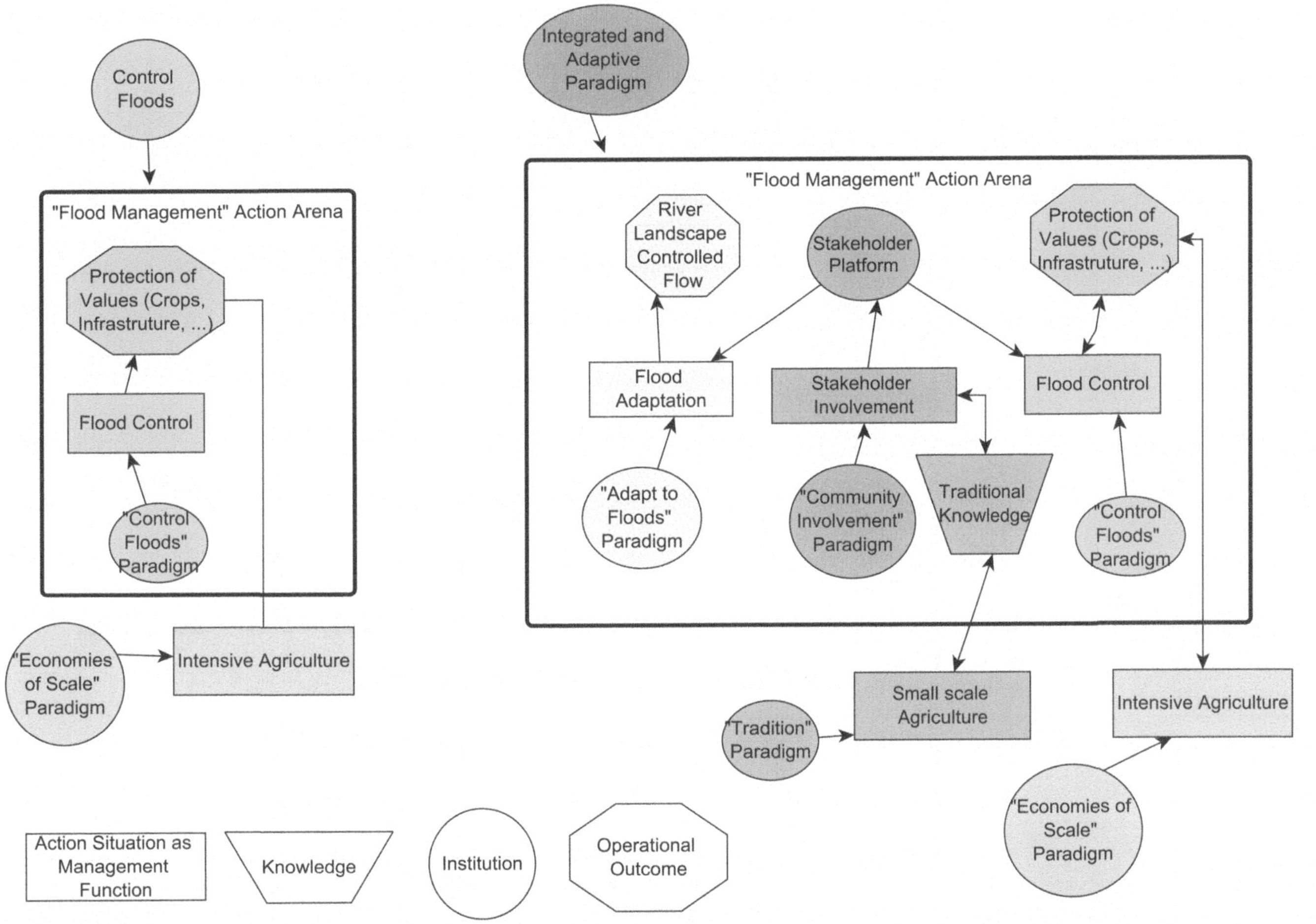

Figure 2 | Functional analysis of the management and governance system (purple colored variables reflect the new 'integrated and adaptive paradigm'; other colors refer to sub-system paradigms, cf., Figure 1). Please refer to the online version of this paper to see this figure in color: http://www.iwaponline.com/wst/toc.htm.

Step 3: 'Visioning of pathways towards sustainable water management'

Based on the functional diagnosis of the status quo and the desired future of the management and governance system (cf., Step 2), necessary measures through time are defined in Step 3 to achieve the desired changes. The transformation towards sustainable water management and governance usually requires substantial investments of resources and reform of institutions. The MTF can be applied to explore these required changes in a holistic manner, and specify a temporal succession of action situations in order to achieve the desired future. The outcome from this task is the design of a concrete pathway towards a sustainable water management and governance system. The resulting pathways include requirements for institutional change, dissemination or production of knowledge, physical interventions (i.e., 'operational outcomes'), as well as participation of actors.

Figure 3 shows an example pathway towards an integrated and adaptive flood management system for the Hungarian reaches of the Tisza River Basin that was developed based on literature review and expert opinion. The pathway starts from a present action situation comprising the reinterpretation of a flood management policy called VTT2 (in Hungarian: 'Vásárhelyi TervTovábbfejlesztése') (cf., Sendzimir *et al.* 2010), which reflects the mindset of the dominating 'Control Floods' paradigm. The transformation process ends with an 'Integrated and Adaptive Flood Management' action situation that delivers the outcome of 'integrated and adaptive flood-land-agriculture management' (see Figure 3). In the example pathway depicted in Figure 3, two action situations are envisioned to lead to the desired outcome. First, a round table is advised under participation of the Ministry of Environment and Water, responsible engineers and water managers, and representatives of the 'Living Tisza Alliance', which is formed of local activists,

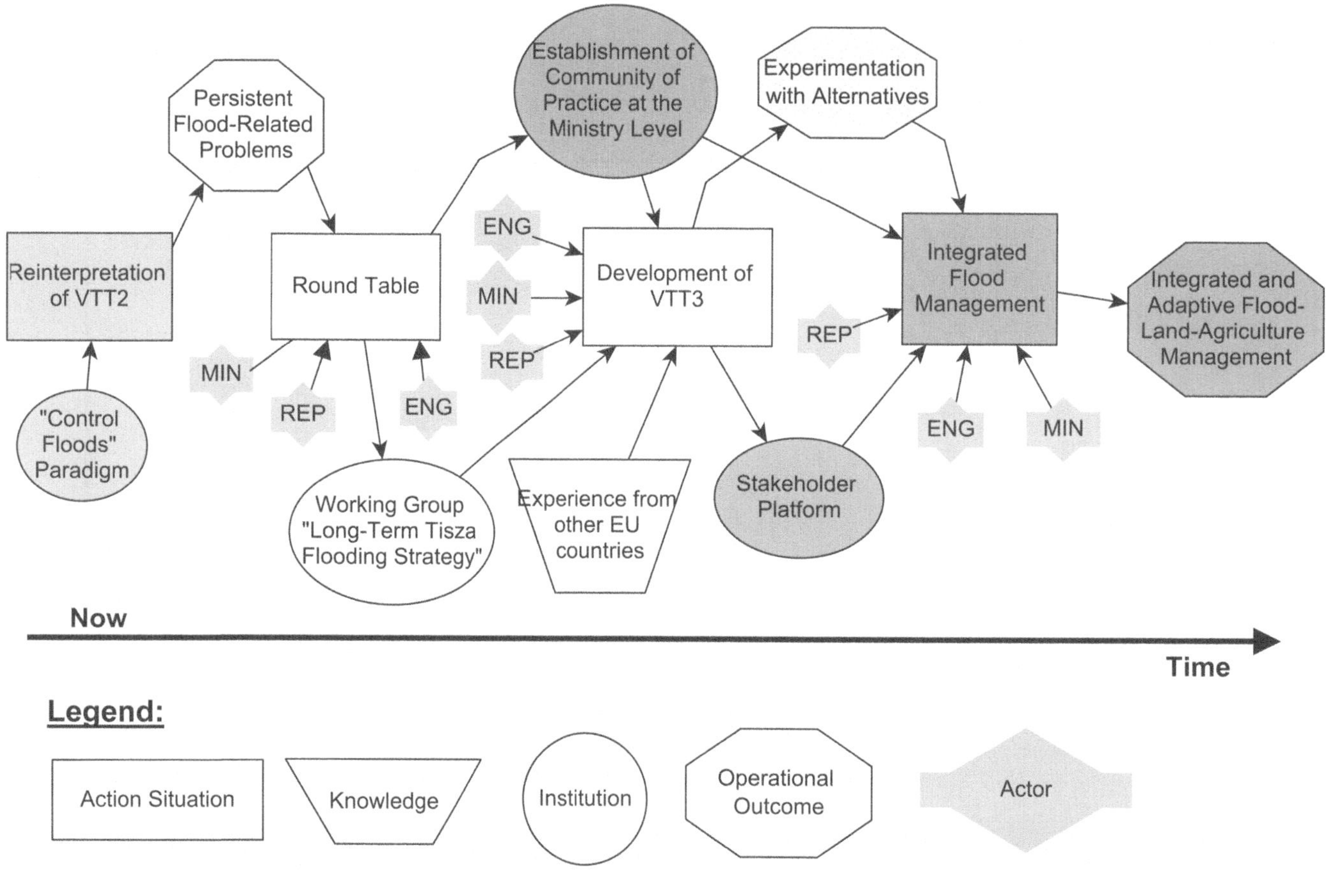

Figure 3 | Example of a possible pathway towards a sustainable water management and governance system in the Tisza River Basin (purple colored variables reflect the new 'integrated and adaptive paradigm'; other colors refer to the respective management paradigm, cf., Figure 1). Abbreviations of actors mean the following: REP = representatives of 'Living Tisza Alliance'; ENG = Responsible Engineers and Water Managers; MIN = Ministry of Environment and Water. Please refer to the online version of this paper to see this figure in color: http://www.iwaponline.com/wst/toc.htm.

national non-governmental organizations and the Village Municipalities Association (cf., Sendzimir *et al.* 2007). This round table sets the rules for a working group for a long-term flood strategy (institution) that includes all participating actors, and facilitates the institutionalization of a community of practice at the ministry level. The community of practice is supposed to link different departments and facilitate a continuous deliberative process on innovative solutions for the flood management problems in the Tisza Basin. In the past, this was temporarily achieved by the leadership of an individual at the ministry level, before his departure from the parliamentary committee stopped this process. A community of practice could lead to a more sustainable network that is less vulnerable to change of personnel. These activities together with experience from other EU countries with innovative flood management could set the basis for the development of a new flood management policy (named VTT3) that could lead to the institutionalization of a stakeholder platform at the local and regional scales as well as more experimentation with alternative approaches (e.g., through pilot studies). The established institutions and the experimental approach would result in the 'Integrated and Adaptive Flood Management' action situation that brings together the ministerial 'community of practice' and the stakeholder platform. This leads to river-landscape controlled flows (operational outcome), which reflects a transformation of the overall management and governance system towards a more 'integrated and adaptive' overall-system paradigm.

The example application of the proposed methodology demonstrates a structured approach to analyze water management and governance systems in an integrated way, and envision transformation processes towards IWRM. Through the use of the proposed methodology, hidden assumptions about the management and governance system, as well as requirements for integrated and adaptive solution strategies, are elicited. This approach can be particularly useful in transboundary river basins as paradigms might vary considerably between international water authorities due to different historical and cultural developments. While the Tisza case was confined to the Hungarian reach of the Tisza, a basin-wide study would be very useful to reveal differences in management paradigms in order to explore opportunities to deal with them constructively. The proposed methodology builds on straightforward methods and is therefore particularly suitable to be applied in collaborative management processes. At the same time, the proposed methodology composed of participatory modeling and the MTF allows for in-depth analysis of the water system. In this way, the linkage of the MTF and participatory modeling allows for the inclusion of scientific insights on water governance into management practice, and vice versa.

In future studies stemming from this research, empirical research will aim to determine concrete sets of management paradigms that facilitate IWRM. In particular, the relationships of overall-system and sub-system management paradigms require additional empirical research to identify which sets of sub-system paradigms are supported or inhibited by overall-system paradigms. In addition, the applicability of the proposed methodology to design concrete pathways towards IWRM will be evaluated through further case-study research in other basins with different socio-economic and hydrological characteristics (especially transboundary basins).

CONCLUSIONS

The analysis of management and governance systems is a complex task. Methods and tools have to deal with this complexity in order to avoid resorting to simplistic solutions or panaceas. In addition, effective science-policy-engineering dialogues need to be initiated to transfer findings between research and policy-making, and facilitate implementation on the ground.

The methodology presented in this paper complies with these requirements by building on participatory model building and the MTF. Participatory model building is a suitable and widely tested method to structure complex problems and elicit different perspectives held by stakeholders. Participatory model building supports the analysis of objective and subjective dimensions of resource issues. The MTF allows for the integration of elicited knowledge into an overall system perspective, and supports subsequent discussions of pathways towards sustainable management and governance systems. The MTF provides a formalization of complex management and governance systems and a 'common language' that can be used in participatory processes.

Management paradigms are proposed in this paper as suitable concepts to analyze the interlinkages between resource, management and governance systems in a comprehensive way. The consideration of sub-system specific management paradigms acknowledges that the effectiveness of paradigms depends on the respective application area or sub-system. However, an overall-system paradigm usually emerges that reflects the common mindset for the overall system. Such an overall-system paradigm can support a variety of management paradigms at the sub-system level (e.g., an adaptive and integrated management paradigm), or hamper variety by being incompatible with other paradigms (e.g., a predict and control paradigm).

A case study in the Tisza River Basin, Hungary, demonstrated the application of the proposed methodology. Further empirical research will be conducted in the future to evaluate the applicability and effectiveness of the proposed methodology in different water management contexts. In addition, concrete sets of management paradigms will be examined that support integrated and adaptive water resources management.

REFERENCES

Brugnach, M., Dewulf, A., Pahl-Wostl, C. & Taillieu, T. 2008 Toward a relational concept of uncertainty: About knowing too little, knowing too differently, and accepting not to know. *Ecology and Society* **13** (2), 30.

Cashman 2009 Alternative manifestations of actor responses to urban flooding: Case studies from Bradford and Glasgow. *Water Science and Technology* **60** (1), 77–85.

Costanza, R. & Ruth, M. 1998 Using dynamic modeling to scope environmental problems and build consensus. *Environmental Management* **22** (2), 183–195.

Folke, C. 2006 Resilience: The emergence of a perspective for social–ecological systems analyses. *Global Environmental Change* **16**, 253–267.

Galaz, V. 2007 Water governance, resilience and global environmental change – a reassessment of integrated water resources management (IWRM). *Water Science and Technology* **56** (4), 1–9.

Grabs, W., Tyagi, A. C. & Hyodo, M. 2006 Integrated flood management. *Water Science and Technology* **56** (4), 97–103.

Halbe, J., Pahl-Wostl, C. & Adamowski, J. (accepted) A framework to support the initiation, planning, and institutionalization of participatory modeling processes in water resources management. *Water Resources Research*.

Holling, C. S. 1978 *Adaptive Environmental Assessment and Management*. Wiley, Chichester, UK.

Jeffrey, P. & Gearey, M. 2006 Integrated water resources management: Lost on the road from ambition to realisation? *Water Science and Technology* **53** (1), 1–8.

Jonsson, A., Andersson, L., Alkan-Olsson, J. & Arheimer, B. 2007 How participatory can participatory modeling be? Degrees of influence of stakeholder and expert perspectives in six dimensions of participatory modeling. *Water Science and Technology* **56** (1), 207–214.

Knieper, C., Holtz, G., Kastens, B. & Pahl-Wostl, C. 2010 Analysing water governance in heterogeneous case studies – experiences with a database approach. *Environmental Science and Policy* **13** (7), 592–603.

Medema, W., McIntosh, B. S. & Jeffrey, P. J. 2008 From premise to practice: A critical assessment of integrated water resources management and adaptive management approaches in the water sector. *Ecology and Society* **13** (2), 29.

Metcalf, S. S., Wheeler, E., BenDor, T., Lubinski, K. S. & Hannon, B. M. 2010 Sharing the floodplain: Mediated modeling for environmental management. *Environmental Modelling and Software* **25** (11), 1282–1290.

Ostrom, E. 2005 *Understanding Institutional Diversity*. Princeton University Press, New Jersey, USA.

Pahl-Wostl, C. 2007 Transitions towards adaptive management of water facing climate and global change. *Water Resources Management* **21**, 49–62.

Pahl-Wostl, C. 2009 A conceptual framework for analysing adaptive capacity and multi-level learning processes in resource governance regimes. *Global Environmental Change* **19**, 354–365.

Pahl-Wostl, C., Craps, M., Dewulf, A., Mostert, E., Tabara, D. & Taillieu, T. 2007 Social learning and water resources management. *Ecology and Society* **12** (2), 5.

Pahl-Wostl, C., Holtz, G., Kastens, B. & Knieper, C. 2010 Analysing complex water governance regimes: The management and transition framework. *Environmental Science and Policy* **13** (7), 571–581.

Pahl-Wostl, C., Jeffrey, P., Isendahl, N. & Brugnach, M. 2011 Maturing the new water management paradigm: Progressing from aspiration to practice. *Water Resources Management* **25** (3), 837–856.

Renger, M., Kolfschoten, G. L. & de Vreede, G.-J. 2008 Challenges in collaborative modeling: A literature review. *International Journal of Simulation, and Process Modelling* **4**, 248–263.

Schlüter, M., Hirsch, D. & Pahl-Wostl, C. 2010 Coping with change: Responses of the Uzbek water management regime to socio-economic transition and global change. *Environmental Science and Policy* **13** (7), 620–636.

Sendzimir, J., Flachner, Z., Pahl-Wostl, C. & Knieper, C. 2010 Stalled regime transition in the upper Tisza river basin: The dynamics of linked action situations. *Environmental Science and Policy* **13** (7), 604–619.

Sendzimir, J., Magnuszewski, P., Flachner, Z., Balogh, P., Molnar, G., Sarvari, A. & Nagy, Z. 2007 Assessing the resilience of a river management regime: Informal learning in a shadow network in the Tisza River Basin. *Ecology and Society* **13** (1), 11.

Tidwell, V. C., Passell, H. D., Conrad, S. H. & Thomas, R. P. 2004 System dynamics modeling for community-based water planning: Application to the Middle Rio Grande. *Aquatic Science* **66** (4), 357–372.

van den Belt, M. 2004 *Mediated Modeling – A System Dynamics Approach to Environmental Consensus Building*. Island Press, Washington.

Voinov, A. & Bousquet, F. 2010 Modelling with stakeholders. *Environmental Modelling and Software* **25**, 1268–1281.

Walker, W. E., Harremoës, P., Rotmans, J., van der Sluis, J. P., van Asselt, M. B. A., Janssen, P. & Krayer von Kraus, M. P. 2003 Defining uncertainty: A conceptual basis for uncertainty management in model-based decision support. *Integrated Assessment* **4** (1), 5–17.

First received 28 September 2012; accepted in revised form 6 February 2013

Index

IWA Publishing's authorised EU representative for General Product Safety Regulations is Diane D'Arras, 15 rue Duret, 75116 Paris, France, e-mail: safety@iwap.co.uk.

Printed and bound by CPI Group (UK) Ltd, Croydon, CR0 4YY
12/05/2026
02108874-0007